Power Systems

More information about this series at http://www.springer.com/series/4622

Farhad Shahnia · Ali Arefi
Gerard Ledwich
Editors

Electric Distribution Network Planning

Springer

Editors
Farhad Shahnia
School of Engineering and Information Technology
Murdoch University
Perth, WA
Australia

Gerard Ledwich
Faculty of Science and Engineering
Queensland University of Technology
Brisbane, QLD
Australia

Ali Arefi
School of Engineering and Information Technology
Murdoch University
Perth, WA
Australia

ISSN 1612-1287 ISSN 1860-4676 (electronic)
Power Systems
ISBN 978-981-13-3911-0 ISBN 978-981-10-7056-3 (eBook)
https://doi.org/10.1007/978-981-10-7056-3

Printed on acid-free paper

This Springer imprint is published by the registered company Springer Nature Singapore Pte Ltd.
part of Springer Nature
The registered company address is: 152 Beach Road, #21-01/04 Gateway East, Singapore 189721, Singapore

Preface

Electric distribution networks are critical parts of power delivery systems. In recent years, many new technologies and distributed energy resources have been integrated into these networks. To provide electricity at the possible lowest cost and at required quality, long-term planning is essential for these networks. In distribution planning, optimal location and size of necessary upgrades are determined to satisfy the demand and the technical requirements of the loads and to tackle uncertainties associated with load and distributed energy resources. To this aim, an optimization algorithm is utilized to find the optimal net present cost of augmentation over the planning period. The distribution network is usually formulated as a mixed-integer nonlinear programming problem, which is solved using various approaches including mathematic and heuristic-based algorithms.

Over the last decades, several researches have been carried out around the world on electric distribution planning, whose results are available as journal articles, conference papers or technical reports. However, to the best of the editors' knowledge, no single book has covered the different aspects of distribution networks' planning so far. The interested readers had to search among several hundreds of papers on this topic through various databases in order to build up their knowledge on the subject. This book is the first one entirely focused on the distribution networks planning and is an effort to provide a research-oriented and a coherent book on the subject for postgraduate students and researchers.

This book is benefited from the inputs and comments of a large number of researchers and experts from the academia and industry. It contains 13 chapters. The breakdown of the chapters is as follows:

- Chapter 1 reviews the multi-stage expansion planning problem of distribution networks where investments in the distribution network and distributed generations are jointly considered;
- Chapter 2 presents static and dynamic models for the planning of distribution networks;
- Chapter 3 discusses the mathematical formulations of unbalance networks, for operation optimization analysis to support decision-making processes;

- Chapter 4 presents an integrated distributed generation and primary–secondary expansion planning in the presence of wholesale and retail markets;
- Chapter 5 discusses a new planning tool based on the concept of a multi-agent system;
- Chapter 6 presents an efficient method for sizing and siting distributed generations in distribution networks;
- Chapter 7 describes probabilistic and possibilistic-based planning methodologies of battery energy storage systems in distribution networks;
- Chapter 8 introduces an optimally distributed generation placement problem towards power and energy loss minimization;
- Chapter 9 presents a hybrid methodology based on a local search algorithm and a genetic algorithm, for the multi-objective and multi-stage distribution expansion planning problem;
- Chapter 10 introduces simultaneous optimization concept of distribution network reconfiguration and distributed generation sizing;
- Chapter 11 studies the implementation of optimal incentive plans for plug-in electric vehicle aggregator to participate in the energy market;
- Chapter 12 presents a survey of optimization techniques used to find the optimal sizes and locations of compensators; and
- Chapter 13 discusses a methodology for the allocation of automatic reclosers within electric distribution networks.

As the editors of the book, we would like to thank all the contributors for their support and hard work. We also would like to thank the reviewers who provided valuable comments for improving the quality of the book. Also, we are grateful to the publisher Springer Nature for agreeing to publish this book. Last but not least, we would like to thank our families—Farhad thanks his parents (Nahideh and Ali) and his spouse (Negar), Ali thanks his wife and son (Behnaz and Amin), and Gerard thanks his family for their continuous encouragement and support.

Perth, Australia — Farhad Shahnia
Perth, Australia — Ali Arefi
Brisbane, Australia — Gerard Ledwich
January 2018

Reviewers

Aggelos S. Bouhouras, Aristotle University of Thessaloniki, Greece; University of Applied Sciences, Greece
Ali Ahmadian, K. N. Toosi University of Technology, Iran
Ali Elkamel, University of Waterloo, Canada
Alexandre M. F. Dias, University of Lisbon and INESC-ID, Portugal
Almoataz Y. Abdelaziz, Ain Shams University, Egypt
Alireza Heidari, University of New South Wales, Australia
Carlos F. M. Almeida, University of Sao Paulo, Brazil
Carlos F. Sabillón, São Paulo State University, Brazil
David Pozo, Skolkovo Institute of Science and Technology, Russia
Dimitris P. Labridis, Aristotle University of Thessaloniki, Greece
Gabriel Quiroga, University of Sao Paulo, Brazil
Gregorio Muñoz-Delgado, Universidad de Castilla-La Mancha, Spain
Hazlie Mokhlis, University of Malaya, Malaysia
Henrique Kagan, University of Sao Paulo, Brazil
Jan Kays, Amprion GmbH, Dortmund, Germany
Javier Contreras, Universidad de Castilla-La Mancha, Spain
John F. Franco, São Paulo State University, Brazil
José M. Arroyo, Universidad de Castilla-La Mancha, Spain
Julio López, University of Cuenca, Cuenca, Ecuador
Karar Mahmoud, Aswan University, Egypt
Kazem Zare, University of Tabriz, Iran
Mehdi Rahmani-Andebili, Sharif University of Technology, Iran
Mahdi Sedghi, K. N. Toosi University of Technology, Iran
Mahmood Reza Haghifam, Tarbiat Modares University, Iran
Mahmud Fotuhi Firuzabad, Sharif University of Technology, Iran
Marcos J. Rider, University of Campinas, Brazil
Masoud Aliakbar Golkar, K. N. Toosi University of Technology, Iran
Mehrdad Setayesh Nazar, Shahid Beheshti University, Iran
Michael Fowler, University of Waterloo, Canada
Moein Moeini-Aghtaie, Sharif University of Technology, Iran

Mohamed Ebeed, Aswan University, Egypt
Nelson Kagan, University of Sao Paulo, Brazil
Paschalis A. Gkaidatzis, Aristotle University of Thessaloniki, Greece
Pedro M. S. Carvalho, University of Lisbon and INESC-ID, Portugal
Rubén Romero, São Paulo State University, Brazil
Salah Kamel, Aswan University, Egypt
Shady H. E. Abdel Aleem, Higher Institute of Engineering, Cairo, Egypt
Wardiah Mohd Dahalan, Universiti Kuala Lumpur, Malaysia
Yorino Naoto, Hiroshima University, Japan

Contents

About the Editors

Farhad Shahnia received his Ph.D. in Electrical Engineering from Queensland University of Technology, Brisbane, Australia, in 2011. Currently, he is a Senior Lecturer of Power Engineering at Murdoch University, Perth, Australia. His professional experience includes three years at Research Office-Eastern Azarbayjan Electric Power Distribution Company, Tabriz, Iran. Prior to joining Curtin University, he was a Research Fellow at the Queensland University of Technology, Brisbane, Australia, and a Lecturer at Curtin University, Perth, Australia. He has authored 1 book, 9 book chapters and over 120 research articles as well as editing 4 books.

Ali Arefi received his Ph.D. in Electrical Engineering in 2011. He is currently a Senior Lecturer of Power Engineering at Murdoch University, Perth, Australia. Prior to that, he was a Lecturer and Research Fellow at the Queensland University of Technology. He also has six years' experience with electric distribution industry and has been the supervisor for five industry-funded research projects. His research interests are in the areas of electric delivery planning, state estimation, power quality and energy efficiency. He has published more than 80 research articles.

Gerard Ledwich received his Ph.D. in Electrical Engineering from University of Newcastle, Newcastle, Australia, in 1976. He has been the Chair Professor in Power Engineering at the Queensland University of Technology, Brisbane, Australia, since 1998. Previously, he was with the University of Queensland, Brisbane, Australia, from 1976 to 1994. His research interests are in the areas of power system operation and control. He is the supervisor of more than 35 Ph.D. graduates and has published 1 book, 3 chapters and over 350 research articles.

Abbreviations

3P3W	Three-phase three-wire
3P4W	Three-phase four-wire
ABC	Artificial bee colony
AI	Artificial intelligence
ALO	Ant lion optimization
AR	Automatic recloser
BES	Battery energy storage
BFOA	Bacterial foraging optimization algorithm
CAMG	Customers' active microgrid
CDF	Cumulative distribution function
CDMG	Customer's dispatchable microgrid
CF	Capacity factor
CNDMG	Customer's non-dispatchable microgrid
CS	Cuckoo search
DA	Distribution automation
D-FACTS	Distributed flexible AC transmission system
DFIG	Doubly fed induction generator
DG	Distributed generation
DisCo	Distribution company
DL	Dispatchable load
DNUDG	Dispatchable non-utility distributed generation
DOD	Depth of discharge
DR	Demand response
DSM	Demand side management
DSO	Distribution system operator
DSTATCOM	Distributed static compensator
D-SVC	Distribution static VAR compensator
DVS	Decision variable set
EDNEP	Distribution network expansion planning
ELF	Exhaustive load flow

EM	Energy management
ENS	Energy not supplied
EP	Evolutionary programming
EPSO	Evolutionary particle swarm optimization
ESS	Energy storage system
EV	Electric vehicle
EVCC	Electric vehicle charging coordination
FCB	Fixed capacitor bank
FLF	Fuzzy load flow
G2V	Grid-to-vehicle
GA	Genetic algorithm
GAP	Genetic algorithm population
GEV	Generalized extreme value
GIS	Georeferenced information system
GOA	Grasshopper optimization algorithm
GPSO	Global particle swarm optimization
GRASP	Greedy randomized adaptive search procedure
GS	Gaussian search
GUPFC	Generalized unified power flow controller
GWO	Grey wolf optimization
HIC	Hours of interrupted customer
HP	Hydro-plant
HS	Harmony search
HV	High voltage
IA	Improved analytical
IC	Interrupted customers
ICT	Information and communications technology
IDGNEP	Integrated distributed generation and primary–secondary network expansion planning
IG	Induction generator
IGBT	Insulated gate bipolar transistor
IPFC	Interline power flow controller
LP	Linear programming
LPSO	Local particle swarm optimization
LS	Load smoothing
LSF	Loss sensitivity factor
LV	Low voltage
MAS	Multi-agent system
MCS	Monte Carlo simulation
MGA	Modified genetic algorithm
MI	Merit index
MICQP	Mixed-integer conic quadratic programming
MILP	Mixed-integer linear programming
MINLC	Mixed-integer-nonlinear constrained
MINLP	Mixed-integer nonlinear programming

MPSO	Modified particle swarm optimization
MU	Monetary unit
MV	Medium voltage
NC	Normally closed
NC-AR	Normally closed automatic recloser
NDL	Non-dispatchable load
NDNUDG	Non-dispatchable non-utility distributed generation
NLP	Nonlinear programming
NO	Normally opened
NO-AR	Normally opened automatic recloser
NR	Network reconfiguration
NUDG	Non-utility distributed generation
NURMP	Non-utility retail market participants
O&M	Operation and maintenance
ODGP	Optimal distributed generation placement
OESSP	Optimal energy storage system placement
OLTC	On-load tap changer
OMS	Outage management system
OPF	Optimal power flow
ORESP	Optimal renewable energy sources placement
PCC	Point of coupling connection
PDF	Probability density function
PEM	Point estimate method
PEV	Plug-in electric vehicle
PLF	Probabilistic/possibilistic load flow
POPF	Probabilistic/possibilistic optimal power flow
PRF	Price factor
PSO	Particle swarm optimization
PV	Photovoltaic
PWM	Pulse-width modulation
RES	Renewable energy source
RPF	Reverse power flow
SABC	Satisfied artificial bee colony
SCA	Sine cosine algorithm
SCB	Switched capacitor bank
SiG	Synchronous generator
SoC	State of charge
SSSC	Static synchronous series compensator
SVC	Static VAR compensator
TCSC	Thyristor controlled series capacitor
UDG	Utility distributed generation
UPQC	Unified power quality conditioner
UPS	Uninterruptible power supply
UPSO	Unified particle swarm optimization
V2G	Vehicle-to-grid

VR	Voltage regulator
VSC	Voltage source converter
VSS	Value of the stochastic solution
VVC	Volt-VAR control
WRI	Weighted reliability index
WT	Wind turbine

Chapter 1
Distribution System Expansion Planning

Gregorio Muñoz-Delgado, Javier Contreras and José M. Arroyo

Abstract The widespread growth of distributed generation (DG), mainly due to its numerous operational and planning benefits and to the penetration of renewable energy, inevitably requires the inclusion of this kind of generation in distribution planning models. This chapter addresses the multistage expansion planning problem of a distribution system where investments in the distribution network and in DG are jointly considered. The optimal expansion plan identifies the best alternative, location, and installation time for the candidate assets. The incorporation of DG in distribution system expansion planning drastically increases the complexity of the optimization process. In order to shed light on the modeling difficulties associated with the co-optimized planning problem, a deterministic model is presented first. The model is driven by the minimization of the net present value of the total cost including the costs related to investment, maintenance, production, losses, and unserved energy. As a relevant feature, radiality conditions are specifically tailored to accommodate the presence of DG in order to avoid the islanding of distributed generators and the issues associated with transfer nodes. Since a large portion of DG relies on non-dispatchable renewable-based technologies, the uncertainty associated with the high variability of the corresponding energy sources needs to be properly characterized in the planning models. Based on the previous deterministic model, uncertainty is incorporated using a stochastic programming framework. Within such a context, the uncertainty featured by renewable-based generation and demand is characterized through a set of scenarios that explicitly

G. Muñoz-Delgado (✉) · J. Contreras · J. M. Arroyo
E.T.S. de Ingenieros Industriales,
Universidad de Castilla-La Mancha, Ciudad Real, Spain
e-mail: gregorio.munoz.delgado@gmail.com

J. Contreras
e-mail: Javier.Contreras@uclm.es

J. M. Arroyo
e-mail: JoseManuel.Arroyo@uclm.es

F. Shahnia et al. (eds.), *Electric Distribution Network Planning*,
Power Systems, https://doi.org/10.1007/978-981-10-7056-3_1

capture the correlation between uncertainty sources. The resulting stochastic program is driven by the minimization of the total expected cost. Both deterministic and stochastic optimization problems are formulated as mixed-integer linear programs for which finite convergence to optimality is guaranteed and efficient off-the-shelf software is available. Numerical results illustrate the effective performance of the approaches presented in this chapter.

Keywords Distributed generation · Distribution system planning
Multistage · Network expansion · Stochastic programming · Uncertainty

1.1 Introduction

Power systems are used to produce, transport, and distribute electric power from generation centers to consumers. A power system usually consists of generation units, transmission and subtransmission networks, distribution networks, consumption centers, system protection devices, and control equipment [1]. Distribution networks are an important part of the electric system since they supply energy from distribution substations to end users. The main components of the distribution network are the substations and the feeder branches. Distribution substations are fed by one or several subtransmission networks, although sometimes they can be directly connected to the transmission network. Distribution substations reduce the voltage level from high voltage to medium voltage ranging between 4.16 and 34.5 kV through the use of transformers. Moreover, from distribution substations energy is injected into the distribution grid by means of one or several primary feeders [2].

Regardless of their topologies, which may be either meshed or radial, most distribution networks are operated in a radial way since it is the cheapest and simplest method from the viewpoints of planning, design, and system protection. Traditionally, these networks have been designed with wide operating ranges, which allows them to be passively operated, thereby resulting in more economical management. However, investment in distribution networks is several times more costly than investment in transmission grids [1], which reveals the economic importance of distribution system planning. Moreover, a viable investment plan must not only be economically driven, but it must also satisfy several criteria and guidelines related to design, components, layout, or performance [3].

From a centralized standpoint, distribution companies are responsible for the operation and planning of distribution networks so that the growing demand is continuously satisfied with quality standards and in a secure fashion. Therefore, planning models are used to obtain an optimal investment plan at minimum cost while meeting the security and quality requirements. Traditionally, these planning models determine the optimal expansion decisions related to the reinforcement and installation of branches, substations, and transformers [3, 4]. However, the widespread growth of distributed generation (DG), mainly due to its numerous operational and planning benefits [5] and to the impetus of renewable energy, inevitably requires the inclusion of this kind of generation in distribution planning models [4, 6]. This new context where DG comes into play calls for changes in the way distribution systems are operated and planned [3].

DG comprises small-scale power units located close to consumption centers. Manifold technologies are currently used for DG including wind turbines, photovoltaic (PV) plants, mini hydro plants, fuel cells, cogeneration plants, micro gas turbines, internal combustion engines, and energy storage devices such as batteries [5, 7]. The use of DG has numerous advantages related to system planning and operation [8], such as

- Reduction in energy losses
- Control of the voltage profile
- Improvement of power quality
- Increase in system reliability
- Reduction or deferral of the network expansion
- Decrease in the emissions of CO_2
- Short lead time
- Low investment risk
- Modularity
- Reduced physical size
- Availability of a wide range of DG technologies.

The presence of DG may have a significant impact on power flows, voltage profiles, system efficiency, and protection devices. As a consequence, the traditionally passive distribution networks are transitioning to a new paradigm where an active role is played. The operational impact of DG depends on many factors such as the type, size, and location of generation units; the types of control equipment; and the characteristics of branches and loads, among others.

The increasing penetration of non-dispatchable renewable-based technologies for DG, such as wind and photovoltaic energy, requires the consideration of the uncertainty associated with the high variability of these energy sources. Furthermore, load demand is another source of uncertainty with a huge impact on generation. Although many tools have been successfully developed to forecast demand as well as wind and PV energy production, the incorporation of such uncertainty sources in planning models is still challenging.

In the technical literature, many works have addressed the joint expansion planning of distribution network assets and DG [4]. However, under a dynamic or multistage framework, only few works have considered a complete expansion model accounting for topological changes due to the connection of new load nodes through the installation of new branches. Relevant works addressing this problem can be classified in two groups, namely those disregarding uncertainty [9–13], and those considering uncertainty [14–19].

In [9], a genetic algorithm combined with an optimal power flow is applied to solve the deterministic co-optimized expansion planning problem driven by the minimization of the overall cost. Reference [10] uses a heuristic method based on particle swarm optimization and shuffled frog leaping to solve a multiobjective version of the optimization problem where two objective functions are minimized, namely cost and reliability. Reference [11] applies a modified particle swarm optimization algorithm to solve this combinatorial optimization problem. Reference [12] presents a multiobjective reliability-based distribution expansion planning model. A hybrid self-adaptive global-based harmony-search algorithm and an optimal power flow are used whereas a fuzzy satisfying method is applied in order to obtain the best solution. Reference [13] presents a mixed-integer linear programming model to solve the joint multistage expansion planning problem.

Reference [14] proposes a genetic algorithm to solve the distribution system expansion planning problem under uncertainty using a multiobjective optimization framework. Reference [15] applies a genetic algorithm combined with an optimal power flow. The uncertainty of demand, electricity prices, and wind is represented through scenarios generated on the basis of their corresponding probability density functions. Reference [16] uses particle swarm optimization to address the planning problem considering load and price uncertainties under an electricity market environment. These uncertainties are modeled through probability density functions and Monte Carlo simulation. In [17], a multiobjective model is formulated to represent the different objectives of a distribution company and private DG investors while considering the uncertainty of demand. The concept of system of systems is proposed to model the expansion of DG owned by private developers. Particle swarm optimization is applied to solve the proposed model. Finally, in [18, 19], a multistage and stochastic mixed-integer linear programming model is developed to support the decision-making process of distribution system planners.

This chapter is focused on multistage expansion planning of distribution network assets and DG [20, 21]. In this problem, the optimal location, alternative, and installation time for each candidate asset is provided, thereby constituting a dynamic approach. Moreover, the connection of new load nodes is considered. Thus, topological changes due to the installation of new branches to supply the demand at those new load nodes are explicitly accounted for. First, a deterministic model for co-optimized expansion planning is described. This model is driven by the minimization of the present value of the total cost, which comprises costs related to investment, maintenance, production, energy losses, and unserved energy. As a consequence of this co-optimization, the radial operation of the system needs to be specifically imposed in order to avoid the islanding of distributed generators and the issues associated with transfer nodes. Second, in order to represent the uncertainty associated with demand and renewable-based generation, a stochastic programming model relying on the previous deterministic model is presented. Uncertainty is modeled through different scenarios that explicitly capture the correlation between uncertainty sources. This model is driven by the minimization of the present value of the total expected cost. The resulting optimization problems are formulated as mixed-integer nonlinear programs, which, using some well-known linearization schemes, are recast as mixed-integer linear programs.

We recognize that the use of the models described in this chapter leads to results that may be optimistic and that a complete study would require the consideration of a more sophisticated operational model. This consideration would, however, render the problem essentially intractable through optimization and would have to be solved by heuristics or repeated simulations. These modeling limitations notwithstanding, the use of our models is acceptable for distribution planning purposes and provides the planner with a first estimate of a cost-effective expansion plan.

Additionally, it is worth mentioning that the use of mixed-integer linear programming features three advantages of utmost importance from a practical perspective, namely (1) the guaranteed finite convergence to optimality, (2) the availability of a measure of the distance to the global optimum along the solution process, and (3) the ready availability of efficient off-the-shelf software.

Finally, note that regardless of the ownership of DG assets, the models described hereinafter provide valuable information about the best investment plan in terms of economics. In the case of DG owned by independent producers, such information may be used to devise appropriate incentive strategies, which are beyond the scope of this chapter.

1.2 Deterministic Model

In this section, a deterministic optimization model for the joint expansion planning of distribution network assets and DG is presented. Built on the distribution network expansion planning models described in [7, 22, 23], our model considers a centralized framework. Thus, the planner is responsible for expanding the existing distribution network in order to meet the growing demand at minimum cost over a planning horizon comprising several stages. To that end, the planner has the possibility of installing new branches, transformers, substations, and generators for which several investment alternatives exist. For each stage, demand, wind speed, and solar irradiation profiles are divided into several time blocks. Figure 1.1 shows an example with a 4-time-block discretization. To that end, first, historical hourly data of system demand, wind speed, and solar irradiation throughout a year are expressed as per-unit factors by dividing each data by the corresponding maximum level. Hence, each set of factors represents the per-unit demand, wind speed, and solar irradiation profile. Second, triplets of hourly factors for demand, wind speed, and solar irradiation are sorted by demand in descending order. Next, the factor curves are discretized into n_B time blocks characterized by a pre-specified number of hours. For each time block, the average values of the ordered demand, wind speed, and solar irradiation factors are calculated.

Moreover, as done in [22, 23], radial operation of the system is explicitly imposed and a lossless approximate network model is used. In addition, as done in [23], the costs of losses are included in the objective function. Next, the deterministic model is described in detail.

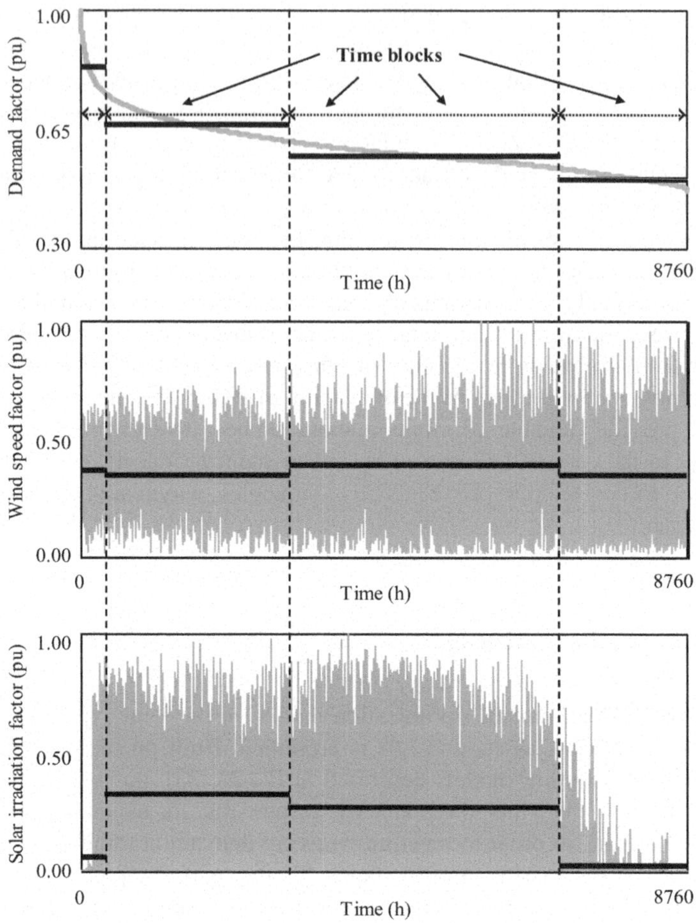

Fig. 1.1 Time block discretization of demand, wind speed, and solar irradiation factors for the deterministic model

1.2.1 Objective Function and Cost-Related Terms

The goal of the distribution planner is to address the estimated demand growth by installing the required system components in an economic and secure fashion. To that end, an optimization problem is presented wherein the sum of investment and operating costs is minimized along the planning horizon. The investment cost is related to the purchase and installation of distributed generation and distribution network assets such as branches, transformers, and substations. The operating cost consists of four terms related to maintenance, production, energy losses, and unserved energy. The maintenance cost represents the cost associated with keeping all system components in good condition through regular inspection and eventual repair if needed. The production cost is related to the energy purchased at substations and the energy produced by generators. The cost of energy losses comprises the energy lost as heat in branches and transformers due to the Joule effect. Finally, the unserved energy cost is a penalty for the demand not supplied under normal operation.

The objective function (1.1) to be minimized represents the present value of the total cost.

$$
\begin{aligned}
c^{TPV} = & \sum_{t \in \mathcal{T}} \frac{(1+I)^{-t}}{I} c_t^I + \sum_{t \in \mathcal{T}} \left[(1+I)^{-t} \left(c_t^M + c_t^E + c_t^R + c_t^U \right) \right] \\
& + \frac{(1+I)^{-n_T}}{I} \left(c_{n_T}^M + c_{n_T}^E + c_{n_T}^R + c_{n_T}^U \right).
\end{aligned}
\tag{1.1}
$$

As formulated in [23], the total cost consists of three terms. In the first term, the present worth value of the investment cost is represented under the assumption of a perpetual or infinite planning horizon [24]. The second term characterizes the present value of the sum of the operating costs including maintenance, production, energy losses, and unserved energy costs. Finally, in the third term, the present value of the sum of the operating costs incurred after the last time stage is modeled. Note that such a term depends on the values of those costs at the last time stage while also assuming a perpetual planning horizon. The cost terms in (1.1) are formulated as

$$
\begin{aligned}
c_t^I = & \sum_{l \in \{NRB, NAB\}} RR^l \sum_{k \in K^l} \sum_{(i,j) \in \Upsilon^l} C_k^{I,l} \ell_{ij} x_{ijkt}^l + RR^{SS} \sum_{i \in \Psi^{SS}} C_i^{I,SS} x_{it}^{SS} + RR^{NT} \sum_{k \in K^{NT}} \sum_{i \in \Psi^{SS}} C_k^{I,NT} x_{ikt}^{NT} \\
& + \sum_{p \in \mathcal{P}} RR^p \sum_{k \in K^p} \sum_{i \in \Psi^p} C_k^{I,p} pf \overline{G}_k^p x_{ikt}^p; \quad \forall t \in \mathcal{T}
\end{aligned}
\tag{1.2}
$$

$$
\begin{aligned}
c_t^M = & \sum_{l \in \mathcal{L}} \sum_{k \in K^l} \sum_{(i,j) \in \Upsilon^l} C_k^{M,l} \left(y_{ijkt}^l + y_{jikt}^l \right) + \sum_{tr \in TR} \sum_{k \in K^{tr}} \sum_{i \in \Psi^{SS}} C_k^{M,tr} y_{ikt}^{tr} \\
& + \sum_{p \in \mathcal{P}} \sum_{k \in K^p} \sum_{i \in \Psi^p} C_k^{M,p} y_{ikt}^p; \quad \forall t \in \mathcal{T}
\end{aligned}
\tag{1.3}
$$

$$
c_t^E = \sum_{b \in \mathcal{B}} \Delta_b pf \left(\sum_{tr \in TR} \sum_{k \in K^{tr}} \sum_{i \in \Psi^{SS}} C_b^{SS} g_{iktb}^{tr} + \sum_{p \in \mathcal{P}} \sum_{k \in K^p} \sum_{i \in \Psi^p} C_k^{E,p} g_{iktb}^p \right); \quad \forall t \in \mathcal{T} \tag{1.4}
$$

$$c_t^R = \sum_{b\in\mathcal{B}} \Delta_b C_b^{SS} pf \left(\sum_{tr\in TR} \sum_{k\in K^{tr}} \sum_{i\in\Psi^{SS}} Z_k^{tr} \left(g_{iktb}^{tr}\right)^2 + \sum_{l\in\mathcal{L}} \sum_{k\in K^l} \sum_{(i,j)\in\Upsilon^l} Z_k^l \ell_{ij} \left(f_{ijktb}^l + f_{jiktb}^l \right)^2 \right); \quad \forall t\in\mathcal{T} \tag{1.5}$$

$$c_t^U = \sum_{b\in\mathcal{B}} \sum_{i\in\Psi_t^{LN}} \Delta_b C^U pf\, d_{itb}^U; \quad \forall t\in\mathcal{T} \tag{1.6}$$

where capital recovery rates are computed as $RR^l = \frac{I(1+I)^{\eta^l}}{(1+I)^{\eta^l}-1}, \forall l \in \{NRB, NAB\}$; $RR^{NT} = \frac{I(1+I)^{\eta^{NT}}}{(1+I)^{\eta^{NT}}-1}$; $RR^p = \frac{I(1+I)^{\eta^p}}{(1+I)^{\eta^p}-1}, \forall p \in \mathcal{P}$; and $RR^{SS} = \frac{I(1+I)^{\eta^{SS}}}{(1+I)^{\eta^{SS}}-1}$.

It is worth pointing out that, for each time stage, one set of binary variables per branch, x_{ijkt}^l, is used to model the associated investment decisions. In contrast, two sets of binary variables, y_{ijkt}^l and y_{jikt}^l, as well as two sets of continuous variables, f_{ijktb}^l and f_{jiktb}^l, are associated with each branch in order to respectively model the direction and magnitude of the corresponding current flow. Note that f_{ijktb}^l is positive and equal to the branch current flow between nodes i and j measured at node i only when the current flows from i to j, being 0 otherwise.

Expressions (1.2) represent the amortized investment cost at each stage, which is formulated as the sum of the costs associated with the replacement and addition of branches, the reinforcement and construction of substations, the installation of new transformers, and the installation of DG. Expressions (1.3) model the maintenance costs of existing and newly added branches, transformers, and generators. In (1.4), the production costs associated with substations and generators are characterized. Expressions (1.5) represent the cost of energy losses in branches and transformers, which, as done in [23], are modeled as quadratic terms. Such nonlinearities can be accurately approximated by a set of tangent lines, as explained in Sect. 1.2.5. This approximation yields piecewise linear functions, which, for practical purposes, are indistinguishable from nonlinear models if enough segments are used. Finally, expressions (1.6) correspond to the penalty cost imposed for the unserved energy.

1.2.2 Kirchhoff's Laws and Operational Limits

System operation is an important factor in expansion planning models since investment and operating decisions have a great impact on each other. The constraints associated with system operation are formulated as

$$\underline{V} \leq v_{itb} \leq \overline{V}; \quad \forall i \in \Psi^N, \forall t \in \mathcal{T}, \forall b \in \mathcal{B} \tag{1.7}$$

$$0 \leq f^l_{ijktb} \leq y^l_{ijkt} \overline{F}^l_k; \quad \forall l \in \mathcal{L}, \forall i \in \Psi^l_j, \forall j \in \Psi^N, \forall k \in K^l, \forall t \in \mathcal{T}, \forall b \in \mathcal{B} \tag{1.8}$$

$$0 \leq g^{tr}_{iktb} \leq y^{tr}_{ikt} \overline{G}^{tr}_k; \quad \forall tr \in TR, \forall i \in \Psi^{SS}, \forall k \in K^{tr}, \forall t \in \mathcal{T}, \forall b \in \mathcal{B} \tag{1.9}$$

$$0 \leq d^U_{itb} \leq \mu^D_b D_{it}; \quad \forall i \in \Psi^{LN}_t, \forall t \in \mathcal{T}, \forall b \in \mathcal{B} \tag{1.10}$$

$$0 \leq g^C_{iktb} \leq y^C_{ikt} \overline{G}^C_k; \quad \forall i \in \Psi^C, \forall k \in K^C, \forall t \in \mathcal{T}, \forall b \in \mathcal{B} \tag{1.11}$$

$$0 \leq g^p_{iktb} \leq y^p_{ikt} \hat{G}^p_{ikb}; \quad \forall p \in \{W, \Theta\}, \forall i \in \Psi^p, \forall k \in K^p, \forall t \in \mathcal{T}, \forall b \in \mathcal{B} \tag{1.12}$$

$$\sum_{p \in \mathcal{P}} \sum_{k \in K^p} \sum_{i \in \Psi^p} g^p_{iktb} \leq \xi \sum_{i \in \Psi^{LN}_t} \mu^D_b D_{it}; \quad \forall t \in \mathcal{T}, \forall b \in \mathcal{B} \tag{1.13}$$

$$\begin{aligned} &\sum_{l \in \mathcal{L}} \sum_{k \in K^l} \sum_{(i,j) \in \Upsilon^l} \left[f^l_{ijktb} - f^l_{jiktb} \right] \\ &= \sum_{tr \in TR} \sum_{k \in K^{tr}} g^{tr}_{iktb} + \sum_{p \in \mathcal{P}} \sum_{k \in K^p} g^p_{iktb} - \mu^D_b D_{it} + d^U_{itb}; \quad \forall i \in \Psi^N, \forall t \in \mathcal{T}, \forall b \in \mathcal{B} \end{aligned} \tag{1.14}$$

$$\begin{aligned} &y^l_{ijkt} \left[Z^l_k \ell_{ij} f^l_{ijktb} - \left(v_{itb} - v_{jtb} \right) \right] = 0; \\ &\forall l \in \mathcal{L}, \forall i \in \Psi^l_j, \forall j \in \Psi^N, \forall k \in K^l, \forall t \in \mathcal{T}, \forall b \in \mathcal{B}. \end{aligned} \tag{1.15}$$

Constraints (1.7) limit nodal voltage magnitudes by setting upper and lower bounds. Similarly, constraints (1.8) set the bounds on branch current flows. Note that, if a branch is not used, i.e., $y^l_{ijkt} = 0$, the corresponding current flow is 0. Analogously, current injection levels of transformers are restricted in constraints (1.9), whereby, if a transformer is not used, i.e., $y^{tr}_{ikt} = 0$, the corresponding current injection is 0. Constraints (1.10) model the variables associated with nodal unserved energy as continuous and nonnegative, with maximum levels equal to the corresponding level of nodal demand. Production limits for conventional and renewable-based generation are formulated in constraints (1.11) and (1.12), respectively. Note that, as done for branches and transformers, the maximum level of generation is imposed through the use of binary utilization variables. Moreover, the upper bound for the production of each conventional generator is the corresponding rated generation capacity whereas the upper generation bound for each renewable generator is the available power associated with the corresponding generation technology. Constraints (1.13) limit the level of penetration of DG to a fraction, ξ, of the demand.

Finally, the effect of the distribution network is characterized through expressions (1.14) and (1.15), which represent Kirchhoff's laws using the linearized network model proposed in [22]. As described in [22], the linearized network model is an adapted version of the per-unit dc model used for the transmission network that is based on three assumptions: (1) nodal voltages lie in a narrow range around the rated value used as base voltage, (2) all per-unit branch current flows and nodal power injections have the same power factor, and (3) the per-unit voltage drop across a branch is equal to the difference between the per-unit magnitudes of the nodal voltages at both ends of the branch.

As per assumption (1), per-unit values of nodal apparent power injections and nodal current injections are identical. Assumption (2) allows representing per-unit branch current flows and nodal power injections, which are complex numbers, through their magnitudes. Hence, Kirchhoff's current law can be expressed as a set of linear scalar equalities in terms of per-unit magnitudes of branch current flows and nodal power injections, giving rise to nodal balance equations (1.14). In addition, assumption (3) allows formulating Kirchhoff's voltage law for each branch in use as a linear expression relating the per-unit magnitudes of branch current flows, nodal voltages, and branch impedances. Constraints (1.15) extend this result to account for the utilization state of all branches. Note that constraints (1.15) include nonlinearities involving the products of binary variables and continuous variables, for which a linear equivalent is formulated in Sect. 1.2.5.

1.2.3 Investment and Utilization Constraints

The investment and utilization decisions are constrained according to

$$x^{l}_{ijkt} \in \{0,1\}; \quad \forall l \in \{NRB, NAB\}, \forall (i,j) \in \Upsilon^{l}, \forall k \in K^{l}, \forall t \in \mathcal{T} \tag{1.16}$$

$$x^{SS}_{it} \in \{0,1\}; \quad \forall i \in \Psi^{SS}, \forall t \in \mathcal{T} \tag{1.17}$$

$$x^{NT}_{ikt} \in \{0,1\}; \quad \forall i \in \Psi^{SS}, \forall k \in K^{NT}, \forall t \in \mathcal{T} \tag{1.18}$$

$$x^{p}_{ikt} \in \{0,1\}; \quad \forall p \in \mathcal{P}, \forall i \in \Psi^{p}, \forall k \in K^{p}, \forall t \in \mathcal{T} \tag{1.19}$$

$$y^{l}_{ijkt} \in \{0,1\}; \quad \forall l \in \mathcal{L}, \forall i \in \Psi^{l}_{j}, \forall j \in \Psi^{N}, \forall k \in K^{l}, \forall t \in \mathcal{T} \tag{1.20}$$

$$y^{tr}_{ikt} \in \{0,1\}; \quad \forall tr \in TR, \forall i \in \Psi^{SS}, \forall k \in K^{tr}, \forall t \in \mathcal{T} \tag{1.21}$$

$$y^{p}_{ikt} \in \{0,1\}; \quad \forall p \in \mathcal{P}, \forall i \in \Psi^{p}, \forall k \in K^{p}, \forall t \in \mathcal{T} \tag{1.22}$$

$$\sum_{t \in \mathcal{T}} \sum_{k \in K^{l}} x^{l}_{ijkt} \leq 1; \quad \forall l \in \{NRB, NAB\}, \forall (i,j) \in \Upsilon^{l} \tag{1.23}$$

$$\sum_{t \in \mathcal{T}} x_{it}^{SS} \leq 1; \quad \forall i \in \Psi^{SS} \tag{1.24}$$

$$\sum_{t \in \mathcal{T}} \sum_{k \in K^{NT}} x_{ikt}^{NT} \leq 1; \quad \forall i \in \Psi^{SS} \tag{1.25}$$

$$\sum_{t \in \mathcal{T}} \sum_{k \in K^{p}} x_{ikt}^{p} \leq 1; \quad \forall p \in \mathcal{P}, \forall i \in \Psi^{p} \tag{1.26}$$

$$x_{ikt}^{NT} \leq \sum_{\tau=1}^{t} x_{i\tau}^{SS}; \quad \forall i \in \Psi^{SS}, \forall k \in K^{NT}, \forall t \in \mathcal{T} \tag{1.27}$$

$$y_{ijkt}^{EFB} + y_{jikt}^{EFB} \leq 1; \quad \forall (i,j) \in \Upsilon^{SW,EFB}, \forall k \in K^{EFB}, \forall t \in \mathcal{T} \tag{1.28}$$

$$y_{ijkt}^{l} + y_{jikt}^{l} \leq \sum_{\tau=1}^{t} x_{ijk\tau}^{l}; \quad \forall l \in \{NRB, NAB\}, \forall (i,j) \in \Upsilon^{SW,l}, \forall k \in K^{l}, \forall t \in \mathcal{T} \tag{1.29}$$

$$y_{ijkt}^{ERB} + y_{jikt}^{ERB} \leq 1 - \sum_{\tau=1}^{t} \sum_{\kappa \in K^{NRB}} x_{ij\kappa\tau}^{NRB}; \quad \forall (i,j) \in \Upsilon^{SW,ERB}, \forall k \in K^{ERB}, \forall t \in \mathcal{T} \tag{1.30}$$

$$y_{ijkt}^{EFB} + y_{jikt}^{EFB} = 1; \quad \forall (i,j) \in \left(\Upsilon^{EFB} \backslash \Upsilon^{SW,EFB}\right), \forall k \in K^{EFB}, \forall t \in \mathcal{T} \tag{1.31}$$

$$y_{ijkt}^{l} + y_{jikt}^{l} = \sum_{\tau=1}^{t} x_{ijk\tau}^{l}; \quad \forall l \in \{NRB, NAB\}, \forall (i,j) \in \left(\Upsilon^{l} \backslash \Upsilon^{SW,l}\right), \forall k \in K^{l}, \forall t \in \mathcal{T} \tag{1.32}$$

$$y_{ijkt}^{ERB} + y_{jikt}^{ERB} = 1 - \sum_{\tau=1}^{t} \sum_{\kappa \in K^{NRB}} x_{ij\kappa\tau}^{NRB}; \quad \forall (i,j) \in \left(\Upsilon^{ERB} \backslash \Upsilon^{SW,ERB}\right), \forall k \in K^{ERB}, \forall t \in \mathcal{T} \tag{1.33}$$

$$y_{ikt}^{NT} \leq \sum_{\tau=1}^{t} x_{ik\tau}^{NT}; \quad \forall i \in \Psi^{SS}, \forall k \in K^{NT}, \forall t \in \mathcal{T} \tag{1.34}$$

$$y_{ikt}^{p} \leq \sum_{\tau=1}^{t} x_{ik\tau}^{p}; \quad \forall p \in \mathcal{P}, \forall i \in \Psi^{p}, \forall k \in K^{p}, \forall t \in \mathcal{T} \tag{1.35}$$

$$\sum_{l \in \{NRB,NAB\}} \sum_{k \in K^l} \sum_{(i,j) \in \Upsilon^l} C_k^{I,l} \ell_{ij} x_{ijkt}^l$$
$$+ \sum_{i \in \Psi^{SS}} C_i^{I,SS} x_{it}^{SS} + \sum_{k \in K^{NT}} \sum_{i \in \Psi^{SS}} C_k^{I,NT} x_{ikt}^{NT} + \sum_{p \in \mathcal{P}} \sum_{k \in K^p} \sum_{i \in \Psi^p} C_k^{I,p} pf \bar{G}_k^p x_{ikt}^p \leq IB_t; \forall t \in \mathcal{T}. \tag{1.36}$$

Expressions (1.16)–(1.22) set the binary nature of investment and utilization variables. Constraints (1.23)–(1.27) are related to investment decisions. As per constraints (1.23), a maximum of one investment can be made in each branch along the planning horizon. Constraints (1.24) impose that investment at substation nodes can only be made once along the planning horizon. One new transformer at most can be installed at each substation node along the time span, as modeled in constraints (1.25). The installation of generators at each candidate node throughout the planning horizon is limited to one by constraints (1.26). Constraints (1.27) guarantee that new transformers can only be added in substations that have been previously expanded or built.

Expressions (1.28)–(1.35) are related to the utilization of existing and newly added components. Note that binary utilization variables are related to investment variables so that a system component cannot be used if it has not been previously installed. Constraints (1.28)–(1.30) model the utilization of switchable branches while explicitly characterizing the direction of current flows. Switching of those branches is considered under normal operation, thereby allowing for network reconfiguration. Analogously, constraints (1.31)–(1.33) are associated with the utilization of branches that are not reconfigurable under normal operation. Constraints (1.34) and (1.35) model the utilization of new transformers and distributed generators, respectively. Finally, constraints (1.36) set the budget for investments at each stage.

1.2.4 Radiality Constraints

The radial operation is modeled by

$$\sum_{l \in \mathcal{L}} \sum_{i \in \Psi_j^l} \sum_{k \in K^l} y_{ijkt}^l = 1; \quad \forall j \in \Psi_t^{LN}, \forall t \in \mathcal{T} \tag{1.37}$$

$$\sum_{l \in \mathcal{L}} \sum_{i \in \Psi_j^l} \sum_{k \in K^l} y_{ijkt}^l \leq 1; \quad \forall j \notin \Psi_t^{LN}, \forall t \in \mathcal{T} \tag{1.38}$$

$$\sum_{l \in \mathcal{L}} \sum_{k \in K^l} \sum_{j \in \Psi_i^l} \left(\tilde{f}_{ijkt}^l - \tilde{f}_{jikt}^l \right) = \tilde{g}_{it}^{SS} - \tilde{D}_{it}; \quad \forall i \in \Psi^N, \forall t \in \mathcal{T} \tag{1.39}$$

$$0 \leq \tilde{f}^{EFB}_{ijkt} \leq n_{DG}; \quad \forall i \in \Psi^{EFB}_j, \forall j \in \Psi^N, \forall k \in K^{EFB}, \forall t \in \mathcal{T} \tag{1.40}$$

$$0 \leq \tilde{f}^{ERB}_{ijkt} \leq n_{DG}\left(1 - \sum_{\tau=1}^{t} \sum_{\kappa \in K^{NRB}} x^{NRB}_{ij\kappa\tau}\right); \quad \forall (i,j) \in \Upsilon^{ERB}, \forall k \in K^{ERB}, \forall t \in \mathcal{T} \tag{1.41}$$

$$0 \leq \tilde{f}^{ERB}_{jikt} \leq n_{DG}\left(1 - \sum_{\tau=1}^{t} \sum_{\kappa \in K^{NRB}} x^{NRB}_{ij\kappa\tau}\right); \quad \forall (i,j) \in \Upsilon^{ERB}, \forall k \in K^{ERB}, \forall t \in \mathcal{T} \tag{1.42}$$

$$0 \leq \tilde{f}^{l}_{ijkt} \leq n_{DG} \sum_{\tau=1}^{t} x^{l}_{ij\kappa\tau}; \quad \forall l \in \{NRB, NAB\}, \forall (i,j) \in \Upsilon^{l}, \forall k \in K^{l}, \forall t \in \mathcal{T} \tag{1.43}$$

$$0 \leq \tilde{f}^{l}_{jikt} \leq n_{DG} \sum_{\tau=1}^{t} x^{l}_{ij\kappa\tau}; \quad \forall l \in \{NRB, NAB\}, \forall (i,j) \in \Upsilon^{l}, \forall k \in K^{l}, \forall t \in \mathcal{T} \tag{1.44}$$

$$0 \leq \tilde{g}^{SS}_{it} \leq n_{DG}; \quad \forall i \in \Psi^{SS}, \forall t \in \mathcal{T} \tag{1.45}$$

where

$$\tilde{D}_{it} = \begin{cases} 1; & \forall i \in \left(\left(\Psi^C \cup \Psi^W \cup \Psi^\Theta\right) \cap \Psi^{LN}_t\right), \forall t \in \mathcal{T} \\ 0; & \forall i \notin \left(\left(\Psi^C \cup \Psi^W \cup \Psi^\Theta\right) \cap \Psi^{LN}_t\right), \forall t \in \mathcal{T}. \end{cases} \tag{1.46}$$

Radiality is modeled through traditional constraints (1.37) and (1.38) [23, 25] in conjunction with constraints (1.39)–(1.45), which were recently presented in [26]. Constraints (1.37) impose load nodes to have a single incoming flow while expressions (1.38) set a maximum of one incoming flow for the remaining nodes. The new radiality constraints (1.39)–(1.45) avoid the issues with transfer nodes and islanded DG that would arise should traditional radiality constraints be solely used in the co-optimized expansion planning problem. The idea behind these new radiality constraints is to set a fictitious demand in those load nodes that could be islanded due to the installation of DG units. Fictitious nodal demands can only be supplied by fictitious substations located at the original substation nodes, where fictitious energy flowing through the branches of the system is injected. As a consequence, the islanding of areas with load demand is prevented under normal operation. Constraints (1.39) represent the nodal fictitious current balance equations. Constraints (1.40)–(1.44) bound the fictitious flows through branches. Finally, constraints (1.45) set the limits for the fictitious currents injected by the fictitious substations.

1.2.5 Mixed-Integer Linear Formulation

The deterministic model for the joint expansion planning of DG and distribution network assets is a mixed-integer nonlinear program for which no exact method is currently available. This issue motivates us to recast the original problem as an instance of mixed-integer linear programming by replacing nonlinear expressions (1.5) and (1.15) with linear terms. As a consequence, finite convergence to the optimum is guaranteed while providing a measure of the distance to optimality along the solution process. Moreover, efficient off-the-shelf software based on the state-of-the-art branch-and-cut algorithm is available.

Based on [27], a piecewise linear approximation is used for the quadratic terms in (1.5). Thus, expressions (1.5) are replaced with

$$\begin{aligned} c_t^R = &\sum_{b\in\mathcal{B}} \Delta_b C_b^{SS} pf \Bigg(\sum_{tr\in TR} \sum_{k\in K^{tr}} \sum_{i\in\Psi^{SS}} \sum_{h=1}^{n_H} M_{kh}^{tr} \delta_{iktbh}^{tr} \\ &+ \sum_{l\in\mathcal{L}} \sum_{k\in K^l} \sum_{(i,j)\in\Upsilon^l} \sum_{h=1}^{n_H} M_{kh}^{l} \ell_{ij} \left(\delta_{ijktbh}^{l} + \delta_{jiktbh}^{l} \right) \Bigg); \quad \forall t\in\mathcal{T} \end{aligned} \tag{1.47}$$

$$g_{iktb}^{tr} = \sum_{h=1}^{n_H} \delta_{iktbh}^{tr}; \quad \forall tr\in TR, \forall i\in\Psi^{SS}, \forall k\in K^{tr}, \forall t\in\mathcal{T}, \forall b\in\mathcal{B} \tag{1.48}$$

$$0\le \delta_{iktbh}^{tr} \le A_{kh}^{tr}; \quad \forall h=1\ldots n_H, \forall tr\in TR, \forall i\in\Psi^{SS}, \forall k\in K^{tr}, \forall t\in\mathcal{T}, \forall b\in\mathcal{B} \tag{1.49}$$

$$f_{ijktb}^{l} = \sum_{h=1}^{n_H} \delta_{ijktbh}^{l}; \quad \forall l\in\mathcal{L}, \forall i\in\Psi_j^l, \forall j\in\Psi^N, \forall k\in K^l, \forall t\in\mathcal{T}, \forall b\in\mathcal{B} \tag{1.50}$$

$$\begin{aligned} &0\le \delta_{ijktbh}^{l} \le A_{kh}^{l}; \forall h=1\ldots n_H, \\ &\forall l\in\mathcal{L}, \forall i\in\Psi_j^l, \forall j\in\Psi^N, \forall k\in K^l, \forall t\in\mathcal{T}, \forall b\in\mathcal{B} \end{aligned} \tag{1.51}$$

where expressions (1.47) are the linearized costs of energy losses, while (1.48)–(1.49) and (1.50)–(1.51) are related to the linearization of energy losses in transformers and branches, respectively.

In addition, using the disjunctive-constraint-based transformation described in [28], nonlinear expressions (1.15) have the linear equivalent of

$$\begin{aligned} &-J\left(1-y_{ijkt}^{l}\right) \le Z_k^l \ell_{ij} f_{ijktb}^{l} - \left(v_{itb} - v_{jtb}\right) \le J\left(1-y_{ijkt}^{l}\right); \\ &\forall l\in\mathcal{L}, \forall i\in\Psi_j^l, \forall j\in\Psi^N, \forall k\in K^l, \forall t\in\mathcal{T}, \forall b\in\mathcal{B}. \end{aligned} \tag{1.52}$$

If y^l_{ijkt} is equal to 1, the corresponding constraint (1.52) becomes $0 \le Z^l_k \ell_{ij} f^l_{ijktb} - \left(v_{itb} - v_{jtb}\right) \le 0$, which is identical to the condition $Z^l_k \ell_{ij} f^l_{ijktb} - \left(v_{itb} - v_{jtb}\right) = 0$ resulting from (1.15). Conversely, if y^l_{ijkt} is equal to 0, the corresponding constraint (1.52) yields $-J \le -\left(v_{itb} - v_{jtb}\right) \le J$, i.e., $\left|v_{itb} - v_{jtb}\right| \le J$, since f^l_{ijktb} is equal to 0 as per (1.8). Thus, for a sufficiently large positive value for parameter J, no relation between nodal voltage magnitudes v_{itb} and v_{jtb} is imposed, as modeled in (1.15) for y^l_{ijkt} equal to 0. Since nodal voltage magnitudes are bounded by $\overline{V}$ and $\underline{V}$ in (1.7), the largest possible value for $\left|v_{itb} - v_{jtb}\right|$ is $\overline{V} - \underline{V}$, which is thus the minimum value for J.

Thus, the resulting mixed-integer linear program is formulated as

$$\begin{aligned} \underset{\Xi^{DT}}{\text{Minimize}} \quad c^{TPV} = & \sum_{t \in \mathcal{T}} \frac{(1+I)^{-t}}{I} c^I_t + \sum_{t \in \mathcal{T}} \left[(1+I)^{-t}\left(c^M_t + c^E_t + c^R_t + c^U_t\right)\right] \\ & + \frac{(1+I)^{-n_T}}{I}\left(c^M_{n_T} + c^E_{n_T} + c^R_{n_T} + c^U_{n_T}\right) \end{aligned} \tag{1.53}$$

subject to

Constraints (1.2)–(1.4), (1.6)–(1.14), (1.16)–(1.45), and (1.47)–(1.52) (1.54)

where

$$\begin{aligned} \Xi^{DT} = \Big\{ & c^E_t, c^I_t, c^M_t, c^R_t, c^U_t, c^{TPV}, d^U_{itb}, f^l_{ijktb}, \tilde{f}^l_{ijkt}, g^p_{iktb}, g^{tr}_{iktb}, \tilde{g}^{SS}_{it}, v_{itb}, \\ & x^l_{ijkt}, x^{NT}_{ikt}, x^p_{ikt}, x^{SS}_{it}, y^l_{ijkt}, y^p_{ikt}, y^{tr}_{ikt}, \delta^l_{ijktbh}, \delta^{tr}_{iktbh} \Big\} \end{aligned}$$

1.3 Stochastic Programming Model

In this section, based on the previous deterministic model, the uncertainty related to renewable-based generation and demand is incorporated using a scenario-based stochastic programming framework [29]. To that end, uncertainty is characterized through a set of scenarios that explicitly capture the correlation between uncertainty sources.

Next, the procedure for generating scenarios of demand, wind speed, and solar irradiation is presented. The objective function and constraints of the stochastic model are subsequently described in detail. This section concludes with the formulation of the resulting mixed-integer linear program.

1.3.1 Uncertainty Modeling

For investment planning, the increasing penetration of stochastic resources in electric distribution networks requires the accurate modeling of the associated uncertainty. Based on the methodology described in [30], a set of scenarios representing the uncertainty associated with demand, wind speed, and solar irradiation is generated using historical data. The scenario-generation procedure comprises six steps that are described as follows:

Step (1) Historical hourly data of system demand, wind speed, and solar irradiation throughout a year are expressed as per-unit factors by dividing each data by the corresponding maximum level. Hence, each set of factors represents the per-unit demand, wind speed, and solar irradiation profile.
Step (2) Triplets of hourly factors for demand, wind speed, and solar irradiation are sorted by demand factor in descending order. Figure 1.2 shows an ordered demand factor curve and the corresponding profiles of wind speed and solar irradiation factors.
Step (3) The factor curves resulting from step (2) are discretized into n_B time blocks. In order to accurately model the peak demand, which usually has a big influence on investment decisions, a relatively small time block related to such peak demand is defined. For each time block, the corresponding wind speed and solar irradiation factors are sorted in descending order. An example with a 4-time-block discretization is depicted in Fig. 1.3.
Step (4) For each time block, the cumulative distribution functions (cdf) of the ordered demand, wind speed, and solar irradiation factors are built. As an example, Fig. 1.4 the cdf corresponding to the curves depicted in Fig. 1.3.
Step (5) The cdfs are divided into segments with their corresponding probabilities. The pre-specified numbers of segments are denoted by n_S^D, n_S^W, and n_S^Θ for demand, wind speed, and solar irradiation factor curves, respectively. In addition to its pre-specified probability, each segment s is characterized by an average factor equal to the average value of the factors within such a segment. As a consequence, pairs probability-average factor are generated for demand, namely $\pi_{sb}^D - \mu_{sb}^D$, for wind speed, namely $\pi_{sb}^W - \mu_{sb}^W$, and for solar irradiation, namely $\pi_{sb}^\Theta - \mu_{sb}^\Theta$. In Fig. 1.5, the cdf of the first time block for the demand factor curve is divided into three segments with probabilities, π_{11}^D, π_{21}^D, and π_{31}^D, equal to 0.4, 0.5, and 0.1, respectively. The demand factors associated with those segments respectively lie in the ranges [0.00, 0.70], (0.70, 0.86], and (0.86, 1.00]. As can be seen in Fig. 1.5, the corresponding average demand factors, μ_{11}^D, μ_{21}^D, and μ_{31}^D, are equal to 0.67, 0.76, and 0.90, respectively.

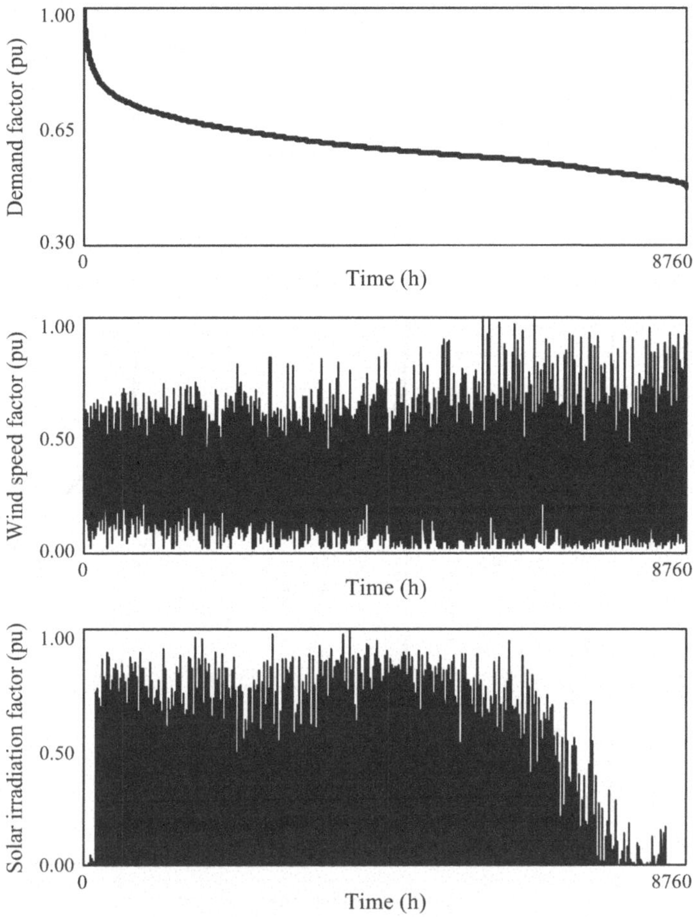

Fig. 1.2 Ordered demand factor curve and the corresponding profiles of wind speed and solar irradiation factors

Step (6) Scenarios, indexed by ω, for each time block result from combining all pairs $\pi_{sb}^{D}-\mu_{sb}^{D}$, $\pi_{sb}^{W}-\mu_{sb}^{W}$, and $\pi_{sb}^{\Theta}-\mu_{sb}^{\Theta}$. Hence, nodal demands in each scenario are equal to the product of the forecasted values and the corresponding factor $\mu_{b}^{D}(\omega)$. For each scenario, average factors μ_{sb}^{W} and μ_{sb}^{Θ} are converted to wind speed and solar irradiation levels so that the maximum levels of wind and photovoltaic power generation, $\hat{G}_{ikb}^{W}(\omega)$ and $\hat{G}_{ikb}^{\Theta}(\omega)$, are determined. Thus, for each time block b,

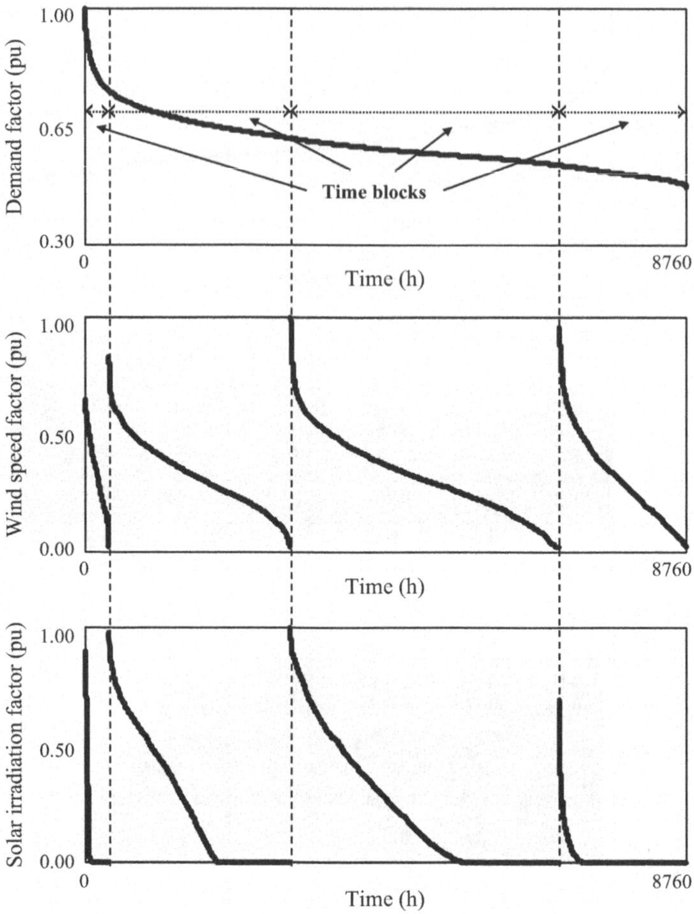

Fig. 1.3 Time block discretization of demand, wind speed, and solar irradiation factors

scenario ω comprises an average demand factor, $\mu_b^D(\omega)$, a vector of maximum levels of wind power generation, $\hat{G}_{ikb}^W(\omega)$, and a vector of maximum levels of photovoltaic power generation, $\hat{G}_{ikb}^\Theta(\omega)$. Mathematically, the set of scenarios Ω_b is formulated as

$$\Omega_b = \left\{ \mu_b^D(\omega), \left\{ \hat{G}_{ikb}^W(\omega) \right\}_{\forall i \in \Psi^W, \forall k \in K^W}, \left\{ \hat{G}_{ikb}^\Theta(\omega) \right\}_{\forall i \in \Psi^\Theta, \forall k \in K^\Theta} \right\}_{\forall \omega = 1 \ldots n_\Omega}; \quad \forall b \in \mathcal{B}. \tag{1.55}$$

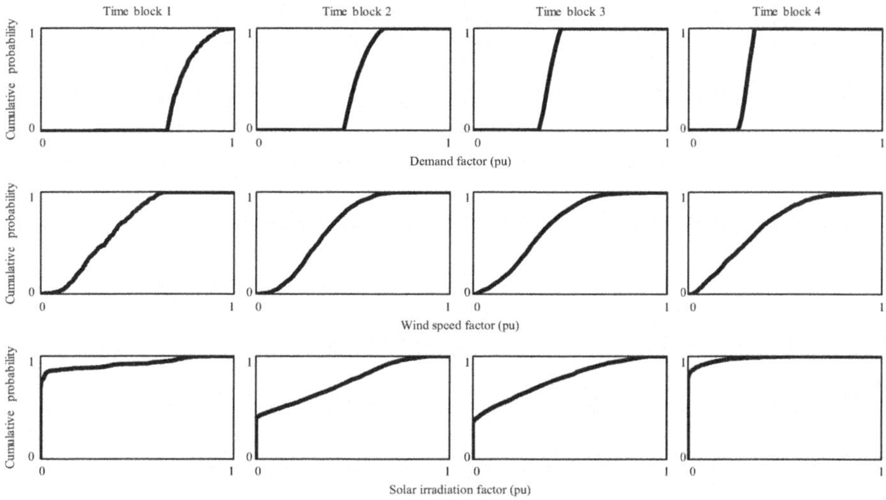

Fig. 1.4 Cumulative distribution functions of demand, wind speed, and solar irradiation factors

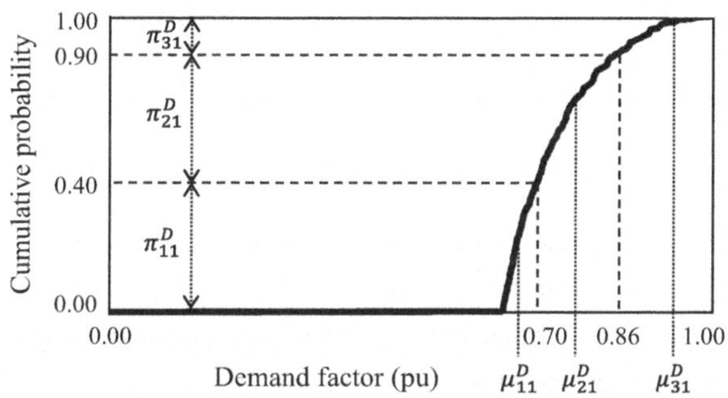

Fig. 1.5 Cumulative distribution function for the first block of the demand factor curve

In addition, the probability of each scenario, $\pi_b(\omega)$, is equal to the corresponding product $\pi_{sb}^D \pi_{sb}^W \pi_{sb}^\Theta$. The number of scenarios per time block, n_Ω, is equal to $n_S^D n_S^W n_S^\Theta$, whereas the number of scenarios or operating conditions is equal to $n_B n_\Omega$.

It should be noted that if data for more than one year were available, the scenario-generation procedure could be applied for each year in order to create different operating conditions at each stage.

1.3.2 Objective Function and Cost-Related Terms

The objective function to be minimized is the net present value of the expected total cost, which can be also represented through expression (1.1) described for the deterministic model. Expected investment and maintenance costs are modeled as in (1.2) and (1.3), respectively. The other cost terms, i.e., the expected costs of production, energy losses, and unserved energy, are modeled as

$$\begin{aligned} c_t^E = \sum_{b \in \mathcal{B}} \sum_{\omega=1}^{n_\Omega} \pi_b(\omega) \Delta_b pf \Bigg(\sum_{tr \in TR} \sum_{k \in K^{tr}} \sum_{i \in \Psi^{SS}} C_b^{SS} g_{iktb}^{tr}(\omega) \\ + \sum_{p \in \mathcal{P}} \sum_{k \in K^p} \sum_{i \in \Psi^p} C_k^{E,p} g_{iktb}^{p}(\omega) \Bigg); \quad \forall t \in \mathcal{T} \end{aligned} \tag{1.56}$$

$$\begin{aligned} c_t^R = \sum_{b \in \mathcal{B}} \sum_{\omega=1}^{n_\Omega} \pi_b(\omega) \Delta_b C_b^{SS} pf \Bigg(\sum_{tr \in TR} \sum_{k \in K^{tr}} \sum_{i \in \Psi^{SS}} Z_k^{tr} \left(g_{iktb}^{tr}(\omega) \right)^2 \\ + \sum_{l \in \mathcal{L}} \sum_{k \in K^l} \sum_{(i,j) \in \Upsilon^l} Z_k^l \ell_{ij} \left(f_{ijktb}^l(\omega) + f_{jiktb}^l(\omega) \right)^2 \Bigg); \quad \forall t \in \mathcal{T} \end{aligned} \tag{1.57}$$

$$c_t^U = \sum_{b \in \mathcal{B}} \sum_{\omega=1}^{n_\Omega} \sum_{i \in \Psi_t^{LN}} \pi_b(\omega) \Delta_b C^U pf \, d_{itb}^U(\omega); \quad \forall t \in \mathcal{T}. \tag{1.58}$$

Similar to the deterministic model, it is worth pointing out that, for each time stage, one set of binary variables per branch, x_{ijkt}^l, is used to model the associated investment decisions. In contrast, two sets of binary variables, y_{ijkt}^l and y_{jikt}^l, as well as two sets of continuous variables, $f_{ijktb}^l(\omega)$ and $f_{jiktb}^l(\omega)$, are associated with each branch in order to respectively model the direction and magnitude of the corresponding current flow. Note that $f_{ijktb}^l(\omega)$ is positive and equal to the branch current flow between nodes i and j measured at node i only when the current flows from i to j, being 0 otherwise.

The three expected costs modeled in (1.56)–(1.58) respectively are the stochastic counterparts of the costs formulated in (1.4)–(1.6) for the deterministic model. Expressions (1.56)–(1.58) mainly differ from (1.4)–(1.6) in the consideration of scenarios and their probabilities in order to model uncertainty.

1.3.3 Kirchhoff's Laws and Operational Limits

For the stochastic model, the constraints associated with system operation are formulated as

$$\underline{V} \leq v_{itb}(\omega) \leq \overline{V}; \quad \forall i \in \Psi^N, \forall t \in \mathcal{T}, \forall b \in \mathcal{B}, \forall \omega = 1 \ldots n_\Omega \tag{1.59}$$

$$\begin{aligned} &0 \leq f^l_{ijktb}(\omega) \leq y^l_{ijkt} \bar{F}^l_k; \\ &\forall l \in \mathcal{L}, \forall i \in \Psi^l_j, \forall j \in \Psi^N, \forall k \in K^l, \forall t \in \mathcal{T}, \forall b \in \mathcal{B}, \forall \omega = 1 \ldots n_\Omega \end{aligned} \tag{1.60}$$

$$\begin{aligned} &0 \leq g^{tr}_{iktb}(\omega) \leq y^{tr}_{ikt} \bar{G}^{tr}_k; \\ &\forall tr \in TR, \forall i \in \Psi^{SS}, \forall k \in K^{tr}, \forall t \in \mathcal{T}, \forall b \in \mathcal{B}, \forall \omega = 1 \ldots n_\Omega \end{aligned} \tag{1.61}$$

$$0 \leq d^U_{itb}(\omega) \leq \mu^D_b(\omega) D_{it}; \quad \forall i \in \Psi^{LN}_t, \forall t \in \mathcal{T}, \forall b \in \mathcal{B}, \forall \omega = 1 \ldots n_\Omega \tag{1.62}$$

$$0 \leq g^C_{iktb}(\omega) \leq y^C_{ikt} \overline{G}^C_k; \quad \forall i \in \Psi^C, \forall k \in K^C, \forall t \in \mathcal{T}, \forall b \in \mathcal{B}, \forall \omega = 1 \ldots n_\Omega \tag{1.63}$$

$$\begin{aligned} &0 \leq g^p_{iktb}(\omega) \leq y^p_{ikt} \hat{G}^p_{ikb}(\omega); \\ &\forall p \in \{W, \Theta\}, \forall i \in \Psi^p, \forall k \in K^p, \forall t \in \mathcal{T}, \forall b \in \mathcal{B}, \forall \omega = 1 \ldots n_\Omega \end{aligned} \tag{1.64}$$

$$\sum_{p \in \mathcal{P}} \sum_{k \in K^p} \sum_{i \in \Psi^p} g^p_{iktb}(\omega) \leq \xi \sum_{i \in \Psi^{LN}_t} \mu^D_b(\omega) D_{it}; \quad \forall t \in \mathcal{T}, \forall b \in \mathcal{B}, \forall \omega = 1 \ldots n_\Omega \tag{1.65}$$

$$\begin{aligned} &\sum_{l \in \mathcal{L}} \sum_{k \in K^l} \sum_{(i,j) \in \Upsilon^l} \left[f^l_{ijktb}(\omega) - f^l_{jiktb}(\omega) \right] \\ &= \sum_{tr \in TR} \sum_{k \in K^{tr}} g^{tr}_{iktb}(\omega) + \sum_{p \in \mathcal{P}} \sum_{k \in K^p} g^p_{iktb}(\omega) \\ &- \mu^D_b(\omega) D_{it} + d^U_{itb}(\omega); \forall i \in \Psi^N, \forall t \in \mathcal{T}, \forall b \in \mathcal{B}, \forall \omega = 1 \ldots n_\Omega \end{aligned} \tag{1.66}$$

$$\begin{aligned} &y^l_{ijkt} \left[Z^l_k \ell_{ij} f^l_{ijktb}(\omega) - \left(v_{itb}(\omega) - v_{jtb}(\omega) \right) \right] = 0; \\ &\forall l \in \mathcal{L}, \forall i \in \Psi^l_j, \forall j \in \Psi^N, \forall k \in K^l, \forall t \in \mathcal{T}, \forall b \in \mathcal{B}, \forall \omega = 1 \ldots n_\Omega. \end{aligned} \tag{1.67}$$

It is worth mentioning that expressions (1.59)–(1.67) correspond to the adaptation of expressions (1.7)–(1.15) in the deterministic model, respectively, to the scenario-based setting. Thus, for the description for (1.59)–(1.67), the interested reader is referred to Sect. 1.2.2.

1.3.4 Investment and Utilization Constraints

For the stochastic model, investment- and utilization-related decisions for distribution network assets and distributed generators along with the budgetary limit for investments at each stage are modeled as in (1.16)–(1.36), which are described in Sect. 1.2.3.

1.3.5 Radiality Constraints

Radiality constraints for the stochastic model are identical to those described for the deterministic model in Sect. 1.2.4, namely expressions (1.37)–(1.46).

1.3.6 Mixed-Integer Linear Formulation

The stochastic model is formulated as a mixed-integer nonlinear program, where nonlinearities are related to the quadratic energy losses in (1.57) and to the bilinear terms in (1.67) involving the product of a binary variable and a continuous variable.

As done for the deterministic model, the quadratic energy losses in (1.57) are recast as piecewise linear expressions as [27]

$$
\begin{aligned}
c_t^R = & \sum_{b\in\mathcal{B}} \sum_{\omega=1}^{n_\Omega} \pi_b(\omega)\Delta_b C_b^{SS} pf \left(\sum_{tr\in TR} \sum_{k\in K^{tr}} \sum_{i\in\Psi^{SS}} \sum_{h=1}^{n_H} M_{kh}^{tr} \delta_{iktbh}^{tr}(\omega) \right. \\
& \left. + \sum_{l\in\mathcal{L}} \sum_{k\in K^l} \sum_{(i,j)\in\Upsilon^l} \sum_{h=1}^{n_H} M_{kh}^{l} \ell_{ij} \left(\delta_{ijktbh}^{l}(\omega) + \delta_{jiktbh}^{l}(\omega) \right) \right); \quad \forall t \in \mathcal{T}
\end{aligned}
\tag{1.68}
$$

$$
\begin{aligned}
& g_{iktb}^{tr}(\omega) = \sum_{h=1}^{n_H} \delta_{iktbh}^{tr}(\omega); \\
& \forall tr \in TR, \forall i \in \Psi^{SS}, \forall k \in K^{tr}, \forall t \in \mathcal{T}, \forall b \in \mathcal{B}, \forall \omega = 1 \ldots n_\Omega
\end{aligned}
\tag{1.69}
$$

$$
\begin{aligned}
& 0 \le \delta_{iktbh}^{tr}(\omega) \le A_{kh}^{tr}; \\
& \forall h = 1 \ldots n_H, \forall tr \in TR, \forall i \in \Psi^{SS}, \forall k \in K^{tr}, \forall t \in \mathcal{T}, \forall b \in \mathcal{B}, \forall \omega = 1 \ldots n_\Omega
\end{aligned}
\tag{1.70}
$$

$$
\begin{aligned}
& f_{ijktb}^{l}(\omega) = \sum_{h=1}^{n_H} \delta_{ijktbh}^{l}(\omega); \\
& \forall l \in \mathcal{L}, \forall i \in \Psi_j^l, \forall j \in \Psi^N, \forall k \in K^l, \forall t \in \mathcal{T}, \forall b \in \mathcal{B}, \forall \omega = 1 \ldots n_\Omega
\end{aligned}
\tag{1.71}
$$

$$0 \leq \delta^{l}_{ijktbh}(\omega) \leq A^{l}_{kh};$$
$$\forall h = 1 \ldots n_H, \forall l \in \mathcal{L}, \forall i \in \Psi^{l}_{j}, \forall j \in \Psi^{N}, \forall k \in K^{l}, \forall t \in \mathcal{T}, \forall b \in \mathcal{B}, \forall \omega = 1 \ldots n_{\Omega}. \tag{1.72}$$

Using the disjunctive-constraint-based transformation described in [28], nonlinear expressions (1.67) have the linear equivalent of

$$\begin{aligned} & -J\left(1 - y^{l}_{ijkt}\right) \leq Z^{l}_{k} \ell_{ij} f^{l}_{ijktb}(\omega) - \left[v_{itb}(\omega) - v_{jtb}(\omega)\right] \leq J\left(1 - y^{l}_{ijkt}\right); \\ & \quad \forall l \in \mathcal{L}, \forall i \in \Psi^{l}_{j}, \forall j \in \Psi^{N}, \forall k \in K^{l}, \forall t \in \mathcal{T}, \forall b \in \mathcal{B}, \forall \omega = 1 \ldots n_{\Omega}. \end{aligned} \tag{1.73}$$

Note that expressions (1.68)–(1.73) respectively represent the stochastic counterparts of expressions (1.47)–(1.52) in the deterministic model.

Thus, the deterministic equivalent associated with the stochastic model can be formulated as a mixed-integer linear program suitable for commercially available software as

$$\begin{aligned} \underset{\Xi^{ST}}{\text{Minimize}} \quad c^{TPV} = & \sum_{t \in \mathcal{T}} \frac{(1+I)^{-t}}{I} c^{I}_{t} + \sum_{t \in \mathcal{T}} \left[(1+I)^{-t}\left(c^{M}_{t} + c^{E}_{t} + c^{R}_{t} + c^{U}_{t}\right)\right] \\ & + \frac{(1+I)^{-n_T}}{I}\left(c^{M}_{n_T} + c^{E}_{n_T} + c^{R}_{n_T} + c^{U}_{n_T}\right) \end{aligned} \tag{1.74}$$

subject to:

Constraints (1.2), (1.3), (1.16)–(1.45), (1.56), (1.58)–(1.66), and (1.68)–(1.73) (1. 75)

where

$$\begin{aligned} \Xi^{ST} = \Big\{ & c^{E}_{t}, c^{I}_{t}, c^{M}_{t}, c^{R}_{t}, c^{U}_{t}, c^{TPV}, d^{U}_{itb}(\omega), f^{l}_{ijktb}(\omega), \tilde{f}^{l}_{ijkt}, g^{P}_{iktb}(\omega), \\ & g^{tr}_{iktb}(\omega), \tilde{g}^{SS}_{it}, v_{itb}(\omega), x^{l}_{ijkt}, x^{NT}_{ikt}, x^{P}_{ikt}, x^{SS}_{it}, y^{l}_{ijkt}, y^{P}_{ikt}, y^{tr}_{ikt}, \delta^{l}_{ijktbh}(\omega), \delta^{tr}_{iktbh}(\omega) \Big\}. \end{aligned}$$

1.4 Numerical Results

Both models have been applied to a distribution system based on the benchmark presented in [31]. As shown in Fig. 1.6, the system comprises 50 load nodes, represented by circles; 4 substation nodes, depicted as squares; and 63 branches, indicated by lines. Base power and base voltage are 1 MVA and 13.5 kV, respectively. Upper and lower bounds for voltages at load nodes are equal to 1.05 and 0.95 pu, respectively. The system power factor *pf* is set at 0.9 and a three-block piecewise linearization is used to approximate energy losses.

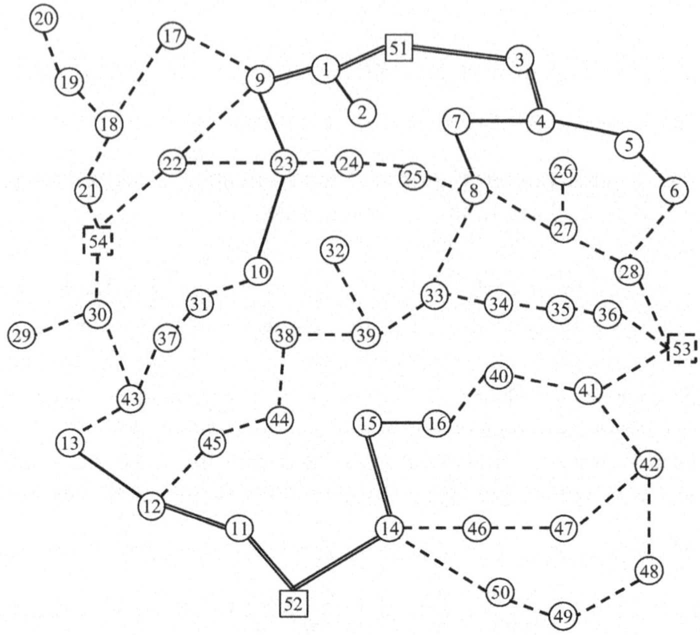

Fig. 1.6 One-line diagram of the test system

The currency used in the simulations is U.S. dollars, hereinafter denoted by \$. Investment decisions are made over a ten-year planning horizon divided into yearly stages considering a 10% interest rate and an investment budget equal to \$3.5 million per year. Nodal peak demands are presented in Table 1.1.

The uncertainty characterization of demand, wind speed, and solar irradiation relies on historical data of year 2012 for an actual site with maximum levels of wind speed and solar irradiation equal to 17.08 m/s and 1114.21 W/m^2, respectively. As done in [32], the same demand, wind speed, and solar irradiation profiles are used for all nodes. Moreover, four time blocks are considered, with durations equal to 350, 2650, 3900, and 1860 h/year, respectively. For the deterministic model, average values for demand, wind speed, and solar irradiation within each time block are used, which are presented in Table 1.2. For the stochastic model, the cumulative distribution functions for demand, wind speed, and solar irradiation are divided into three equiprobable segments. Thus, as reported in Table 1.3, for each time block, three different conditions of demand, wind speed, and solar irradiation are considered according to the procedure described in Sect. 1.3.1. As a result, 27 equiprobable scenarios are generated for each time block, thereby totaling 108 scenarios for each stage.

In Fig. 1.6, existing branches not subject to modification are represented by solid lines, existing branches that can undergo replacement are depicted as solid double lines, and non-existing branches that are candidates for the installation are drawn as

Table 1.1 Nodal peak demands (MVA)

Node	Stage									
	1	2	3	4	5	6	7	8	9	10
1	2.0500	2.1115	2.1730	2.2345	2.2960	2.3575	2.4190	2.4805	2.5420	2.6035
2	0.7800	0.8034	0.8268	0.8502	0.8736	0.8970	0.9204	0.9438	0.9672	0.9906
3	1.5800	1.6274	1.6748	1.7222	1.7696	1.8170	1.8644	1.9118	1.9592	2.0066
4	0.3200	0.3296	0.3392	0.3488	0.3584	0.3680	0.3776	0.3872	0.3968	0.4064
5	0.2800	0.2884	0.2968	0.3052	0.3136	0.3220	0.3304	0.3388	0.3472	0.3556
6	1.1700	1.2051	1.2402	1.2753	1.3104	1.3455	1.3806	1.4157	1.4508	1.4859
7	1.2400	1.2772	1.3144	1.3516	1.3888	1.4260	1.4632	1.5004	1.5376	1.5748
8	0.7200	0.7416	0.7632	0.7848	0.8064	0.8280	0.8496	0.8712	0.8928	0.9144
9	1.1400	1.1742	1.2084	1.2426	1.2768	1.3110	1.3452	1.3794	1.4136	1.4478
10	1.5600	1.6068	1.6536	1.7004	1.7472	1.7940	1.8408	1.8876	1.9344	1.9812
11	1.9100	1.9673	2.0246	2.0819	2.1392	2.1965	2.2538	2.3111	2.3684	2.4257
12	0.9300	0.9579	0.9858	1.0137	1.0416	1.0695	1.0974	1.1253	1.1532	1.1811
13	1.1500	1.1845	1.2190	1.2535	1.2880	1.3225	1.3570	1.3915	1.4260	1.4605
14	1.3500	1.3905	1.4310	1.4715	1.5120	1.5525	1.5930	1.6335	1.6740	1.7145
15	1.6200	1.6686	1.7172	1.7658	1.8144	1.8630	1.9116	1.9602	2.0088	2.0574
16	2.1600	2.2248	2.2896	2.3544	2.4192	2.4840	2.5488	2.6136	2.6784	2.7432
17	1.4000	1.4420	1.4840	1.5260	1.5680	1.6100	1.6520	1.6940	1.7360	1.7780
18	2.1000	2.1630	2.2260	2.2890	2.3520	2.4150	2.4780	2.5410	2.6040	2.6670
19	1.8100	1.8643	1.9186	1.9729	2.0272	2.0815	2.1358	2.1901	2.2444	2.2987
20	0.0000	1.2900	1.3287	1.3674	1.4061	1.4448	1.4835	1.5222	1.5609	1.5996
21	0.0000	0.1600	0.1648	0.1696	0.1744	0.1792	0.1840	0.1888	0.1936	0.1984

(continued)

Table 1.1 (continued)

Node	Stage									
	1	2	3	4	5	6	7	8	9	10
22	0.0000	1.6300	1.6789	1.7278	1.7767	1.8256	1.8745	1.9234	1.9723	2.0212
23	0.0000	0.0000	0.3400	0.3502	0.3604	0.3706	0.3808	0.3910	0.4012	0.4114
24	0.0000	0.0000	2.5100	2.5853	2.6606	2.7359	2.8112	2.8865	2.9618	3.0371
25	0.0000	0.0000	1.7200	1.7716	1.8232	1.8748	1.9264	1.9780	2.0296	2.0812
26	0.0000	0.0000	0.0000	1.4300	1.4729	1.5158	1.5587	1.6016	1.6445	1.6874
27	0.0000	0.0000	0.0000	1.6300	1.6789	1.7278	1.7767	1.8256	1.8745	1.9234
28	0.0000	0.0000	0.0000	1.2200	1.2566	1.2932	1.3298	1.3664	1.4030	1.4396
29	0.0000	0.0000	0.0000	0.0000	0.1600	0.1648	0.1696	0.1744	0.1792	0.1840
30	0.0000	0.0000	0.0000	0.0000	1.3500	1.3905	1.4310	1.4715	1.5120	1.5525
31	0.0000	0.0000	0.0000	0.0000	1.7900	1.8437	1.8974	1.9511	2.0048	2.0585
32	0.0000	0.0000	0.0000	0.0000	0.2300	0.2369	0.2438	0.2507	0.2576	0.2645
33	0.0000	0.0000	0.0000	0.0000	0.0000	1.4700	1.5141	1.5582	1.6023	1.6464
34	0.0000	0.0000	0.0000	0.0000	0.0000	1.6700	1.7201	1.7702	1.8203	1.8704
35	0.0000	0.0000	0.0000	0.0000	0.0000	2.0700	2.1321	2.1942	2.2563	2.3184
36	0.0000	0.0000	0.0000	0.0000	0.0000	1.2300	1.2669	1.3038	1.3407	1.3776
37	0.0000	0.0000	0.0000	0.0000	0.0000	0.0000	0.8200	0.8446	0.8692	0.8938
38	0.0000	0.0000	0.0000	0.0000	0.0000	0.0000	1.7100	1.7613	1.8126	1.8639
39	0.0000	0.0000	0.0000	0.0000	0.0000	0.0000	0.2400	0.2472	0.2544	0.2616
40	0.0000	0.0000	0.0000	0.0000	0.0000	0.0000	0.0000	0.9400	0.9682	0.9964
41	0.0000	0.0000	0.0000	0.0000	0.0000	0.0000	0.0000	1.3000	1.3390	1.3780
42	0.0000	0.0000	0.0000	0.0000	0.0000	0.0000	0.0000	1.7400	1.7922	1.8444

(continued)

Table 1.1 (continued)

Node	Stage									
	1	2	3	4	5	6	7	8	9	10
43	0.0000	0.0000	0.0000	0.0000	0.0000	0.0000	0.0000	1.0900	1.1227	1.1554
44	0.0000	0.0000	0.0000	0.0000	0.0000	0.0000	0.0000	0.0000	0.4400	0.4532
45	0.0000	0.0000	0.0000	0.0000	0.0000	0.0000	0.0000	0.0000	1.1200	1.1536
46	0.0000	0.0000	0.0000	0.0000	0.0000	0.0000	0.0000	0.0000	1.3000	1.3390
47	0.0000	0.0000	0.0000	0.0000	0.0000	0.0000	0.0000	0.0000	0.2000	0.2060
48	0.0000	0.0000	0.0000	0.0000	0.0000	0.0000	0.0000	0.0000	0.0000	1.0300
49	0.0000	0.0000	0.0000	0.0000	0.0000	0.0000	0.0000	0.0000	0.0000	1.1700
50	0.0000	0.0000	0.0000	0.0000	0.0000	0.0000	0.0000	0.0000	0.0000	0.5500

Table 1.2 Operational conditions used for the deterministic model (pu)

Time block	Average demand factor	Average wind speed factor	Average solar irradiation factor
1	0.7418	0.3345	0.0734
2	0.5248	0.3279	0.2441
3	0.3914	0.3236	0.2216
4	0.3011	0.3177	0.0207

Table 1.3 Operational conditions used for the stochastic model (pu)

Time block	Average demand factor	Average wind speed factor	Average solar irradiation factor
1	0.83340	0.50535	0.22026
	0.72168	0.32588	0.00000
	0.67027	0.17218	0.00000
2	0.58940	0.49425	0.58421
	0.51504	0.31629	0.14816
	0.47014	0.17309	0.00000
3	0.42664	0.51211	0.54507
	0.38973	0.31093	0.11964
	0.35800	0.14763	0.00000
4	0.32606	0.53766	0.06212
	0.30166	0.29549	0.00000
	0.27546	0.11993	0.00000

dashed lines. Branch lengths are presented in Table 1.4. Network reconfiguration under normal operation can be implemented by switching branches 10-31, 13-43, 23-24, 33-39, and 38-44. The capacity and unitary impedance of existing branches are 6.28 MVA and 0.557 Ω/km, respectively. Table 1.5 presents the data for candidate conductors in branches subject to replacement and in non-existing branches. Note that two conductor alternatives are available per branch. Based on [23], all branches are characterized by a maintenance cost equal to $450/year and a 25-year lifetime.

Existing substations, which are located at nodes 51 and 52 (Fig. 1.6), include a 12-MVA transformer characterized by an impedance equal to 0.16 Ω and a maintenance cost equal to $2000. Nodes 53 and 54 represent the locations of candidate substations. Voltages at substation nodes are set at 1.05 pu. Based on the cost data reported in [33], the costs of the energy supplied by all substations, C_b^{SS}, are identical and equal to $225.33/MWh, $182.72/MWh, $154.43/MWh, and $81.62/MWh, for time blocks 1–4, respectively. The cost of unserved energy, C^U, is $1000/MWh. Investment decisions consist in (i) expanding existing substations

Table 1.4 Branch lengths (km)

From	To	Length	From	To	Length	From	To	Length
1	2	0.66	12	45	1.33	30	43	1.47
1	9	0.86	13	43	1.07	30	54	1.02
1	51	1.11	14	15	1.81	31	37	0.45
3	4	0.90	14	46	1.31	32	39	1.46
3	51	2.06	14	50	2.25	33	34	0.81
4	5	1.45	14	52	2.21	33	39	1.19
4	7	1.24	15	16	0.91	34	35	0.76
5	6	0.81	16	40	1.29	35	36	0.45
6	28	1.55	17	18	1.83	36	53	1.28
7	8	1.00	18	19	0.68	37	43	1.01
8	25	0.79	18	21	0.98	38	39	1.19
8	27	1.60	19	20	0.96	38	44	1.27
8	33	1.92	21	54	0.58	40	41	1.39
9	17	1.61	22	23	1.85	41	42	1.52
9	22	2.08	22	54	1.89	41	53	1.73
9	23	1.36	23	24	0.82	42	47	1.82
10	23	1.89	24	25	0.89	42	48	1.77
10	31	0.92	26	27	0.68	44	45	1.02
11	12	1.42	27	28	1.15	46	47	1.29
11	52	1.50	28	53	1.64	48	49	1.58
12	13	1.70	29	30	1.17	49	50	0.92

Table 1.5 Data for candidate conductors

Type	Alternative 1			Alternative 2		
l	Upper limit (MVA)	Impedance (Ω/km)	Investment cost ($/km)	Upper limit (MVA)	Impedance (Ω/km)	Investment cost ($/km)
NRB	9.00	0.478	19140	12.00	0.423	29,870
NAB	6.28	0.557	15020	9.00	0.478	25,030

by adding a new transformer and (ii) building a new substation from scratch. Expansion costs of substations at nodes 51–54 are $100000, $100000, $150000, and $150000, respectively. Data for candidate transformers are listed in Table 1.6, where the same two alternatives are available for each substation. The lifetime of all candidate transformers is 15 years. Moreover, it is assumed that η^{SS} is considerably larger than the lifetimes of the other distribution assets. Thus, $RR^{SS} = I$.

For the sake of simplicity, investment in renewable-based DG is only allowed, with the penetration limit, ξ, set at 25%. Candidate nodes for installation of wind generators are 3, 15, 23, 35, and 42. Candidate nodes for installation of PV generators are 4, 12, 24, 36, and 43. The economic and technical features of candidate

Table 1.6 Data for candidate transformers

Alternative 1				Alternative 2			
Rated capacity (MVA)	Impedance (Ω)	Maintenance cost ($)	Investment cost ($)	Rated capacity (MVA)	Impedance (Ω)	Maintenance cost ($)	Investment cost ($)
12	0.16	2000	750,000	15	0.13	3000	950,000

Table 1.7 Data for candidate DG units

Alternative k	Wind generators			Photovoltaic generators		
	Rated capacity (MVA)	Investment cost ($/MW)	Cost of energy supplied ($/MWh)	Rated capacity (MVA)	Investment cost ($/MW)	Cost of energy supplied ($/MWh)
1	0.91	185,000	0	0.70	172,000	0
2	2.05	184,000	0	1.65	171,000	0

DG units are presented in Table 1.7, where two alternatives are considered for each technology. Maintenance costs for DG units are set so that $C_k^{M,p} = 0.05 C_k^{I,p} pf \overline{G}_k^p$. Wind generation limits for the deterministic model, $\hat{G}_{ikb}^W$, and for the stochastic model, $\hat{G}_{ikb}^W(\omega)$, associated with wind speed levels, are determined using the data available in [34] related to wind generators *E44* and *E82*, which correspond to alternatives 1 and 2, respectively. PV generation limits for the deterministic model, $\hat{G}_{ikb}^\Theta$, and for the stochastic model, $\hat{G}_{ikb}^\Theta(\omega)$, associated with solar irradiation levels, are determined using the data available in [35] related to 5000 solar panels of types *KD100-36* and *KD300-80* within *F Series*, which correspond to alternatives 1 and 2, respectively. A 20-year lifetime is considered for all units.

Both models have been implemented on a Dell PowerEdge R920X64 with four Intel Xeon E7-4820 processors at 2.00 GHz and 768 GB of RAM using CPLEX 12.6 [36] and GAMS 24.8 [37]. The stopping criterion for the branch-and-cut algorithm of CPLEX is based on an optimality gap equal to 1%. Under this stopping criterion, computing times were equal to 3.73 min for the deterministic model and 9.88 h for the stochastic model.

The solutions to both cases are depicted in Fig. 1.7. As can be observed, both expansion plans differ in both investment decisions and topology. The solution provided by the stochastic model features a radial topology. In contrast, a meshed solution was identified by the deterministic model, which is compliant with the consideration of network reconfiguration under normal operation. Thus, both solutions topologically differ in the branches connecting new load nodes to the distribution system. For the solution attained by the deterministic model, the demands at nodes 22 and 24 are supplied by substation 51 through branches 9-22 and 24-25, respectively. For the solution provided by the stochastic model, those demands are respectively fed by substation 51 through branch 23-24 and by substation 54 through branch 22-54. Additionally, for the solution to the deterministic

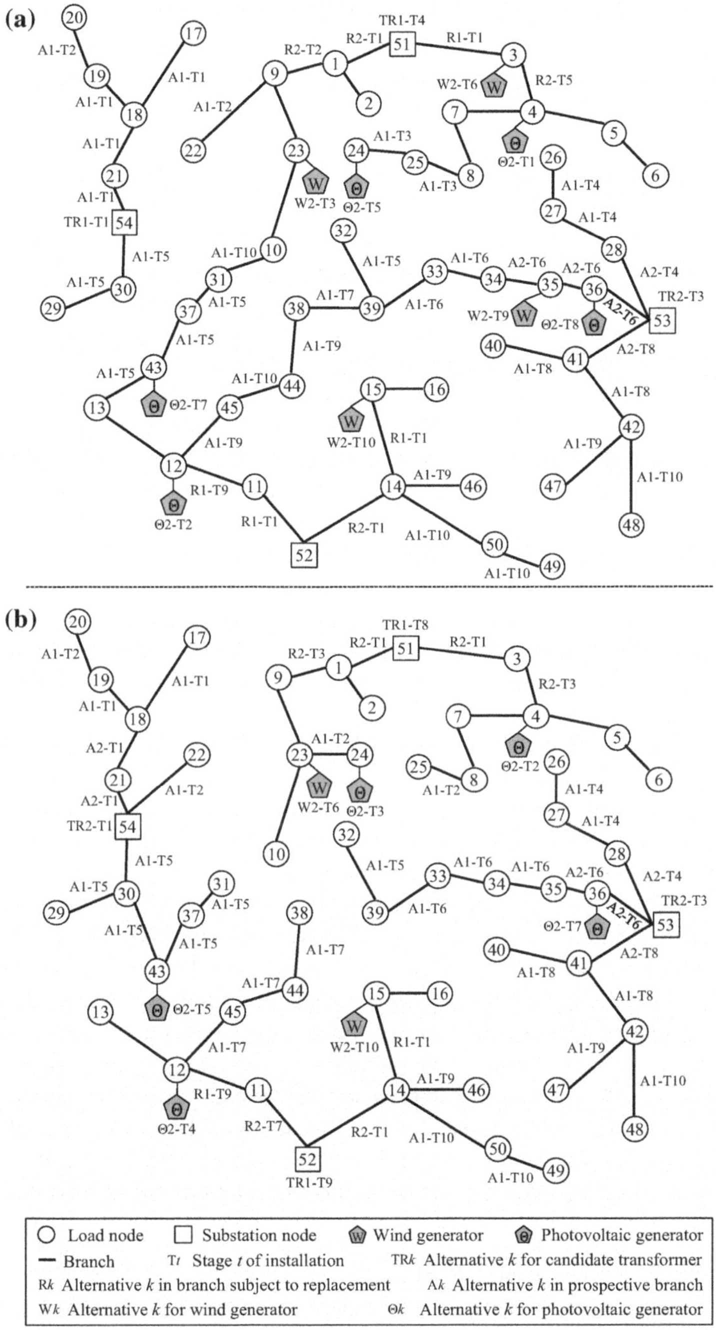

Fig. 1.7 Solutions: **a** Deterministic, **b** Stochastic

model, substation 52 supplies the demands at nodes 31, 37, and 43 at stages 5–9 through branches 13-43, 37-43, and 31-37, while substation 51 supplies those demands at stage 10 through branches 10-31, 31-37, and 37-43. However, for the solution to the stochastic model, those demands are supplied by substation 54 through branches 30-43, 37-43, and 31-37 at stages 5–10. Finally, additional differences arise between both expansion plans related to the demand supply at load nodes 32, 38, and 39.

Moreover, the installation of transformers at substation nodes differs in three aspects. The installation of a transformer at substation node 51 takes place at different stages. At substation node 52, a transformer is installed for the solution attained by the stochastic model, while no transformer is installed for the deterministic model. Furthermore, at substation node 54, another difference arises in the transformer alternative being installed. Regarding DG investment, the expansion plan obtained by the deterministic model includes the same generator types, alternatives, and locations as those determined by the stochastic model with the addition of a wind generator and a photovoltaic generator at nodes 3 and 35, respectively. The installation times of generators also differ except for that installed at node 15.

Finally, it is worth mentioning that the number of generators installed for the solution obtained by the deterministic model is higher than that for the stochastic solution. On the contrary, the solution provided by the stochastic model includes a larger number of transformers with bigger capacity than those of the solution obtained by the deterministic model.

Table 1.8 lists the present values of the different costs associated with the solutions to both models. As can be seen, the consideration of uncertainty yields an expected production cost that is greater than the production cost incurred by the deterministic solution. However, this production cost increase is offset by the reduction in costs related to investment, maintenance, losses, and unserved energy.

This case study is also useful to substantiate the use of both the approximate network model and the piecewise linear approximation for quadratic energy losses. In this regard, load flow results for the high load level at the last time stage provided by the deterministic approach have been compared with those achieved with a full ac load flow model. The average errors obtained for branch current flows, injections at substations, and nodal voltage magnitudes are 0.72, −0.26, and 0.49%, respectively. These results, which are consistent with the experience reported in [22], validate the suitability of the linearized network model. Moreover, the quality of the piecewise linearization for losses has been assessed by solving the deterministic problem with quadratic losses wherein binary investment and utilization variables

Table 1.8 Present values of costs (10^6 \$)

Case	Investment	Maintenance	Production	Losses	Unserved energy	c^{TPV}
Deterministic	21.08	8.88	266.87	9.43	1.45	307.72
Stochastic	17.38	6.80	273.46	8.07	1.13	306.84

were fixed to the values attained by the mixed-integer linear program with a three-piece linearization. For this particular case study, the total cost only differs by 0.87%, thereby corroborating the appropriateness of the piecewise linear approximation in terms of solution quality.

Additionally, the convenience of using a stochastic solution approach rather than a simpler deterministic one is analyzed. To that end, a widely-used metric, namely the value of the stochastic solution (VSS), is calculated [29]. For a minimization problem, the VSS is defined as the difference between two cost terms. The first term, denoted by c^{DP}, represents the value of the objective function obtained from the stochastic model by fixing decision variables not depending on scenarios to the values resulting from solving the associated deterministic problem. The second term, denoted by c^{SP}, represents the value of the objective function resulting from the stochastic model. Thus, the VSS quantifies the potential gain associated with the stochastic solution.

For the problem under consideration, c^{DP} is equal to the value of c^{TPV} obtained from solving the stochastic model (1.74) and (1.75) by fixing decision variables x^{l}_{ijkt}, x^{NT}_{ikt}, x^{p}_{ikt}, x^{SS}_{it}, y^{l}_{ijkt}, y^{p}_{ikt}, and y^{tr}_{ikt} to the values resulting from solving the deterministic model (1.53) and (1.54). For this particular case study, c^{DP} is equal to \$311.49 million, whereas c^{SP} is equal to the value of c^{TPV} provided by the stochastic model (1.74) and (1.75), which, as can be seen in Table 1.8, is equal to \$306.84 million. Thus, the VSS is equal to \$4.56 million, which represents a potential 1.5% improvement upon the deterministic solution.

Appendix

The symbols used throughout this chapter are listed below:

Indices

b	Index for time blocks.
h	Index for the blocks used in the piecewise linearization of energy losses.
i,j	Indices for nodes.
k, κ	Indices for available investment alternatives.
l	Index for branch types.
p	Index for generator types.
s	Index for segments of the cumulative distribution functions.
t, τ	Indices for time stages.
tr	Index for transformer types.
ω	Index for scenarios.

Sets

$\mathcal{B}$ Index set of time blocks.

K^l Index set of available alternatives for branches of type l.

K^p Index set of available alternatives for generators of type p.

K^{tr} Index set of available alternatives for transformers of type tr.

$\mathcal{L}$ Set of branch types. $\mathcal{L} = \{EFB, ERB, NRB, NAB\}$ where EFB, ERB, NRB, and NAB denote existing fixed branch, existing replaceable branch, new replacement branch, and newly added branch, respectively.

$\mathcal{P}$ Set of generator types. $\mathcal{P} = \{C, W, \Theta\}$ where C, W, and Θ stand for conventional, wind power, and photovoltaic generation, respectively.

$\mathcal{T}$ Index set of time stages.

TR Set of transformer types. $TR = \{ET, NT\}$ where ET and NT denote existing transformer and new transformer, respectively.

Ξ^{DT} Set of variables associated with the deterministic model.

Ξ^{ST} Set of variables associated with the stochastic model.

Υ^l Index set of branches of type l.

$\Upsilon^{SW,l}$ Subset of Υ^l comprising those branches that are switchable under normal operation.

Ψ_i^l Index set of nodes connected to node i by a branch of type l.

Ψ_t^{LN} Index set of load nodes at stage t.

Ψ^N Index set of system nodes.

Ψ^p Index set of candidate nodes for the installation of generators of type p.

Ψ^{SS} Index set of substation nodes.

Ω_b Set of scenarios for time block b.

Parameters

A_{kh}^l Width of block h of the piecewise linear energy losses for alternative k for branches of type l.

A_{kh}^{tr} Width of block h of the piecewise linear energy losses for alternative k for transformers of type tr.

$C_k^{E,p}$ Cost coefficient for the energy supplied by alternative k for generators of type p.

$C_k^{I,l}$ Investment cost coefficient for alternative k for branches of type l.

$C_k^{I,NT}$ Investment cost coefficient for alternative k for new transformers.

$C_k^{I,p}$ Investment cost coefficient for alternative k for generators of type p.

$C_i^{I,SS}$ Investment cost coefficient for the substation at node i.

$C_k^{M,l}$ Maintenance cost coefficient for alternative k for branches of type l.

$C_k^{M,p}$ Maintenance cost coefficient for alternative k for generators of type p.

$C_k^{M,tr}$ Maintenance cost coefficient for alternative k for transformers of type tr.

C_b^{SS} Cost coefficient for the energy supplied by substations for time block b.
C^U Cost coefficient for unserved energy.
D_{it} Actual peak demand at node i and stage t.
$\tilde{D}_{it}$ Fictitious peak demand at node i and stage t.
$\overline{F}_k^l$ Upper limit for the actual current flow through alternative k for branches of type l.
$\overline{G}_k^p$ Rated capacity for alternative k for generators of type p.
$\hat{G}_{ikb}^p$ Maximum power availability for alternative k for generators of type p at node i and time block b.
$\hat{G}_{ikb}^p(\omega)$ Maximum power availability for alternative k for generators of type p at node i, time block b, and scenario ω.
$\overline{G}_k^{tr}$ Rated capacity of alternative k for transformers of type tr.
I Annual interest rate.
IB_t Investment budget for stage t.
J Sufficiently large positive constant.
ℓ_{ij} Length of the branch connecting nodes i and j.
M_{kh}^l Slope of block h of the piecewise linear energy losses for alternative k for branches of type l.
M_{kh}^{tr} Slope of block h of the piecewise linear energy losses for alternative k for transformers of type tr.
n_B Number of time blocks.
n_{DG} Number of candidate nodes for installation of distributed generation.
n_H Number of blocks of the piecewise linear energy losses.
n_T Number of time stages.
n_Ω Number of scenarios per time block.
n_S^D Number of segments for demand factors at each time block.
n_S^p Number of segments for factors for generation of type p at each time block.
pf System power factor.
RR^l Capital recovery rate for investment in branches of type l.
RR^{NT} Capital recovery rate for investment in new transformers.
RR^p Capital recovery rate for investment in generators of type p.
RR^{SS} Capital recovery rate for investment in substations.
$\underline{V}$ Lower bound for nodal voltages.
$\overline{V}$ Upper bound for nodal voltages.
Z_k^l Unitary impedance magnitude for alternative k for branches of type l.
Z_k^{tr} Impedance magnitude for alternative k for transformers of type tr.
Δ_b Duration of time block b.
η^l Lifetime of branches of type l.
η^{NT} Lifetime of new transformers.
η^p Lifetime of generators of type p.
η^{SS} Lifetime of substation assets other than transformers.
μ_b^D Average demand factor of time block b.

μ_{sb}^{D} Average factor for demand in segment s of time block b.
μ_{sb}^{p} Average factor for generation of type p in segment s of time block b.
$\mu_{b}^{D}(\omega)$ Average demand factor of time block b and scenario ω.
ξ Penetration limit for distributed generation.
π_{sb}^{D} Probability of the average factor for demand in segment s of time block b.
π_{sb}^{p} Probability of the average factor for generation of type p in segment s of time block b.
$\pi_{b}(\omega)$ Probability of scenario ω of time block b.

Variables

c_{t}^{E} Production cost at stage t.
c_{t}^{I} Amortized investment cost at stage t.
c_{t}^{M} Maintenance cost at stage t.
c_{t}^{R} Energy losses cost at stage t.
c_{t}^{U} Unserved energy cost at stage t.
c^{TPV} Present value of the total cost.
d_{itb}^{U} Unserved energy at node i, stage t, and time block b.
$d_{itb}^{U}(\omega)$ Unserved energy at node i, stage t, time block b, and scenario ω.
f_{ijktb}^{l} Actual current flow through alternative k for the branch of type l connecting nodes i and j at stage t and time block b.
$f_{ijktb}^{l}(\omega)$ Actual current flow through alternative k for the branch of type l connecting nodes i and j at stage t, time block b, and scenario ω.
$\tilde{f}_{ijkt}^{l}$ Fictitious current flow through alternative k for the branch of type l connecting nodes i and j at stage t.
g_{iktb}^{p} Current injection at node i for alternative k for the generator of type p at stage t and time block b.
$g_{iktb}^{p}(\omega)$ Current injection at node i for alternative k for the generator of type p at stage t, time block b, and scenario ω.
g_{iktb}^{tr} Actual current injection at substation node i for alternative k for the transformer of type tr at stage t and time block b.
$g_{iktb}^{tr}(\omega)$ Actual current injection at substation node i for alternative k for the transformer of type tr at stage t, time block b, and scenario ω.
$\tilde{g}_{it}^{SS}$ Fictitious current injection at substation node i and stage t.
v_{itb} Voltage magnitude at node i, stage t, and time block b.
$v_{itb}(\omega)$ Voltage magnitude at node i, stage t, time block b, and scenario ω.
x_{ijkt}^{l} Binary investment variable for alternative k for the branch of type l connecting nodes i and j at stage t.
x_{ikt}^{NT} Binary investment variable for alternative k for the new transformer at substation node i and stage t.
x_{ikt}^{p} Binary investment variable for alternative k for the generator of type p at node i and stage t.

x_{it}^{SS} Binary investment variable for the substation at node i and stage t.

y_{ijkt}^{l} Binary utilization variable for alternative k for the branch of type l connecting nodes i and j at stage t.

y_{ikt}^{p} Binary utilization variable for alternative k for the generator of type p at node i and stage t.

y_{ikt}^{tr} Binary utilization variable for alternative k for the transformer of type tr at substation node i and stage t.

δ_{ijktbh}^{l} Current in block h of the piecewise linear energy losses for alternative k for the branch of type l connecting nodes i and j at stage t and time block b.

$\delta_{ijktbh}^{l}(\omega)$ Current in block h of the piecewise linear energy losses for alternative k for the branch of type l connecting nodes i and j at stage t, time block b, and scenario ω.

δ_{iktbh}^{tr} Current injection in block h of the piecewise linear energy losses for alternative k for the transformer of type tr at substation node i, stage t, and time block b.

$\delta_{iktbh}^{tr}(\omega)$ Current injection in block h of the piecewise linear energy losses for alternative k for the transformer of type tr at substation node i, stage t, time block b, and scenario ω.

References

1. A. Gómez-Expósito, A.J. Conejo, C. Cañizares, *Electric Energy Systems. Analysis and Operation* (CRC Press, Boca Raton, FL, USA, 2009)
2. W.H. Kersting, *Distribution System Modeling and Analysis*, 3rd edn. (CRC Press, Boca Raton, FL, USA, 2012)
3. H.L. Willis, *Power Distribution Planning Reference Book*, 2nd edn. (Marcel Dekker Inc, New York, NY, USA, 2004)
4. P.S. Georgilakis, N.D. Hatziargyriou, A review of power distribution planning in the modern power systems era: models, methods and future research. Electr. Power Syst. Res. **121**, 89–100 (2015)
5. N. Jenkins, R. Allan, P. Crossley, D. Kirschen, G. Strbac, *Embedded Generation* (The Institution of Engineering and Technology, London, UK, 2000)
6. A. Keane, L.F. Ochoa, C.L.T. Borges, G.W. Ault, A.D. Alarcon-Rodriguez, R.A.F. Currie, F. Pilo, C. Dent, G.P. Harrison, State-of-the-art techniques and challenges ahead for distributed generation planning and optimization. IEEE Trans. Power Syst. **28**(2), 1493–1502 (2013)
7. W. El-Khattam, Y.G. Hegazy, M.M.A. Salama, An integrated distributed generation optimization model for distribution system planning. IEEE Trans. Power Syst. **20**(2), 1158–1165 (2005)
8. R. Viral, D.K. Khatod, Optimal planning of distributed generation systems in distribution system: a review. Renew. Sust. Energ. Rev. **16**(7), 5146–5165 (2012)
9. H. Falaghi, C. Singh, M.-R. Haghifam, M. Ramezani, DG integrated multistage distribution system expansion planning. Int. J. Electr. Power Energy Syst. **33**(8), 1489–1497 (2011)
10. M. Gitizadeh, A.A. Vahed, J. Aghaei, Multistage distribution system expansion planning considering distributed generation using hybrid evolutionary algorithms. Appl. Energy **101**, 655–666 (2013)

11. M. Sedghi, M. Aliakbar-Golkar, M.-R. Haghifam, Distribution network expansion considering distributed generation and storage units using modified PSO algorithm. Int. J. Electr. Power Energy Syst. **52**, 221–230 (2013)
12. M. Shivaie, M.T. Ameli, M.S. Sepasian, P.D. Weinsier, V. Vahidinasab, A multistage framework for reliability-based distribution expansion planning considering distributed generations by a self-adaptive global-based harmony search algorithm. Reliab. Eng. Syst. Saf. **139**, 68–81 (2015)
13. A. Tabares, J.F. Franco, M. Lavorato, M.J. Rider, Multistage long-term expansion planning of electrical distribution systems considering multiple alternatives. IEEE Trans. Power Syst. **31** (3), 1900–1914 (2016)
14. C.L.T. Borges, V.F. Martins, Multistage expansion planning for active distribution networks under demand and distributed generation uncertainties. Int. J. Electr. Power Energy Syst. **36** (1), 107–116 (2012)
15. A. Bagheri, H. Monsef, H. Lesani, Renewable power generation employed in an integrated dynamic distribution network expansion planning. Electr. Power Syst. Res. **127**, 280–296 (2015)
16. R. Hemmati, R.-A. Hooshmand, N. Taheri, Distribution network expansion planning and DG placement in the presence of uncertainties. Int. J. Electr. Power Energy Syst. **73**, 665–673 (2015)
17. H. Arasteh, M.S. Sepasian, V. Vahidinasab, P. Siano, SoS-based multiobjective distribution system expansion planning. Electr. Power Syst. Res. **141**, 392–406 (2016)
18. S.F. Santos, D.Z. Fitiwi, M. Shafie-Khah, A.W. Bizuayehu, C.M.P. Cabrita, J.P.S. Catalão, New multistage and stochastic mathematical model for maximizing RES hosting capacity–part I: problem formulation. IEEE Trans. Sustain. Energy **8**(1), 304–319 (2017)
19. S.F. Santos, D.Z. Fitiwi, M. Shafie-Khah, A.W. Bizuayehu, C.M.P. Cabrita, J.P.S. Catalão, New multi-stage and stochastic mathematical model for maximizing RES hosting capacity–part II: numerical results. IEEE Trans. Sustain. Energy **8**(1), 320–330 (2017)
20. G. Muñoz-Delgado, J. Contreras, J.M. Arroyo, Joint expansion planning of distributed generation and distribution networks. IEEE Trans. Power Syst. **30**(5), 2579–2590 (2015)
21. G. Muñoz-Delgado, J. Contreras, J.M. Arroyo, Multistage generation and network expansion planning in distribution systems considering uncertainty and reliability. IEEE Trans. Power Syst. **31**(5), 3715–3728 (2016)
22. S. Haffner, L.F.A. Pereira, L.A. Pereira, L.S. Barreto, Multistage model for distribution expansion planning with distributed generation–part I: problem formulation. IEEE Trans. Power Deliv. **23**(2), 915–923 (2008)
23. R.C. Lotero, J. Contreras, Distribution system planning with reliability. IEEE Trans. Power Deliv. **26**(4), 2552–2562 (2011)
24. L. Blank, A. Tarquin, *Engineering Economy*, 7th edn. (McGraw-Hill, New York, NY, USA, 2012)
25. P.C. Paiva, H.M. Khodr, J.A. Domínguez-Navarro, J.M. Yusta, A.J. Urdaneta, Integral planning of primary-secondary distribution systems using mixed integer linear programming. IEEE Trans. Power Syst. **20**(2), 1134–1143 (2005)
26. M. Lavorato, J.F. Franco, M.J. Rider, R. Romero, Imposing radiality constraints in distribution system optimization problems. IEEE Trans. Power Syst. **27**(1), 172–180 (2012)
27. S.P. Bradley, A.C. Hax, T.L. Magnanti, *Applied Mathematical Programming* (Addison-Wesley, Reading, MA, USA, 1977)
28. S. Binato, M.V.F. Pereira, S. Granville, A new Benders decomposition approach to solve power transmission network design problems. IEEE Trans. Power Syst. **16**(2), 235–240 (2001)
29. J.R. Birge, F. Louveaux, *Introduction to Stochastic Programming*, 2nd edn. (Springer, New York, NY, USA, 2011)
30. L. Baringo, A.J. Conejo, Wind power investment within a market environment. Appl. Energy **88**(9), 3239–3247 (2011)

31. V. Miranda, J.V. Ranito, L.M. Proença, Genetic algorithms in optimal multistage distribution network planning. IEEE Trans. Power Syst. **9**(4), 1927–1933 (1994)
32. Y.M. Atwa, E.F. El-Saadany, M.M.A. Salama, R. Seethapathy, Optimal renewable resources mix for distribution system energy loss minimization. IEEE Trans. Power Syst. **25**(1), 360–370 (2010)
33. Red Eléctrica de España (2017) [Online], Available: https://www.esios.ree.es/en
34. ENERCON, ENERCON wind energy converters: products overview (July 2010) [Online], Available: http://www.enercon.de
35. KYOCERA SOLAR Europe (2017) [Online], Available: http://www.kyocerasolar.eu
36. IBM ILOG CPLEX (2017) [Online], Available: https://www.ibm.com/analytics/data-science/prescriptive-analytics/cplex-optimizer
37. GAMS Development Corporation (2017) [Online], Available: http://www.gams.com

Chapter 2
Static and Dynamic Convex Distribution Network Expansion Planning

Julio López, David Pozo and Javier Contreras

Abstract This chapter presents static and dynamic optimization-based models for planning the electric distribution network. Based on a branch flow model, two Mixed-Integer Conic Quadratic Programming (MICQP) convex formulations are proposed to solve the network expansion planning models including high modeling fidelity of the intrinsic interaction of the manifold elements of the networks. The objective of the presented models is to minimize investment and operation costs by optimally deciding on installing new feeders and/or changing existing ones for others with larger capacities, installing new substations or expanding existing ones and, finally, installing capacitor banks and voltage regulators, modifying the network topology. In addition, discrete tap settings of voltage regulators are modeled as a set of mixed-integer linear equations, which are embedded in an ac optimal power flow. The presented MICQP models are convex optimization problems. Therefore globality and convergence are guaranteed. Computational results to verify the efficiency of the proposed methodology are obtained for a 24-node test system. Finally, conclusions are duly drawn.

Keywords Capacitor banks · Convex optimization · Dynamic model
Electric distribution network expansion planning · Static models
Voltage regulators

J. López (✉)
Electrical Engineering School, University of Cuenca, Cuenca, Ecuador
e-mail: julio.lopez@ucuenca.edu.ec

D. Pozo
Center of Energy Systems, Skolkovo Institute of Science and Technology, Moscow, Russia
e-mail: d.pozo@skoltech.ru

J. Contreras
Escuela Técnica Superior de Ingenieros Industriales, Universidad de Castilla – La Mancha, Ciudad Real, Spain
e-mail: Javier.Contreras@uclm.es

F. Shahnia et al. (eds.), *Electric Distribution Network Planning*,
Power Systems, https://doi.org/10.1007/978-981-10-7056-3_2

2.1 Introduction

The rapid increase of renewable power generation connected to electric distribution networks has complicated their operation. Because this operational complexity, it could lead to a high impact on the economic efficiency of networks due to the significant investment costs of new control devices that properly guarantee appropriate levels of security, quality, and reliability at competitive costs. In this vein, optimization tools to solve the Electric Distribution Network Expansion Planning (EDNEP) problem has recently attracted more attention representing a shifting toward feasible-based to optimization-based planning paradigms. It is clear that the use of optimization tools in EDNEP represent substantial gains or savings in planning electric distribution networks. However, it is essential to properly capture the complexity of the non-linear interactions of the manifolds elements and physic laws with high fidelity. This chapter is devoted to this propose presenting convex formulations that could be implemented on off-the-shelf solvers with globality and convergence guarantee.

In its simplest version, the EDNEP problem consists of determining the investments that guarantee an economical and reliable distribution network operation. Technical constraints such as maximum current flows through feeders, maximum power from substation transformers, voltage magnitude limits in nodes and network radiality must be considered [1, 2]. The EDNEP problem can be established as follows: an electric distribution network needs to meet the demands of a fixed number of consumers due to demand growth, hence, it is necessary to carry out expansion planning consisting of: installing new feeders and/or changing existing ones for others with larger capacities, installing new substations or expanding existing ones and finally, installing capacitor banks and voltage regulators. The objective is to minimize the total costs related to those investments and network operation costs, subject to a set of physical, operating and economic constraints [3].

The installation of capacitor banks in electric distribution networks is important, mainly to maintain the voltage magnitude and energy losses within pre-established limits. Their optimal sizes and locations make these improvements feasible [4]. Thus, optimal capacitor bank placement aims at placing and sizing them, minimizing the costs associated with capacitor banks and energy losses. Within this context, another important aspect is the location of voltage regulators [5].

Some works about the EDNEP problem have independently addressed feeder and substation installation [7–13, 17–20], allocation of capacitor banks [5, 21–23], allocation of voltage regulators [24–26], and joint allocation of capacitor banks and voltage regulators [27–30]. However, the EDNEP problem should focus not only on a single technology–or device–based planning. In this context, an integral EDNEP problem considering a co-optimization of all the above control devices could foster benefits to the electric distribution network by reducing operation costs and losses, increasing flexibility and reliability is proposed.

However, this problem becomes a large-scale mixed-integer non-linear programming (MINLP) problem. The EDNEP problem has been solved using different techniques, such as heuristic and meta-heuristic algorithms and classical optimization techniques [6]. Heuristic algorithms have produced solutions with a relatively low computational effort, like branch-exchange in [7, 8] and the constructive heuristic algorithm in [9]. Meta-heuristic algorithms have also been used, like evolutionary algorithms in [10], genetic algorithms in [11] and [12], ant colony algorithms in [13], simulated annealing algorithms in [14, 15] and particle swarm algorithm in [16]. Although metaheuristics are flexible and achieve good results, they also present many problems, such as a high computational demand, the need for adjusting and fine-tuning the parameters and the definition of a stopping criterion. In addition, they cannot guarantee convergence to a global optimum, or indicate the quality of the final solution, because they do not provide a distance indicator to the optimal solution.

2.2 Time Framework

An important aspect to be considered in the optimal EDNEP problem is the decision-making process in the planning horizon. According to [31], the EDNEP problem can be divided into two periods: short-term planning (1 up to 4 years) and long-term planning (5 up to 20 years), leading to two types of EDNEP optimization models, *static* and *dynamic*.

In a *static* model, EDNEP decisions are only made at the beginning of the planning horizon, i.e., at a single point in time, where the load demand data considered remains constant until the end of the planning horizon. This modeling type is known as a single-stage model as well and considers the whole planning horizon in a single period, which is the target period. Since the EDNEP in mainly conditioned by the load demand in the electric distribution network, which usually increases over time, the reference period is usually selected as the last year of the planning horizon.

In a *dynamic* model, EDNEP decisions are made at different points in time. This modeling type is known as multi-period as well and represents the real behavior of the electric distribution network. In this approach, the planning horizon is divided into different time periods, each one comprising a specific number of years [32, 33].

The advantage of using a static approach for the EDNEP problem is that the resulting model is relatively simple. One of the disadvantages is that the EDNEP is solved for the last year of the planning horizon. Another drawback is that, if the EDNEP problem is solved for a long-term planning horizon, the load demand in the electric distribution network will probably be much higher than the load demand in the short-term. Therefore, the EDNEP will probably result in an oversize of the installed components and higher investments.

2.3 AC Power Flow in Electric Distribution Networks

The analysis of an electric distribution network requires the solution of the power flow problem to calculate the state of the system represented by voltage magnitudes in nodes, current flows in feeders, energy losses and other variables of interest. Therefore, power flow models are widely tools used in the steady-state analysis of the networks. Most ac power flow models in electric distribution networks are based on power-mismatch and current-mismatch formulations either in polar or rectangular formats, mainly using Newton-Raphson algorithms [35]. On the other hand, radial networks are characterized by a high R/X ratio. This renders the load flow problem ill-conditioned. Previous research indicates that standard load flow methods fail to converge in ill-conditioned test systems [36, 37].

In this work, the equations that represent the steady-state of radial networks are obtained from the branch flow model proposed in [38–40] as

$$P_k = \sum_{j\in\alpha(k)} \left(P_{km} + R_{km} I_{km}^2\right) - \sum_{j\in\alpha(k)} P_{jk} \quad \forall k \in \mathrm{B} \tag{2.1}$$

$$Q_k = \sum_{j\in\alpha(k)} \left(Q_{km} + X_{km} I_{km}^2\right) - \sum_{j\in\alpha(k)} Q_{jk} \quad \forall k \in \mathrm{B} \tag{2.2}$$

$$V_k^2 - V_m^2 = 2(R_{km}P_{km} + X_{km}Q_{km}) - \left(R_{km}^2 + X_{km}^2\right) I_{km}^2 \quad \forall km \in \mathrm{BR} \tag{2.3}$$

$$V_m^2 I_{km}^2 = P_{km}^2 + Q_{km}^2 \quad \forall km \in \mathrm{BR} \tag{2.4}$$

where constraints (2.1) and (2.2) are the active and reactive power injections; (2.3) describes the forward voltage drop in each line and (2.4) defines apparent power flow injection at the head bus of each line. Equations (2.1)–(2.4) are frequently used in the power flow sweep method of radial networks and can be used to formulate the MINLP model for the EDNEP problem.

Without loss generality, the power flow optimization problem can be formulated using the above steady-state equations of radial networks, including an objective function that minimizes real power loss [27]. The compact form of the non-linear ac power flow problem can be expressed as

$$\begin{aligned}
&\min \sum_{j\in\alpha(k)} R_{km} I_{km}^2 \quad \forall km \in \mathrm{BR}\\
&\text{subject to:}\\
&\text{constraints } (2.1) - (2.4)\\
&P_k = P_k^{SE} - P_k^{D} \quad \forall km \in \mathrm{B}\\
&Q_k = Q_k^{SE} - Q_k^{D} \quad \forall km \in \mathrm{B}
\end{aligned} \tag{2.5}$$

where the two last equations are the nodal active and reactive power balancing conditions, respectively.

The optimization problem (2.5) is a non-linear and non-convex in the ac power flow for radial networks due to the quadratic terms in their constraints and objective function. However, the square terms in (2.1)–(2.4) and the objective function in (2.5) can be dropped using auxiliary variables that represent those square terms, $d_k = V_k^2$ and $l_{km} = I_{km}^2$, and (2.4) can be relaxed to a convex constraint by second order conic programming (SOCP) [41], where it is relaxed to an inequality constraint. The relaxed convex ac power flow problem for radial networks can be expressed as

$$\min \sum_{j \in \alpha(k)} R_{km} l_{km} \tag{2.6}$$

$$P_k^{SE} - P_k^D = \sum_{j \in \alpha(k)} (P_{km} + R_{km} l_{km}) - \sum_{j \in \alpha(k)} P_{jk} \quad \forall k \in \mathrm{B} \tag{2.7}$$

$$Q_k^{SE} - Q_k^D = \sum_{j \in \alpha(k)} (Q_{km} + X_{km} l_{km}) - \sum_{j \in \alpha(k)} Q_{jk} \quad \forall k \in \mathrm{B} \tag{2.8}$$

$$d_k - d_m = 2(R_{km} P_{km} + X_{km} Q_{km}) - \left(R_{km}^2 + X_{km}^2\right) l_{km} \quad \forall km \in \mathrm{BR} \tag{2.9}$$

$$d_m l_{km} \geq P_{km}^2 + Q_{km}^2 \quad \forall km \in \mathrm{BR} \tag{2.10}$$

2.4 Convex Model for the EDNEP Problem

In this section, an optimization model is presented that includes the minimization of investment costs by installing new feeders and/or changing existing ones for others with larger capacities, installing new substations or expanding existing ones and finally, installing capacitor banks and voltage regulators, as well as the timing to add new assets or expand existing ones in case of dynamic (multi-stage) planning, including the operating costs associated with energy loss. Equations related to each device considered are depicted to capture the physical laws that govern them. Then, a convex formulation is derived to build a MICQP optimization problem that benefits from the advances of off-the-shelf MICQP solvers. It is worth to clarify that Mixed-Integer Conic Quadratic Programming (MICQP) problems are non-convex because of the integer nature of some decision variables. However, MICQP resolution algorithms iteratively solve optimization problems where integer variables are fixed. In this case, our problem is convex. Additionally, in this case, MICQP theory guarantees global solution of the optimization problem. For further details about Conic Programming theory, interested readers are referred to [42].

2.4.1 Capacitor Bank Model

In electric distribution networks, there are two types of capacitor banks: (1) Fixed Capacitor Banks (FCB) and (2) Switched Capacitor Banks (SCB) [43], as shown in Fig. 2.1. FCBs are capacitor banks composed of units which, after being installed in the planning stage, are always connected throughout all load levels; whereas, SCBs are composed of units which, after being installed at the planning stage, can be fully or partially connected at every load level. According to Fig. 2.1a, expressions (2.11)–(2.13) model the allocation and operation of FCBs in the networks.

$$q_k^{fcb} = Q^{fc} n_k^{fc} \quad \forall k \in \text{FCB} \tag{2.11}$$

$$0 \le n_k^{fc} \le N_k^{fcb} \quad \forall k \in \text{FCB} \tag{2.12}$$

$$n_k^{fc} \in \mathbb{Z}^+ \quad \forall k \in \text{FCB} \tag{2.13}$$

Equation (2.11) represents the reactive power produced by the fixed capacitor banks installed at node k. Constraint (2.12) limits the number of units to be installed

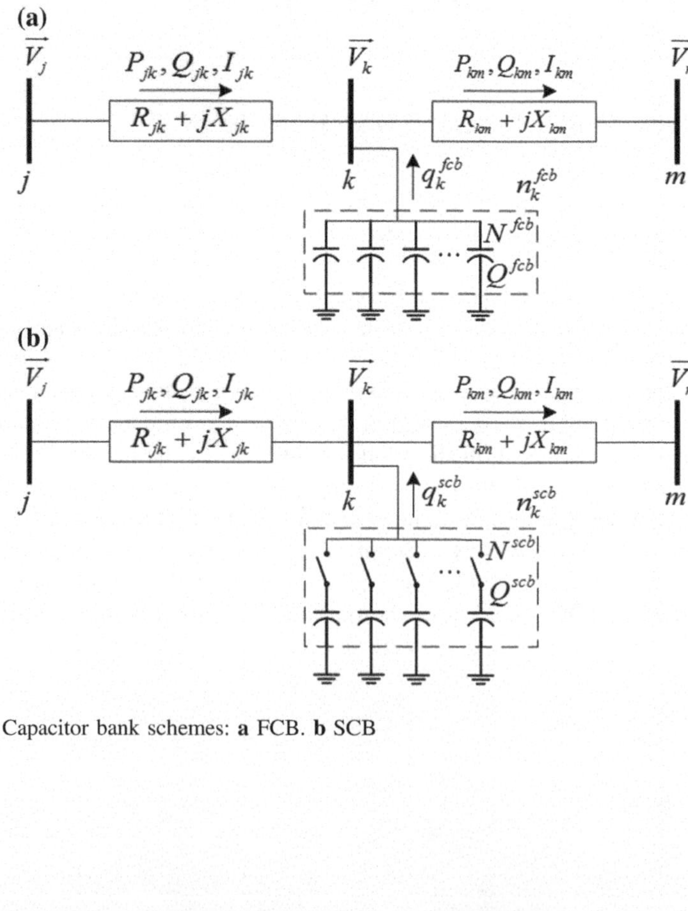

Fig. 2.1 Capacitor bank schemes: **a** FCB. **b** SCB

in a fixed capacitor bank and (2.13) establishes the integrality condition for the number of units installed.

To model SCBs (Fig. 2.1b), it is important to take into account the load variations in a given period of time, therefore, these equations are modeled considering the time period and can only be included in the dynamic model. Expressions (2.14)–(2.16) represent the SCBs installation and operation.

$$q_{k,t}^{scb} = Q^{sc} n_{k,t}^{sc} \quad \forall k \in \mathrm{SCB},\ \forall t \in \mathrm{P} \tag{2.14}$$

$$0 \leq n_{k,t}^{sc} \leq N_k^{scb} \quad \forall k \in \mathrm{SCB},\ \forall t \in \mathrm{P} \tag{2.15}$$

$$n_{k,t}^{sc} \in \mathbb{Z}^+ \quad \forall k \in \mathrm{SCB},\ \forall t \in \mathrm{P} \tag{2.16}$$

where (2.14) represents the reactive power produced by the fixed capacitors banks installed at node *k*. Constraint (2.15) limits the number of units to be installed in a fixed capacitor bank and (2.16) establishes the integrality condition for the number of units installed.

2.4.2 Voltage Regulator Model

To model the voltage regulator, consider an autotransformer with an automatic changing mechanism of the tap position (number of turns) of the series winding to maintain a predetermined level of voltage magnitude along an electric distribution feeder in case of load level variations. Standard voltage regulators contain a reversing switch that enables a regulating range, which determines the tap step-size Δ (+ increases and – decreases the voltage magnitude), taking into account the reference voltage magnitude and the maximum number of steps, *Ns*, as shown in Fig. 2.2, where Δ and *Ns* are known parameters, e.g. 0.00625, 32 steps, respectively, *a* and *tp* are the tap setting and tap position, respectively, which are considered variables in the planning and operation of electric distribution networks.

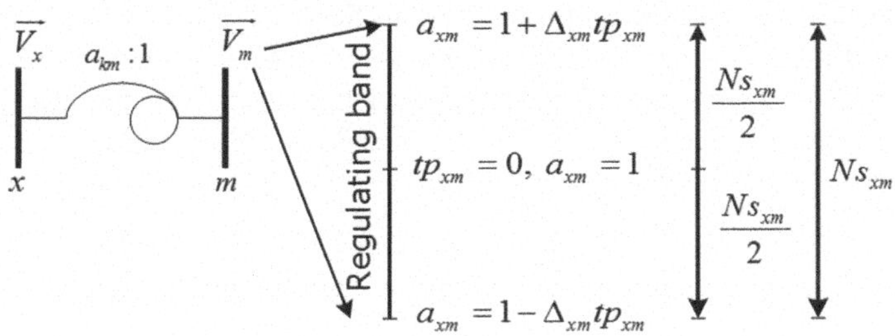

Fig. 2.2 Voltage regulator in operation

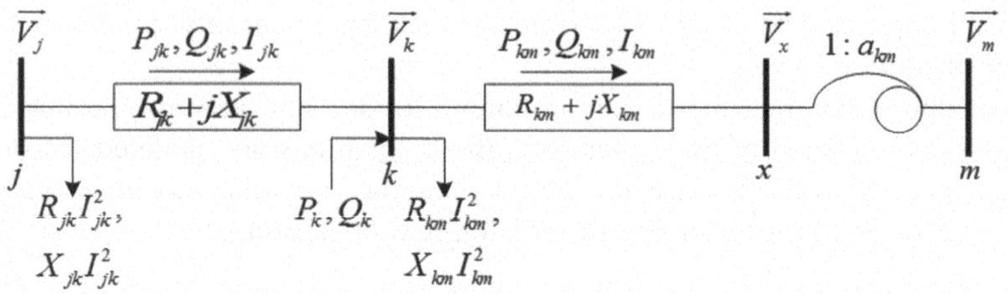

Fig. 2.3 Feeder with a voltage regulator model

Consider the voltage regulator located at node m of branch km, where node x is the non-regulated voltage magnitude node, as shown in Fig. 2.3, where a partial electric distribution feeder with a voltage regulator is shown. The voltage regulator in branch km can be divided into branch kx and branch xm, where branch xm only contains the tap changer and the impedance kx is the same as branch km. With these considerations, similar to DistFlow [38], the injected active and reactive power equations of the voltage regulator can be described as

$$P_k = \sum_{j \in \alpha(k)} \left(P_{km} + R_{km} I_{km}^2\right) - \sum_{j \in \alpha(k)} P_{jk} \quad \forall k \in \mathrm{B} \tag{2.17}$$

$$Q_k = \sum_{j \in \alpha(k)} \left(Q_{km} + X_{km} I_{km}^2\right) - \sum_{j \in \alpha(k)} Q_{jk} \quad \forall k \in \mathrm{B} \tag{2.18}$$

$$V_k^2 - V_x^2 = 2(R_{km}P_{km} + X_{km}Q_{km}) - \left(R_{km}^2 + X_{km}^2\right) I_{km}^2 \quad \forall km \in \mathrm{VR} \tag{2.19}$$

$$V_m^2 I_{km}^2 = P_{km}^2 + Q_{km}^2 \quad \forall km \in \mathrm{VR} \tag{2.20}$$

$$V_m^2 = a_{km}^2 V_x^2 \quad \forall km \in \mathrm{VR} \tag{2.21}$$

where constraints (2.17) and (2.18) are the active and reactive power injections; (2.19) describes the forward voltage drop in each line, (2.20) defines the apparent power flow injection at the head bus of each line and (2.21) is the voltage magnitude regulated by the voltage regulator. Equations (2.17)–(2.20) can be convexified in a similar form to the ac power flow in Sect. 2.3.

Equation (2.21) can be transformed into several mixed integer linear expressions. In order to do so, the voltage regulator tap-changer is formulated as

$$a_{km} = 1 + \Delta_{km} tp_{km} \quad \forall km \in \mathrm{VR} \tag{2.22}$$

$$\Delta_{km} = \frac{\left(a_{km}^{\max} - a_{km}^{\min}\right)}{Ns_{km}} \quad \forall km \in \mathrm{VR} \tag{2.23}$$

$$-\frac{Ns_{km}}{2}\alpha_{km}^{vr} \leq tp_{km} \leq \frac{Ns_{km}}{2}\alpha_{km}^{vr} \quad \forall km \in \mathrm{VR} \tag{2.24}$$

$$tp_{km} \in \mathbb{Z} \quad \forall km \in \mathrm{VR} \tag{2.25}$$

where (2.22) is the expression that determines the tap setting, (2.23) computes the tap step-size, (2.24) limits the number of tap positions and states if it is necessary to install a voltage regulator in the electric distribution network and (2.25) establishes the integrality condition to the number of tap positions.

Integer variable tp_{km} can be expressed by the binary expansion scheme [44] as

$$tp_{km} = \sum_{n=0}^{Ns_{km}} \left[\left(n - \frac{Ns_{km}}{2} \right) \alpha_{kmn} \right] \quad \forall km \in \mathrm{VR} \tag{2.26}$$

$$\sum_{n=0}^{Ns_{km}} \alpha_{kmn} = 1 \quad \forall km \in \mathrm{VR} \tag{2.27}$$

where $\alpha_{kmn} \in \{0/1\}$. Replacing (2.26) in (2.22), yields

$$a_{km} = 1 + \Delta_{km} \sum_{n=0}^{Ns_{km}} \left[\left(n - \frac{Ns_{km}}{2} \right) \alpha_{kmn} \right] \quad km \in \mathrm{VR} \tag{2.28}$$

Using the auxiliary variable $d_k = V_k^2$ in (2.21), one obtains

$$d_m = a_{km}^2 d_x \quad \forall km \in \mathrm{VR} \tag{2.29}$$

Multiplying both sides of (2.28) by d_x, results in

$$a_{km} d_x = d_x + \Delta_{km} \sum_{n=0}^{Ns_{km}} \left[\left(n - \frac{Ns_{km}}{2} \right) \alpha_{kmn} d_x \right] \quad \forall km \in \mathrm{VR} \tag{2.30}$$

Defining auxiliary variables $z_{km} = a_{km} d_x$ and $x_{kmn} = \alpha_{kmn} d_x$, one obtains

$$z_{km} = d_x + \Delta_{km} \sum_{n=0}^{Ns_{km}} \left[\left(n - \frac{Ns_{km}}{2} \right) x_{kmn} \right] \quad \forall km \in \mathrm{VR} \tag{2.31}$$

From (2.29), $d_m = a_{km} z_{km}$, and replacing (2.31) yields

$$d_m = z_{km} + \Delta_{km} \sum_{n=0}^{Ns_{km}} \left[\left(n - \frac{Ns_{km}}{2} \right) \alpha_{kmn} z_{km} \right] \quad \forall km \in \mathrm{VR} \tag{2.32}$$

Defining the auxiliary variable $y_{kmn} = \alpha_{kmn} z_{km}$, one obtains

$$d_m = z_{km} + \Delta_{km} \sum_{n=0}^{Ns_{km}} \left[\left(n - \frac{Ns_{km}}{2} \right) y_{kmn} \right] \quad \forall km \in \text{VR} \tag{2.33}$$

Auxiliary variables x_{kmn} and y_{kmn} are the products of binary and continuous variables, which can be reformulated into an exact linear set of constraints using the big-M method, resulting in

$$-M\alpha_{kmn} \leq x_{kmn} \leq M\alpha_{kmn} \quad \forall km \in \text{VR}, \forall n = 0, 1, 2, \ldots, Ns_{km} \tag{2.34}$$

$$-M(1 - \alpha_{kmn}) \leq d_x - x_{kmn} \leq M(1 - \alpha_{kmn}) \quad \forall km \in \text{VR}, \forall n = 0, 1, 2, \ldots, Ns_{km} \tag{2.35}$$

$$-M\alpha_{kmn} \leq y_{kmn} \leq M\alpha_{kmn} \quad \forall km \in \text{VR}, \forall n = 0, 1, 2, \ldots, Ns_{km} \tag{2.36}$$

$$-M(1 - \alpha_{kmn}) \leq z_{km} - y_{kmn} \leq M(1 - \alpha_{kmn}) \quad \forall km \in \text{VR}, \forall n = 0, 1, 2, \ldots, Ns_{km} \tag{2.37}$$

2.4.3 Static Convex EDNEP Model

As previously explained, the static model considers the whole planning horizon in a single target period. In addition, the steady state, capacitor banks and voltage regulators convex formulations previously developed are taken into account. The static convex EDNEP model can be formulated as

$$\begin{aligned}
\min f = {} & K_L \sum_{km \in \text{BR}} \sum_{c \in \text{C}} C^C_{km,c} \alpha^C_{km,c} L_{km} + K_S \sum_{k \in \text{SE}} \left(C^{SE}_k \alpha^{SE}_k + C^{SER}_k \alpha^{SER}_k \right) \\
& + K_C \sum_{k \in \text{FCB}} C^{FCB}_k n^{fc}_k + K_{VR} \sum_{km \in \text{VR}} C^{VR}_k \alpha^{vr}_{km} \\
& + 8760 \text{C}_{loss} \sum_{k \in \text{B}} \left(P^{SE}_k - P^D_k \right) + 8760 \sum_{k \in \text{B}} C^{OS}_k \left[\left(P^{SE}_k \right)^2 + \left(Q^{SE}_k \right)^2 \right]
\end{aligned} \tag{2.38}$$

$$P^{SE}_k - P^D_k = \sum_{km \in \text{BR}} \sum_{c \in \text{C}} \left(P_{km,c} + R_{km,c} l_{km,c} \right) - \sum_{jk \in \text{BR}} \sum_{c \in \text{C}} P_{jk,c} \quad \forall k \in \text{B} \tag{2.39}$$

$$Q^{SE}_k - Q^D_k + q^{fcb}_k = \sum_{km \in \text{BR}} \sum_{c \in \text{C}} \left(Q_{km,c} + X_{km,c} l_{km,c} \right) - \sum_{jk \in \text{BR}} \sum_{c \in \text{C}} Q_{jk,c} \quad \forall k \in \text{B} \tag{2.40}$$

$$d_k \le M(1-\alpha_{km}) + d_m - \sum_{c \in C}\left[2\left(R_{km,c}P_{km,c} + X_{km,c}Q_{km,c}\right) - \left(R_{km,c}^2 + X_{km,c}^2\right)l_{km,c}\right] \quad \forall km \in BR \tag{2.41}$$

$$d_k \ge -M(1-\alpha_{km}) + d_m - \sum_{c \in C}\left[2\left(R_{km,c}P_{km,c} + X_{km,c}Q_{km,c}\right) - \left(R_{km,c}^2 + X_{km,c}^2\right)l_{km,c}\right] \quad \forall km \in BR \tag{2.42}$$

$$d_m l_{km} \ge P_{km}^2 + Q_{km}^2 \quad \forall km \in BR \tag{2.43}$$

$$\left(V_k^2\right)^{\min} \le d_k \le \left(V_k^2\right)^{\max} \quad \forall km \in B \tag{2.44}$$

$$0 \le l_{km,c} \le \left(I_{km,c}^2\right)^{\max} \alpha_{km,c}^C \quad \forall km \in BR, \forall c \in C \tag{2.45}$$

$$\beta_{km} = \sum_{c \in C} \alpha_{km,c}^C \le 1 \quad \forall km \in BR \tag{2.46}$$

$$\sum_{km \in BR} \beta_{km} = |B| - |SE0| - \sum_{k \in SE} \alpha_k^{SE} \tag{2.47}$$

$$\left(P_k^{SE}\right)^2 + \left(Q_k^{SE}\right)^2 \le \left(S_k^{SE0}\right)^2 + \left[2S_k^{SE0}S_k^{SE} + \left(S_k^{SE}\right)^2\right]\alpha_k^{SE} \quad \forall k \in SE0 \cup SE \tag{2.48}$$

$$-\frac{Ns_{km}}{2}\alpha_{km}^{vr} \le \sum_{n=0}^{Ns_{km}}\left[\left(n - \frac{Ns_{km}}{2}\right)y_{kmn}\right] \le \frac{Ns_{km}}{2}\alpha_{km}^{vr} \quad \forall km \in VR \tag{2.49}$$

$$\text{Constraints } (2.11)-(2.13), (2.23), (2.27), (2.31), (2.33), (2.34)-(2.37) \tag{2.50}$$

$$\alpha_{km,c}^C \in \{0,1\} \quad \forall km \in BR, \forall c \in C \tag{2.51}$$

$$\alpha_k^{SE} \in \{0,1\} \quad \forall km \in SE \tag{2.52}$$

$$\alpha_{km} \in \{0,1\} \quad \forall km \in BR \tag{2.53}$$

$$\alpha_{km}^{vr} \in \{0,1\} \quad \forall km \in VR \tag{2.54}$$

where objective function (2.38) represents the annualized investment and operation costs. The first term represents the costs associated with the new feeders installation and/or changing existing ones, the second term represents the costs of installing or expanding substations, the third and fourth terms represent the costs associated with installing fixed capacitor banks and voltage regulators, respectively, and the fifth and sixth terms represent the costs associated with energy losses and substation operation, respectively.

Equations (2.39) and (2.40) represent the active and reactive power balance nodal equations. Equations (2.41) and (2.42) describe the forward voltage drops in each branch while (2.43) defines the apparent power flow injection at the head bus of each branch. The limits of voltage and current magnitudes are established in (2.44) and (2.45). Equation (2.46) accounts for the case when a line with a specific conductor type is already installed, by setting the corresponding binary variable to 1, or the case when a line is already installed but can have its conductor type changed by replacing the inequality sign in the summation of (2.46) with an equality. Equation (2.47), together with power balance constraints (2.39) and (2.40), ensures that the network has a number of trees equal to the number of existing and installed substations. Equation (2.48) indicates substation installation ($S_k^{SE0} = 0$ and $S_k^{SE} \neq 0$) and substation expansion ($S_k^{SE0} \neq 0$ and $S_k^{SE} \neq 0$) and binary variable α_k^{SE} is related to the substation capacity limit. Equation (2.49) states whether it is necessary to install a voltage regulator in the electric distribution network, i.e.$\alpha_{km}^{vr} = 1$. Finally, constraints (2.51)–(2.54) represent the binary nature of the variables. The static model in (2.38)–(2.54) is a convex MICQP formulation for the EDNEP problem that guarantees global optimum solutions.

2.4.4 Dynamic Convex EDNEP Model

As previously explained, in the EDNEP dynamic model, the decisions are made at different points in time. In order to do so, the proposed static model in (2.38)–(2.54) can be extended to a dynamic (multi-period) model by adding the period index and some constraints, as

$$\begin{aligned}
\min f = {} & K_L \sum_{t \in \mathrm{P}} \sum_{km \in \mathrm{BR}} \sum_{c \in \mathrm{C}} C_{km,c}^{C} \alpha_{km,c,t}^{C} L_{km} + K_S \sum_{t \in \mathrm{P}} \sum_{k \in \mathrm{SE}} \left(C_k^{SE} \alpha_{k,t}^{SE} + C_k^{SER} \alpha_{k,t}^{SER} \right) \\
& + K_C \sum_{t \in \mathrm{P}} \sum_{k \in \mathrm{FCB}} C_k^{FCB} n_{k,t}^{fc} + K_C \sum_{t \in \mathrm{P}} \sum_{k \in \mathrm{FCB}} C_k^{SCB} n_{k,t}^{sc} + K_{VR} \sum_{t \in \mathrm{P}} \sum_{km \in \mathrm{VR}} C_k^{VR} \alpha_{km,t}^{vr} \\
& + \mathrm{C}_{loss} \sum_{t \in \mathrm{P}} \sum_{k \in \mathrm{B}} \left(P_{k,t}^{SE} - P_{k,t}^{D} \right) + \sum_{t \in \mathrm{P}} \sum_{k \in \mathrm{B}} C_k^{OS} \left[\left(P_{k,t}^{SE} \right)^2 + \left(Q_{k,t}^{SE} \right)^2 \right]
\end{aligned} \tag{2.55}$$

$$P_{k,t}^{SE} - P_{k,t}^{D} = \sum_{km \in \mathrm{BR}} \sum_{c \in \mathrm{C}} \left(P_{km,c,t} + R_{km,c} l_{km,c,t} \right) - \sum_{jk \in \mathrm{BR}} \sum_{c \in \mathrm{C}} P_{jk,c,t} \quad \forall k \in \mathrm{B}, \forall t \in \mathrm{P} \tag{2.56}$$

$$Q_{k,t}^{SE} - Q_{k,t}^{D} + q_{k,t}^{fcb} + q_{k,t}^{scb} = \sum_{km \in BR} \sum_{c \in C} \left(Q_{km,c,t} + X_{km,c} l_{km,c,t}\right) - \sum_{jk \in BR} \sum_{c \in C} Q_{jk,c,t} \quad \forall k \in B, \forall t \in P \tag{2.57}$$

$$d_{k,t} \leq M\left(1 - \alpha_{km,t}\right) + d_{m,t} - \sum_{c \in C} \left[2\left(R_{km,c} P_{km,c,t} + X_{km,c} Q_{km,c,t}\right) - \left(R_{km,c}^2 + X_{km,c}^2\right) l_{km,c,t}\right] \quad \forall km \in BR, \forall t \in P \tag{2.58}$$

$$d_{k,t} \geq -M\left(1 - \alpha_{km,t}\right) + d_{m,t} - \sum_{c \in C} \left[2\left(R_{km,c} P_{km,c,t} + X_{km,c} Q_{km,c,t}\right) - \left(R_{km,c}^2 + X_{km,c}^2\right) l_{km,c,t}\right] \quad \forall km \in BR, \forall t \in P \tag{2.59}$$

$$d_{m,t} l_{km,t} \geq P_{km,t}^2 + Q_{km,t}^2 \quad \forall km \in BR, \forall t \in P \tag{2.60}$$

$$\left(V_k^{\min}\right)^2 \leq d_{k,t} \leq \left(V_k^{\max}\right)^2 \quad \forall km \in B, \forall t \in P \tag{2.61}$$

$$0 \leq l_{km,c,t} \leq \left(I_{km,c}^{\max}\right)^2 \alpha_{km,c,t}^{C} \quad \forall km \in BR, \forall c \in C, \forall t \in P \tag{2.62}$$

$$\beta_{km,t} = \sum_{c \in C} \alpha_{km,c,t}^{C} \leq 1 \quad \forall km \in BR, \forall t \in P \tag{2.63}$$

$$\sum_{km \in BR} \beta_{km,t} = |B| - |SE0| - \sum_{k \in SE} \alpha_{k,t}^{SE} \quad \forall t \in P \tag{2.64}$$

$$\left(P_{k,t}^{SE}\right)^2 + \left(Q_{k,t}^{SE}\right)^2 \leq \left(S_k^{SE0}\right)^2 + \left[2 S_k^{SE0} S_k^{SE} + \left(S_k^{SE}\right)^2\right] \alpha_{k,t}^{SE} \quad \forall k \in SE0 \cup SE, \forall t \in P \tag{2.65}$$

$$q_{k,t}^{fcb} = Q^{fc} n_{k,t}^{fc} \quad \forall k \in FCB, \forall t \in P \tag{2.66}$$

$$0 \leq n_{k,t}^{fc} \leq N_k^{fcb} \quad \forall k \in FCB, \forall t \in P \tag{2.67}$$

$$q_{k,t}^{scb} = Q^{sc} n_{k,t}^{sc} \quad \forall k \in SCB, \forall t \in P \tag{2.68}$$

$$0 \leq n_{k,t}^{sc} \leq N_k^{scb} \quad \forall k \in SCB, \forall t \in P \tag{2.69}$$

$$\Delta_{km} = \frac{\left(a_{km}^{\max} - a_{km}^{\min}\right)}{Ns_{km}} \quad \forall km \in VR \tag{2.70}$$

$$z_{km,t} = d_{x,t} + \Delta_{km} \sum_{n=0}^{Ns_{km}} \left[\left(n - \frac{Ns_{km}}{2} \right) x_{kmn,t} \right] \quad \forall km \in \text{VR}, \forall t \in \text{P} \tag{2.71}$$

$$d_{m,t} = z_{km,t} + \Delta_{km} \sum_{n=0}^{Ns_{km}} \left[\left(n - \frac{Ns_{km}}{2} \right) y_{kmn,t} \right] \quad \forall km \in \text{VR}, \ \forall t \in \text{P} \tag{2.72}$$

$$-M\alpha_{kmn,t} \leq x_{kmn,t} \leq M\alpha_{kmn,t} \quad \forall km \in \text{VR}, \forall t \in \text{P}, \forall n = 0, 1, 2, \ldots, Ns_{km} \tag{2.73}$$

$$\begin{gathered} -M(1 - \alpha_{kmn,t}) \leq d_{x,t} - x_{kmn,t} \leq M(1 - \alpha_{kmn,t}) \\ \forall km \in \text{VR}, \forall t \in \text{P}, \forall n = 0, 1, 2, \ldots, Ns_{km} \end{gathered} \tag{2.74}$$

$$-M\alpha_{kmn,t} \leq y_{kmn,t} \leq M\alpha_{kmn,t} \quad \forall km \in \text{VR}, \forall t \in \text{P}, \forall n = 0, 1, 2, \ldots, Ns_{km} \tag{2.75}$$

$$\begin{gathered} -M(1 - \alpha_{kmn,t}) \leq z_{km,t} - y_{kmn,t} \leq M(1 - \alpha_{kmn,t}) \\ \forall km \in \text{VR}, \forall t \in \text{P}, \forall n = 0, 1, 2, \ldots, Ns_{km} \end{gathered} \tag{2.76}$$

$$\sum_{n=0}^{Ns_{km}} \alpha_{kmn,t} = 1 \quad \forall km \in \text{VR}, \forall t \in \text{P} \tag{2.77}$$

$$-\frac{Ns_{km}}{2} \alpha_{km,t}^{vr} \leq \sum_{n=0}^{Ns_{km}} \left[\left(n - \frac{Ns_{km}}{2} \right) y_{kmn,t} \right] \leq \frac{Ns_{km}}{2} \alpha_{km,t}^{vr} \quad \forall km \in \text{VR}, \forall t \in \text{P} \tag{2.78}$$

$$\sum_{t \in \text{P}} \alpha_{km,c,t}^{C} = 1 \quad \forall t \in \text{P} \tag{2.79}$$

$$\sum_{t \in \text{P}} \alpha_{k,t}^{SE} = 1 \quad \forall t \in \text{P} \tag{2.80}$$

$$\sum_{t \in \text{P}} \alpha_{km,t}^{vr} = 1 \quad \forall t \in \text{P} \tag{2.81}$$

$$n_{k,t}^{sc} \in \mathbb{Z}^{+} \quad \forall k \in \text{SCB}, \forall t \in \text{P} \tag{2.82}$$

$$n_{k,t}^{fc} \in \mathbb{Z}^{+} \quad \forall k \in \text{FCB}, \forall t \in \text{P} \tag{2.83}$$

$$\alpha_{km,c,t}^{C} \in \{0, 1\} \quad \forall km \in \text{BR}, \forall c \in \text{C}, \forall t \in \text{P} \tag{2.84}$$

$$\alpha_{k,t}^{SE} \in \{0, 1\} \quad \forall km \in \text{SE}, \forall t \in \text{P} \tag{2.85}$$

$$\alpha_{km,t} \in \{0, 1\} \quad \forall km \in \text{BR}, \forall t \in \text{P} \tag{2.86}$$

$$\alpha_{km,t}^{vr} \in \{0, 1\} \quad \forall km \in \text{VR}, \forall t \in \text{P} \tag{2.87}$$

where (2.55)–(2.87) represent the convex dynamic (multi-period) EDNEP problem, which was extended from a convex static EDNEP problem to a study for all time periods. In this problem, constraints (2.79)–(2.81) has added, limiting the pieces of equipment (feeders, substations and voltage regulators) to be installed in a specific branch or node along the planning horizon.

2.5 Numerical Results

2.5.1 Data Specifications

Two cases are used for the solution of the EDNEP problem using the 24-node electric distribution network [45]: (1) a static test case and (2) a dynamic test case. The system considered has 24 nodes, 4 substations, 20 load demand nodes and 34 branches, operating at 13.8 kV. For the static model, the planning horizon considered is 20 years in only one stage. For the dynamic model, the planning horizon considered is 20 years divided into 4 periods of 5 years each. These models have been solved using the CPLEX optimization solver [46] in AMPL [47], in a Dell PC server with 256 GB of RAM memory and 2.27 GHz.

The base topology of the 24-node network is shown in Fig. 2.4. Investment alternatives for substations and conductors, which were adopted from [20], are shown in Tables 2.1 and 2.2. Load demand data for each period is shown in Table 2.3. The costs and data related to the capacitor banks and voltage regulators are adapted from [27]. The location cost of capacitor banks is US$1500, the cost of each module is equal to US$1000, the capacity of each module is equal to 300 kVar. For other hand, the location cost of voltage regulators is equal to US $8400 and 32 steps. The interest rate is set at 13%. The price of energy is equal to $0.25 /kWh, and finally the minimum and maximum voltage magnitude limits are 0.95 and 1.05 pu, respectively.

2.5.2 Static Test Case

This case is solved in 6 min 45 s, the objective function found is US$118,321,152. The EDNEP problem is solved taking into account the load demand in the last stage (year 20), but the decisions are made in the first year. The solutions obtained are as follows: substations 23 and 24 are built, feeders 2–12, 4–9, 10–16 and 7–19 are built with conductor type 1, feeders 4–15, 15–17, 1–14, 5–24, 7–23, 10–23, 11–23, 17–22, 18 – 24 and 13–20 are built with conductor type 2 and feeders 1–21 and 6–22 are changed with conductor type 1. In addition, FCB with 1800 kVar are

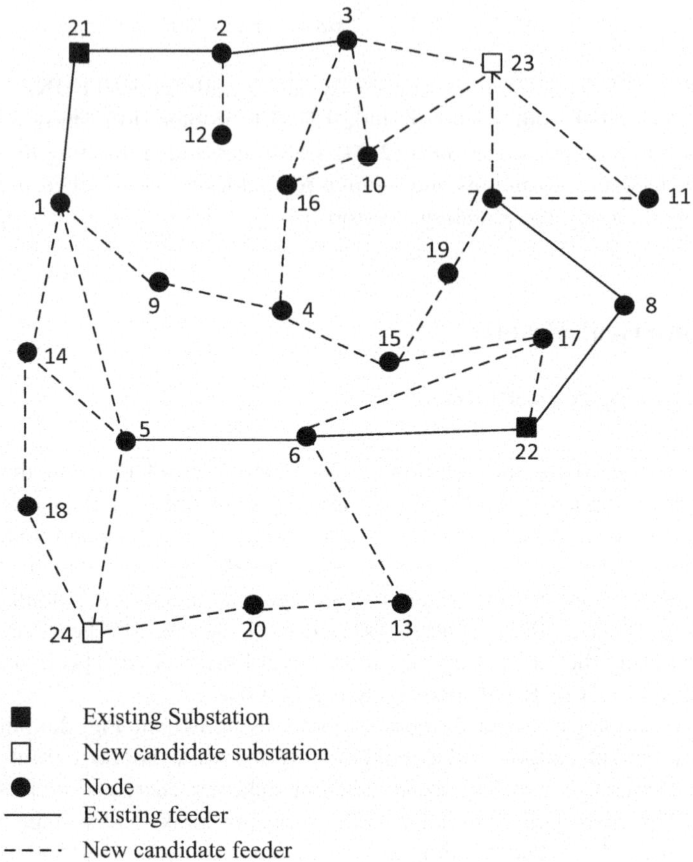

Fig. 2.4 Base topology of the 24-node electric distribution network

Table 2.1 Conductor's data

C	$R_{km,c}(\Omega/\text{km})$	$X_{km,c}(\Omega/\text{km})$	$C^{C}_{km,c}$ (\$/km)
1	0.3655	0.2520	20×10^3
2	0.2921	0.2466	30×10^3
3	0.2359	0.2402	40×10^3

Table 2.2 Substations data

SE	$S_k^{SE}(kVA)$	$S_k^{SER}(kVA)$	$C_k^{SE}(\$)$	$C_k^{SER}(\$)$
21	10,000	8000	0.00	1×10^6
22	15,000	12,000	0.00	1×10^6
23	20,000	0.00	5×10^6	0.00
24	25,000	0.00	8×10^6	0.00

Table 2.3 Load data (kVA)

Bus	$t = 1$	$t = 2$	$t = 3$	$t = 4$	Bus	$t = 1$	$t = 2$	$t = 3$	$t = 4$
1	4050	4658	5356	6160	13	0	1350	1553	1785
2	780	897	1032	1186	14	0	3280	3772	4338
3	2580	2967	3412	3924	15	0	1460	1679	1931
4	320	368	423	487	16	0	0	1530	1760
5	280	322	370	426	17	0	2330	2680	3081
6	1170	1346	1547	1779	18	0	0	2310	2657
7	4040	4646	5343	6144	19	0	0	1750	2013
8	720	828	952	1095	20	0	0	4020	4623
9	1140	1311	1508	1734	21	0	0	0	0
10	1560	1794	2063	2373	22	0	0	0	0
11	0	2000	2300	2645	23	0	0	0	0
12	0	850	978	1124	24	0	0	0	0

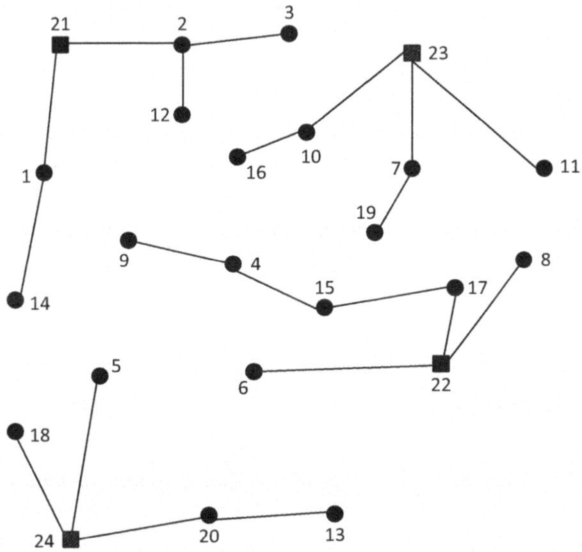

Fig. 2.5 Solution topology of the 24-node electric distribution network in static case

located at nodes 1, 3, 7, 9, 13, 14 and one voltage regulator is located in circuit 4–15 with a tap position of 6. The topology obtained from the optimization process is shown in Fig. 2.5.

2.5.3 Dynamic Test Case

This case is solved in 18 min 23 s, the objective function found is US$55,654,205. In period 1, feeders 4–9, 17–22, 4–16 are built with conductor type 1, while feeders 10–16, 4–15 and 17–22 are built with conductor type 2 and feeder 8–22 is changed with conductor type 1. FCB with 1200 kVar is located at node 1 and FCB with 900 kVar is located at nodes 3, 7, 9 and 5.

In period 2, substation 23 is built and feeders 2–12, 1–14 and 11–23 are built with conductor type 1, while feeders 6–13 and 7–23 are built with conductor type 2 and feeder 1–21 is changed with conductor type 1. FCB with a capacity of 1200 kVar is installed at node 14. In addition, one voltage regulator is installed in branch 4–15 with a tap position of 5.

In period 3, feeder 10–23 is built with conductor type 1 and feeders 7–19, 14–18, and 13–20 are built with conductor type 2. In addition, the voltage regulator installed in branch 4–15 in period 2 changes the tap position to 2.

In period 4, substation 24 is built and feeders 24–18, 24–5 and 24–20 are built with conductor type 1 and feeder 1–9 is built with conductor type 2. In addition, the voltage regulator installed in branch 4–15 in period 2 changes the tap position to 1. The topologies obtained from the optimization process in each period are shown in Fig. 2.6.

The results found for the static and dynamic test cases show that the solutions obtained using the dynamic model have lower costs than those generated using the static model. This can be explained by the appropriate execution of investments in the dynamic model. Note that the cost difference between the static and the dynamic test cases is US$62,666,947.87, which represents a 47.03% reduction in total investment cost.

Appendix

The notations used throughout this chapter are listed below:

Sets

B	Set of nodes.
BR	Set of branches.
C	Set of conductor types.
FCB	Set of fixed capacitor banks.
P	Set of periods.
SE0	Set of existing substations.

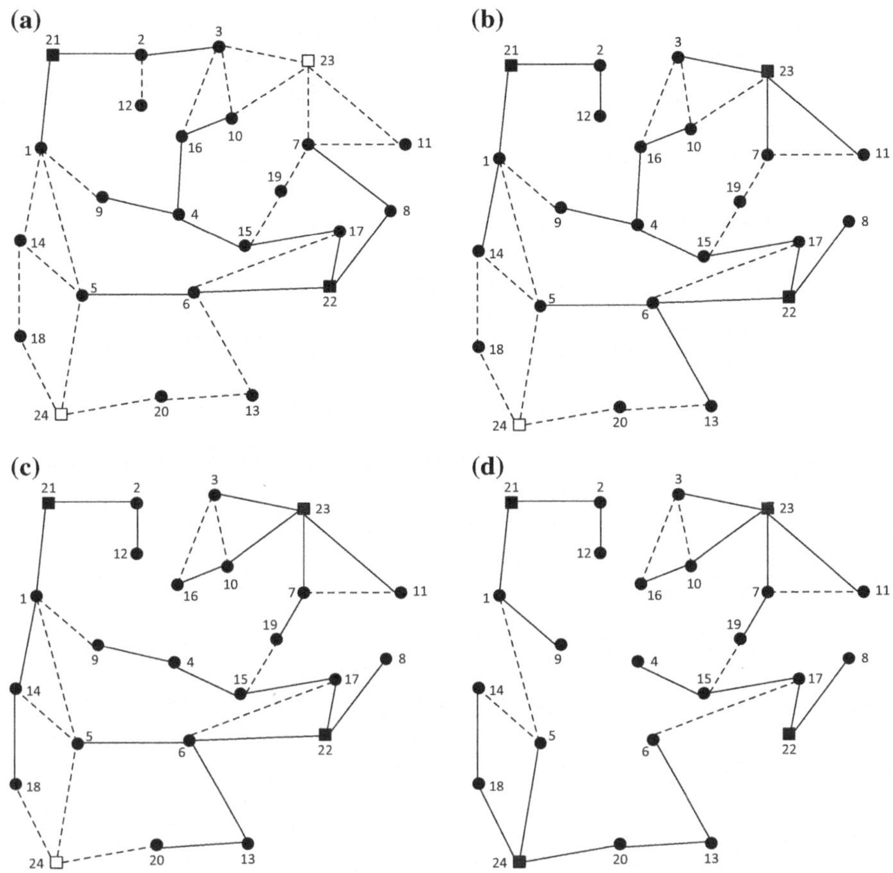

Fig. 2.6 Solution topologies of the 24-node electric distribution network in dynamic case: **a** In period 1. **b** In period 2. **c** In period 3. **d** In period 4

SE	Set of new installed substations.
SCB	Set of switched capacitor banks.
VR	Set of voltage regulators.
$\mathbb{Z}^+$	Set of positive integer numbers.

Parameters

$a_{km}^{\max}, a_{km}^{\min}$	Upper and lower tap settings of voltage regulator in branch *km*.
$C_{km,c}^{C}$	Installation cost of branch *km* using conductor type *c* (US$).
C_{k}^{SE}	Installation cost of new substation *k* (US$).
C_{k}^{SER}	Expanding cost of substation *k* (US$).

C_k^{FCB}	Installation cost of fixed capacitor bank in node k (US$).
C_k^{SCB}	Installation cost of switched capacitor bank in node k (US$).
C_{km}^{VR}	Installation cost of voltage regulator in branch km (US$).
C_{loss}	Real power loss cost (US$).
C_k^{OS}	Operating cost of substation k (US$).
$I_{km,c}^{\max}$	Upper current limit from conductor type c in branch km (pu).
K_L	Interest rate for branches.
K_S	Interest rate for substations.
K_C	Interest rate for capacitor banks.
K_{VR}	Interest rate for voltage regulators.
L_{km}	Length of branch km (km).
M	Large positive number.
N_k^{fcb}	Maximum number of fixed capacitors in installed bank k (pu).
N_k^{scb}	Maximum number of switched capacitors in installed bank k (pu).
Ns_{km}	Total number of steps of voltage regulator in branch km.
$P_{k,t}^D$	Active power demand at node k in period t (pu).
$Q_{k,t}^D$	Reactive power demand at node k in period t (pu).
Q^{fc}	Rating of fixed capacitors (pu).
Q^{sc}	Rating of switched capacitors (pu).
$R_{km,c}$	Resistance per length of conductor type c installed in branch km (pu).
S_k^{SE0}	Maximum apparent power capacity of existing substation k (pu).
S_k^{SE}	Maximum apparent power capacity of newly installed substation k (pu).
$V_k^{\min}, V_k^{\max}$	Lower and upper voltage magnitude limits in substation k (pu).
$X_{km,c}$	Reactance per length of conductor type c installed in branch km (pu).
Δ_{km}	Tap step-size of voltage regulator in km.

Decision Variables

$d_{k,t}$	Auxiliary variable containing $V_{k,t}^2$.
$l_{km,c,t}$	Auxiliary variable containing $I_{km,c,t}^2$.
$n_{k,t}^{fc}$	Integer variable associated with the number of fixed capacitors installed in a bank in node k in period t.
$n_{k,t}^{sc}$	Integer variable associated with the number of switched capacitors installed in a bank in node k in period t.
$P_{k,t}^{SE}$	Active power from substation k in period t (pu).
$P_{km,c,t}$	Active power flow by branch km conductor type c in period t (pu).
$Q_{k,t}^{SE}$	Reactive power from substation k in period t (pu).
$Q_{km,c,t}$	Reactive power flow by branch km conductor type c in period t (pu).

$q^{fcb}_{k,t}$ Reactive power injection from fixed capacitor bank in node k in period t (pu).

$q^{scb}_{k,t}$ Reactive power injection from switched capacitor bank in node k in period t (pu).

$\alpha^{C}_{km,c,t}$ Binary variable associated with installing and/or changing of branch km using conductor type c in period t.

$\alpha^{SE}_{k,t}$ Binary variable associated with installing of a new substation k in period t.

$\alpha^{SER}_{k,t}$ Binary variable associated with expansion of substation k in period t.

$\alpha^{vr}_{km,t}$ Binary variable associated with installing of voltage regulator in branch km in period t.

$\alpha_{km,t}$ Variable used in the calculation of the voltage magnitude drop of branch km in period t.

$\beta_{km,t}$ Binary variable associated with radiality in branch km in period t.

References

1. I.J. Ramirez-Rosado, J.A. Dominguez-Navarro, Possibilistic model based on fuzzy sets for the multiobjective optimal planning of electric power distribution networks. IEEE Trans. Power Syst. **19**(4), 1801–1810 (2004)
2. I.J. Ramirez-Rosado, J.A. Dominguez-Navarro, New multiobjective tabu search algorithm for fuzzy optimal planning of power distribution systems. IEEE Trans. Power Syst. **21**(1), 224–233 (2006)
3. T. Gönen, *Electric Power Distribution Systems Engineering* (McGraw-Hill, New York, 1986)
4. S.F. Mekhamer, M.E. El-Hawary, S.A. Soliman, M.A. Moustafa, M.M. Mansour, New heuristic strategies for reactive power compensation of radial distribution feeders. IEEE Trans. Power Delivery **17**(4), 1128–1135 (2002)
5. J.Y. Park, J.M. Sohn, J.K. Park, Optimal capacitor allocation in a distribution system considering operation costs. IEEE Trans. Power Syst. **24**(1), 462–468 (2009)
6. N.C. Sahoo, S. Ganguly, D. Das, Recent advances on power distribution system planning: a stage-of-the-art survey. Energy Syst. **4**(2), 165–193 (2013)
7. K. Nara, T. Satoh, H. Kuwabara, K. Aoki, M. Kitagawa, T. Ishihara, Distribution systems expansion planning by multi-stage branch exchange. IEEE Trans. Power Syst. **7**(1), 208–214 (1992)
8. E. Míguez, J. Cidrás, E. Díaz-Dorado, J. García-Dornelas, An improved branch-exchange algorithm for large-scale distribution network planning. IEEE Trans. Power Syst. **17**(4), 931–936 (2002)
9. M. Lavorato, M. Rider, A.V. Garcia, R. Romero, A constructive heuristic algorithm for distribution system planning. IEEE Trans. Power Syst. **25**(3), 1734–1742 (2010)
10. G. Yang, Z. Dong, K. Wong, A modified differential evolution algorithm with fitness sharing for power system planning. IEEE Trans. Power Syst. **23**(2), 514–522 (2008)
11. V. Miranda, J.V. Ranito, L.M. Proença, Genetic algorithm in optimal multistage distribution network planning. IEEE Trans. Power Syst. **9**(4), 1927–1933 (1994)
12. I. Ramirez-Rosado, J. Bernal-Agustín, Genetic algorithms applied to the design of large power distribution systems. IEEE Trans. Power Syst. **13**(2), 696–703 (1998)
13. J. Gómez, H. Khodr, P. Oliveira, L. Ocque, J. Yusta, R. Villasana, A. Urdaneta, Ant colony system algorithm for the planning of primary distribution circuits. IEEE Trans. Power Syst. **19** (2), 996–1004 (2004)

14. V. Parada, J. Ferland, M. Arias, K. Daniels, Optimization of electrical distribution feeders using simulated annealing. IEEE Trans. Power Del. **19**(3), 1135–1141 (2004)
15. J.M. Nahman, D.M. Peric, Optimal planning of radial distribution networks by simulated annealing technique. IEEE Trans. Power Syst. **23**(2), 790–795 (2008)
16. S. Ganguly, N. Sahoo, D. Das, Mono- and multi-objective planning of electrical distribution networks using particle swarm optimization. Appl. Soft Comput. **11**(2), 2391–2405 (2011)
17. R. Lotero, J. Contreras, Distribution system planning with reliability. IEEE Trans. Power Del. **26**(4), 2552–2562 (2011)
18. S. Ganguly, N. Sahoo, D. Das, Multi-objective planning of electrical distribution systems using dynamic programming. Int. J. Electr. Power Energy Syst. **46**, 65–78 (2013)
19. S.N. Ravadanegh, R.G. Roshanagh, On optimal multistage electric power distribution networks expansion planning. Electr. Power Energy Syst. **54**, 487–497 (2014)
20. R.A. Jabr, Polyhedral formulations and loop elimination constraints for distribution network expansion planning. IEEE Trans. Power Syst. **28**, 1888–1897 (2013)
21. R. Gallego, J. Monticelli, R. Romero, Optimal capacitor placement in radial distribution networks. IEEE Trans. Power Syst. **16**(4), 630–637 (2001)
22. D.F. Pires, A.G. Martins, C.H. Antunes, A multiobjective model for VAR planning in radial distribution networks based on tabu search. IEEE Trans. Power Syst. **20**(2), 1089–1094 (2005)
23. I.C. Silva Junior, S. Carneiro Junior, E.J. Oliveira, J.S. Costa, J.L.R. Pereira, P.A.N. Garcia, A heuristic constructive algorithm for capacitor placement on distribution system. IEEE Trans. Power Syst. **23**(4), 1619–1626 (2008)
24. C.A.N. Pereira, C.A. Castro, Optimal placement of voltage regulators in distribution systems, in *Proceedings of IEEE Bucharest Power Tech* (Bucharest, Romania, 2009), pp. 1–5
25. A.S. Safigianni, G.J. Salis, Optimum voltage regulator placement in a radial power distribution network. IEEE Trans. Power Syst. **15**(2), 879–886 (2000)
26. J. Mendoza, D. Morales, R. López, J. Vannier, C. Coello, Multiobjetive location of automatic voltage regulators in radial distribution network using a micro genetic algorithm. IEEE Trans. Power Syst. **22**(1), 404–412 (2007)
27. J.F. Franco, M.J. Rider, M. Lavorato, R.A. Romero, A mixed integer LP model for the optimal allocation of voltage regulators and capacitors in radial distribution systems. Electr. Power Energy Syst. **48**, 123–130 (2013)
28. E.P. Madruga, L.N. Canha, Allocation and integrated configuration of capacitor banks and voltage regulators considering multi-objective variables in smart grid distribution system, in *Proceedings of International Conference on Industry Applications* (São Paulo, Brazil, Nov. 2010), pp. 1–6
29. B.A. de Souza, A.M.F. de Almeida, Multiobjective optimization and fuzzy logic applied to planning of the volt/var problem in distributions systems. IEEE Trans. Power Syst. **25**(3), 1274–1281 (2010)
30. J. Sugimoto, R. Yokoyama, Y. Fukuyama, V.V.R. Silva, H. Sasaki, Coordinated allocation and control of voltage regulators based on reactive tabu search, in *IEEE Russian Power Tech* (St. Petersburg, Russia, 27–30 June 2005), pp. 1–6
31. H. Fletcher, K. Strunz, Optimal distribution system horizon planning-part I: formulation. IEEE Trans. Power Syst. **22**(2), 791–799 (2007)
32. S. Haffner, L.F.A. Pereira, L.A. Pereira, L.S. Barreto, Multistage model for distribution expansion planning with distributed generation—part I: problem formulation. IEEE Trans. Power Delivery **23**(2), 915–923 (2008)
33. S. Haffner, L.F.A. Pereira, L.A. Pereira, L.S. Barreto, Multistage model for distribution expansion planning with distributed generation—part II: numerical results. IEEE Trans. Power Delivery **23**(2), 924–929 (2008)
34. A. Sorokin, S. Rebennack, P. Pardalos, N. Iliadis, M. Pereira, *Handbook of Networks in Power Systems I. Energy Systems* (Springer, Berlin, 2012)
35. J. Tate, T. Overbye, A comparison of the optimal multiplier in polar and rectangular coordinates. IEEE Trans. Power Syst. **20**(4), 1667–1674 (2005)

36. S.C. Tripathy, G.D. Prasad, O.P. Malik, G. S. Hope, Load-flow solutions for ill-conditioned power systems by a Newton-like method. IEEE Trans. Power App. Syst. **PAS-101**(10), 3648–3657 (1982)
37. R.A. Jabr, Radial distribution load flow using conic programming. IEEE Trans. Power Syst. **21**, 1458–1459 (2006)
38. M.E. Baran, F.F. Wu, Network reconfiguration in distribution systems for loss reduction and load balancing. IEEE Trans. Power Delivery **4**, 1401–1407 (1989)
39. M.E. Baran, F.F. Wu, Optimal capacitor placement on radial distribution systems. IEEE Trans. Power Delivery **4**, 725–734 (1989)
40. M.E. Baran, F.F. Wu, Optimal sizing of capacitors placed on a radial distribution system. IEEE Trans. Power Delivery **4**, 735–743 (1989)
41. M. Farivar, S.H. Low, Branch flow model: relaxations and convexification—part I. IEEE Trans. Power Syst. **28**(3), 2554–2564 (2013)
42. D. Luenberger, Y. Ye, *Linear and Nonlinear Programming* (Springer, 2008)
43. R.A. Jabr, Optimal placement of capacitors in a radial network using conic and mixed integer linear programming. Electr. Power Syst. Res. **78**, 941–948 (2008)
44. J. López, D. Pozo, J. Contreras, J.R.S. Mantovani, A multiobjective minimax regret robust VAr planning model. IEEE Trans. Power Syst. **32**, 1761–1771 (2017)
45. I. Gönen, I. Ramirez-Rosado, Review of distribution system planning models: a model for optimal multi-stage planning. IEE Proc. Gen. Trans. Dist. **133**(7), 397–408 (1986)
46. IBM, IBM ILOG CPLEX V12.1. User's Manual for CPLEX (2009)
47. R. Fourer, D.M. Gay, B.W. Kernighan, *AMPL: A Modeling Language for Mathematical Programming* (Duxbury Press, 2002)

Chapter 3
Mathematical Optimization of Unbalanced Networks with Smart Grid Devices

Carlos F. Sabillón, John F. Franco, Marcos J. Rider and Rubén Romero

Abstract Electric distribution networks should be prepared to provide an economic and reliable service to all customers, as well as to integrate technologies related to distributed generation, energy storage, and plug-in electric vehicles. A proper representation of the electric distribution network operation, taking into account smart grid technologies, is key to accomplish these goals. This chapter presents mathematical formulations for the steady-state operation of electric distribution networks, which consider the unbalance of three-phase grids. Mathematical models of the operation of smart grid related devices present in networks are discussed (e.g., volt-var control devices, energy storage systems, and plug-in electric vehicles). Furthermore, features related to the voltage dependency of loads, distributed generation, and voltage and thermal limits are also included. These formulations constitute a mathematical framework for optimization analysis of the network operation, which makes it possible to model decision-making processes. Different objectives related to technical and/or economic aspects can be pursued within the framework; in addition, the extension to multi-period and multi-scenario optimization is discussed. The presented models are built based on mixed integer linear programming formulations, avoiding the use of conventional mixed integer nonlinear formulations. The application of the presented framework

C. F. Sabillón (✉) · R. Romero
Department of Electrical Engineering, São Paulo State University (UNESP),
Ilha Solteira, Brazil
e-mail: carlos.sabillon@unesp.br

R. Romero
e-mail: ruben.romero@unesp.br

J. F. Franco
São Paulo State University (UNESP), Rosana, Brazil
e-mail: j.f.franco@ieee.org

M. J. Rider
School of Electrical and Computer Engineering,
University of Campinas, Campinas, Brazil
e-mail: mjrider@dsee.fee.unicamp.br

F. Shahnia et al. (eds.), *Electric Distribution Network Planning*,
Power Systems, https://doi.org/10.1007/978-981-10-7056-3_3

is illustrated throughout control approaches for the voltage control and the plug-in electric vehicle charging coordination problems.

Keywords Distribution network operation · Mathematical optimization Mixed integer linear programming · Smart grids devices · Steady-state operation point

3.1 Introduction

At the present time, high levels of reliability are demanded to power systems, as electricity is required, among others, in industry, communication, lightning, heating and entertainment. Smart grids emerged from earlier attempts of power utilities to use the improvements on electronic technologies to bring a reliable supply of good quality electricity to their customers. Nowadays, smart grids stand out as the current responses in order to cope with the challenges brought by a rising electrical demand.

The electric distribution network is the final stage in the transfer of power to individual consumers. It routes power from small energy sources nearby or power substation fed by transmission lines, to residential, industrial, and commercial customers, through power lines, switches, and transformers [1]. Nowadays, utilities are in charge of operating the distribution network, maintaining a reliable supply of electric power to all costumers connected into the grid. Traditionally, the electrical distribution network has been designed to carry the power from the sources downstream to the consumers, but lately this one-way electricity delivery model has been changing.

With the evolution of the smart grids, the distributed generation (DG) growth, and the introduction of renewable sources and energy storage systems (ESS), the classic distribution network model is evolving. All of these factors impact directly over the planning, engineering, construction, operation, and maintenance of the network. As smart grids technologies continue to strengthen, the current electric distribution network will become a more intelligent and real-time optimized grid; in consequence, the complexity of the planning, operation, and maintenance of the network will increase, as new technologies and distribution practices, which offer greater efficiency, sustainability, and cost savings, are provided. Thereupon, the network must evolve in order to engage all network elements and participants including consumer, generators, and those that do both, in an active management seeking to fulfill technical, economic, and environmental objectives [2].

Since mid-1960s optimization concepts and techniques have been part of the power system planning and operation. The development of strong optimization methods and algorithms and their proper application to the power system depends on a suitable representation of the electric distribution network behavior under smart grid schemes. Hence, mathematical modeling is crucial to achieve an enhanced representation of the network operation, endorsing the decision-making of optimization algorithms.

Once a problem has been properly represented, it is up to the planners/operators to choose the most appropriate method in order to solve it. Heuristic and meta-heuristic techniques, as well as mathematical optimization have foregrounded among those techniques to become the most commonly applied methods to problem solving in electric distribution networks. In the latest years, the accelerated advent of efficient commercial solvers based on mathematical optimization has increased the interest of researchers in the development of complex and realistic mathematical models for optimization problems. Therefore, once the mathematical model is properly defined, the commercial solver finds the best solution; i.e., the planner/operator does not need to take care of the development of the solution method.

According to the nature of the adopted formulation for the optimization problem, the corresponding mathematical model may be classified as

- Linear Programming (LP), where the term 'linear' indicates that all constraints, as well as the objective function, are barren of any nonlinearity.
- Nonlinear Programming (NLP), which aims to deal with problems involving nonlinear constraints and/or objective function.
- Mixed Integer Linear Programming (MILP), which are a special type of LP, where all or some of the decision variables are confined to only integer values.
- Mixed Integer Nonlinear Programming (MINLP), a special type of NLP (analogue to MILP).

For each type of mathematical model there are several well-known optimization techniques. For example, LP problems can be solved by using simplex or interior point algorithms. For NLP problem one can use several traditional optimization techniques (gradient-based techniques, Lagrangian relaxation, Newton's method, successive linear programming, etc.) or an interior point algorithm. To solve MILP problems, a branch and bound algorithm, improved versions of branch and bound such as branch and price or branch and cut, Benders' decomposition, Gomory's cutting planes, among others, might be used. Finally, solving a MINLP problem is a very complicated task and there is few theory related to classical optimization in this regard. Thus, commercial solvers are assumed to solve this type of problems based on branch and bound algorithms, sensitivity and barrier methods, and interior point methods.

In the decade of 2000, commercial solvers based on classical optimization techniques excelled to become extremely efficient, taking advantage of the improvement on resolution techniques. Since then, solvers that target LP and MILP problems such as CPLEX [3], MOSEK [4], GUROBI [5], and similar, had become extremely efficient compared to prior versions. In counterpart, the development of specialized solvers on NLP and MINLP problems is still in progress.

Due to the aforementioned, the interest in the development of mathematical models to represent the operation and planning of electric distribution networks has grown among the researchers. Thereby, commercial solvers such as CPLEX have been used when the problem is represented by a LP or MILP mathematical model. On the other hand, for those cases in which the problems are represented by NLP or

MINLP formulations, these formulations have been transformed into equivalent or approximated LP problems, if possible, or have been solved using solvers for NLP or MINLP problems such as KNITRO [6] or BONMIN [7], even though NLP specialized software are not equally efficient (i.e., they cannot guarantee the global optimum of the problem and they usually demand high computational efforts).

In electric distribution networks, mathematical modeling has become an important tool, as it is widely used in operation and expansion planning problems, especially those including mixed integer programming. This is due to the fact that commonly planners and operators of electric distribution networks have to meet specific goals with limited sources. Hence, a great share of the optimization problems can be classified as 'yes' or 'no' problems, which may be represented by binary variables. Decisions such as

- schedule/not schedule, e.g., electric vehicle charging;
- build/not build, e.g., the construction of a new distribution line;
- K out of N, e.g., the number of capacitors operating in a bank; and
- N-possible values, e.g., the tap-position of a voltage regulator;

can be optimized using mathematical models [1].

Moreover, the inclusion of smart grid devices and technologies into the electric distribution network urge improvements in mathematical formulations previously employed to model the grid. The flexibility offered by smart grid devices demand higher levels of accuracy and resolution in the problem formulations, leading to more realistic but also more complex models. In this regard, the utilization of three-phase representation has become crucial in the solution of problems related to network operation. A three-phase representation, in despite of the commonly used single-phase, takes into account the imbalance in the network and allows the inclusion of mutual coupling effects, conveying to a more accurate determination of the steady-state operation point.

This chapter presents two mathematical formulations to represent the steady-state operation of unbalanced electric distribution networks. The network is initially represented by NLP models; hereupon, LP models are developed throughout linearization techniques and approximations, which are implemented in order to avoid the complexity associated with the solution of the NLP problems. These formulations constitute a framework that can be used by planners and operators as a tool inside optimization methods and algorithms aiming to optimize specific goals, guaranteeing by these means, the feasibility of the solutions found. Besides, constraints related to load conditions, DG features, and voltage and thermal limits are also included in both formulations.

3.2 Mathematical Representation of Unbalanced Electric Distribution Networks

Traditionally, the evaluation of the electric distribution network state has been determined by solving a power flow. The objective of the power flow is to determine, given a set of specific values, the steady-state operation point of the network, i.e., obtain the voltage magnitudes, phase angles in all nodes, and derived quantities (e.g., active and reactive power flows, current magnitudes in the circuits, and power loss). The power flow is a useful tool for the analysis of networks in steady-state, being widely utilized in real time operation, as well as in the planning of expansion and operation. This problem is typically modeled as a system of nonlinear equations, solved through iterative methods [8, 9].

In this section, two approaches are presented aiming to model the operation of an unbalanced electric distribution network. In contradistinction to the iterative trait proper of commonly used power flow methods, these formulations can be solved using mathematical optimization; and they can be extended in order to be used as tools in optimization analysis in order to mathematically formalize decision-making regarding different objectives related to technical and/or economic constraints. Thereupon, in order to determine the steady-state operation point of a network using mathematical optimization, the operation of the grid must be modeled as a conventional mathematical programming problem (3.1). These problems have as a common feature the involvement of optimization. A goal is established and defined as an objective function (f) which has to be maximized or minimized by the setting of a set of control variables (x) and subject to a set of constraints.

$$\begin{aligned} \text{max/min} \quad & f(x) \\ \text{subject to:} \quad & g(x) \leq 0; \\ & h(x) = 0; \end{aligned} \tag{3.1}$$

Although the behavior of an electric distribution network follows a set of nonlinear constraints, it is desired to reach LP representations, which avoid the complexity related to the solution of NLP problems. For this, the nonlinear set of equations is initially presented; later, linearization and approximation techniques are applied to reach an LP model, in each approach. Moreover, in this section all loads are considered as constant power loads.

3.2.1 Current-Based Mathematical Formulation

This formulation is based on the real and imaginary parts of currents through circuits and node voltages in the network. A single branch of an unbalanced network is depicted in Fig. 3.1 Each vector of voltages and currents represent the sum of the corresponding real and imaginary parts, e.g., $\vec{I} = I^{re} + jI^{im}$. Hence, the set of

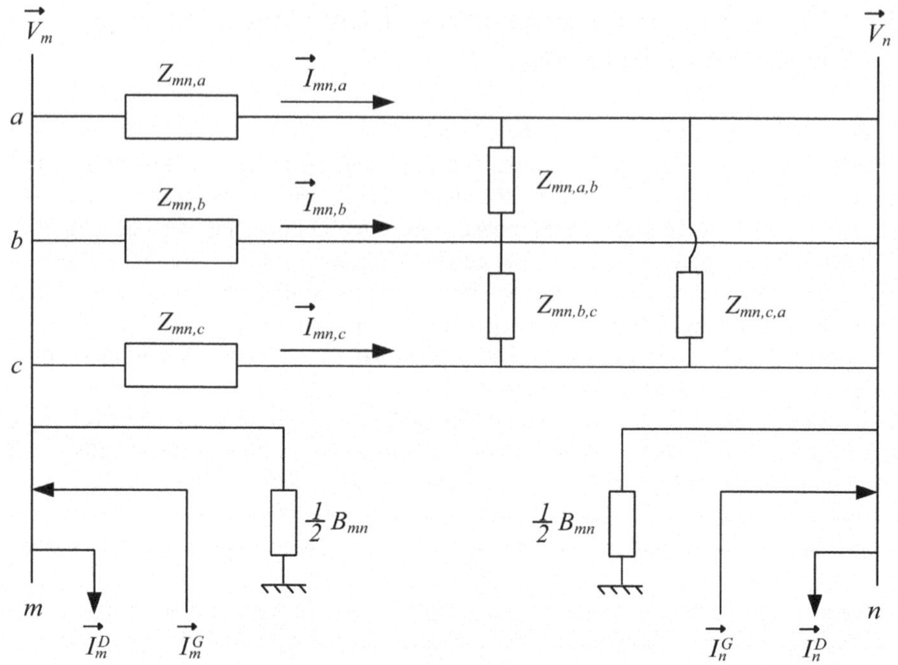

Fig. 3.1 Current-based representation of a single branch of an unbalanced electric distribution network

nonlinear mathematical relationships that represent the steady-state operation of an unbalanced network are written in terms of the current and voltages real and imaginary parts, the active and reactive power demanded by loads, and the circuit impedance, resistance, and reactance.

In the following formulation, consider the sets F, L, and N representing phases, circuits, and nodes, respectively. Furthermore, $V_{n,f}^{re/im}$ are the real/imaginary parts of the voltage at node n and phase f, while $I_{n,f}^{Gre/im}$ and $I_{n,f}^{Dre/im}$ are the real/imaginary parts of the generated and demanded currents, respectively. $P_{n,f}^{G/D}$ and $Q_{n,f}^{G/D}$ are the generated/demanded active and reactive powers. In addition, $I_{mn,f}^{re/im}$ represents the real/imaginary parts of the current through the circuit connecting nodes m and n, while $B_{mn,f}$ represents its shunt susceptance for phase f. Finally, $R_{mn,f,h}$ and $X_{mn,f,h}$ are the resistance and reactance for circuit mn between phases f and h, respectively.

From Fig. 3.1, the voltage drop from node m to node n can be derived as

$$\begin{bmatrix} \Delta\vec{V}_{mn,a} \\ \Delta\vec{V}_{mn,b} \\ \Delta\vec{V}_{mn,c} \end{bmatrix} = \begin{bmatrix} Z_{mn,a} & Z_{mn,a,b} & Z_{mn,a,c} \\ Z_{mn,a,b} & Z_{mn,b} & Z_{mn,b,c} \\ Z_{mn,a,c} & Z_{mn,b,c} & Z_{mn,c} \end{bmatrix} \begin{bmatrix} \vec{I}_{mn,a} \\ \vec{I}_{mn,b} \\ \vec{I}_{mn,c} \end{bmatrix} = \begin{bmatrix} Z_{mn,a}\vec{I}_{mn,a} + Z_{mn,a,b}\vec{I}_{mn,b} + Z_{mn,a,c}\vec{I}_{mn,c} \\ Z_{mn,a,b}\vec{I}_{mn,a} + Z_{mn,b}\vec{I}_{mn,b} + Z_{mn,b,c}\vec{I}_{mn,c} \\ Z_{mn,a,c}\vec{I}_{mn,a} + Z_{mn,b,c}\vec{I}_{mn,b} + Z_{mn,c}\vec{I}_{mn,c} \end{bmatrix} \tag{3.2}$$

Analyzing the individual case for phase a, (3.3) is formulated as

$$\Delta\vec{V}_{mn,a} = Z_{mn,a}\vec{I}_{mn,a} + Z_{mn,a,b}\vec{I}_{mn,b} + Z_{mn,a,c}\vec{I}_{mn,c} \tag{3.3}$$

Extending (3.3), considering that $Z_{mn,f,h} = R_{mn,f,h} + jX_{mn,f,h}$ and separating currents and voltages in their real and imaginary parts, (3.4) and (3.5) are obtained.

$$V^{re}_{m,a} - V^{re}_{n,a} = R_{mn,a}I^{re}_{mn,a} + R_{mn,a,b}I^{re}_{mn,b} + R_{mn,a,c}I^{re}_{mn,c} - X_{mn,a}I^{im}_{mn,a} - X_{mn,a,b}I^{im}_{mn,b} - X_{mn,a,c}I^{im}_{mn,c} \tag{3.4}$$

$$V^{im}_{m,a} - V^{im}_{n,a} = R_{mn,a}I^{im}_{mn,a} + R_{mn,a,b}I^{im}_{mn,b} + R_{mn,a,c}I^{im}_{mn,c} + X_{mn,a}I^{re}_{mn,a} + X_{mn,a,b}I^{re}_{mn,b} + X_{mn,a,c}I^{re}_{mn,c} \tag{3.5}$$

Generalizing these expressions, the voltage drop of an unbalanced network, written in terms of the real and imaginary parts of the voltages and currents, can be expressed by (3.6) and (3.7).

$$V^{re}_{m,f} - V^{re}_{n,f} = \sum_{h \in F} (R_{mn,f,h}I^{re}_{mn,h} - X_{mn,f,h}I^{im}_{mn,h}) \quad \forall mn \in L, f \in F \tag{3.6}$$

$$V^{im}_{m,f} - V^{im}_{n,f} = \sum_{h \in F} (X_{mn,f,h}I^{re}_{mn,h} + R_{mn,f,h}I^{im}_{mn,h}) \quad \forall mn \in L, f \in F \tag{3.7}$$

Furthermore, to model the complete steady-state operation of an unbalanced network, the first Kirchhoff's Law for the real and imaginary parts of the currents in each node is applied, as shown in (3.8) and (3.9). Hereupon, (3.10) and (3.11) establish the relationship between power, voltage and current for the loads.

$$I^{Gre}_{m,f} + \sum_{km \in L} I^{re}_{km,f} - \sum_{mn \in L} I^{re}_{mn} - \left(\sum_{km \in L} B_{km,f} + \sum_{mn \in L} B_{mn,f} \right) \frac{V^{im}_{m,f}}{2} = I^{Dre}_{m,f} \quad \forall m \in N, f \in F \tag{3.8}$$

$$I_{m,f}^{Gim} + \sum_{km \in L} I_{km,f}^{im} - \sum_{mn \in L} I_{mn}^{im} - \left(\sum_{km \in L} B_{km,f} + \sum_{mn \in L} B_{mn,f} \right) \frac{V_{m,f}^{re}}{2} = I_{m,f}^{Dim} \quad \forall m \in N, f \in F \tag{3.9}$$

$$P_{n,f}^{D} = V_{n,f}^{re} I_{n,f}^{Dre} + V_{n,f}^{im} I_{n,f}^{Dim} \quad \forall n \in N, f \in F \tag{3.10}$$

$$Q_{n,f}^{D} = -V_{n,f}^{re} I_{n,f}^{Dim} + V_{n,f}^{im} I_{n,f}^{Dre} \quad \forall n \in N, f \in F \tag{3.11}$$

Equation (3.12) presents the complete NLP formulation developed to determine the steady-state operation point of an unbalanced network.

$$\begin{gathered} \min \alpha \\ \text{subject to : (3.6)–(3.11)} \end{gathered} \tag{3.12}$$

where α is the objective function of the NLP model, and can be designed to minimize or maximize the network operator best interests (e.g., power loss, voltage deviation, or reliability).

Aiming to achieve a LP model based on (3.12), linearization techniques and approximations must be applied to the nonlinearities shown in (3.10) and (3.11). In this regard, [10] proposed the application of Taylor's approximation around an estimated point $(V_{n,f}^{re^*}, V_{n,f}^{im^*})$. Hence, (3.10) and (3.11) are rewritten, expressing the real and imaginary currents in terms of the power and voltages, as shown in (3.13) and (3.14). Those equations represent the nonlinear expressions for the real and imaginary demanded currents as the functions $g(P_{n,f}^{D}, Q_{n,f}^{D}, V_{n,f}^{re}, V_{n,f}^{im})$ and $h(P_{n,f}^{D}, Q_{n,f}^{D}, V_{n,f}^{re}, V_{n,f}^{im})$, respectively.

$$I_{n,f}^{Dre} = \frac{P_{n,f}^{D} V_{n,f}^{re} + Q_{n,f}^{D} V_{n,f}^{im}}{V_{n,f}^{re2} + V_{n,f}^{im2}} = g(P_{n,f}^{D}, Q_{n,f}^{D}, V_{n,f}^{re}, V_{n,f}^{im}) \quad \forall n \in N, f \in F \tag{3.13}$$

$$I_{n,f}^{Dim} = \frac{P_{n,f}^{D} V_{n,f}^{im} - Q_{n,f}^{D} V_{n,f}^{re}}{V_{n,f}^{re2} + V_{n,f}^{im2}} = h(P_{n,f,t}^{D}, Q_{n,f,t}^{D}, V_{n,f,t}^{re}, V_{n,f,t}^{im}) \quad \forall n \in N, f \in F \tag{3.14}$$

Hereupon, taking advantage of the relatively small and limited variation range of the voltage magnitude in a distribution network, (3.15) and (3.16) present the Taylor's approximation used to linearize (3.10) and (3.11).

$$I_{n,f}^{Dre} = g^* + \frac{\partial g}{\partial V^{re}}\Big|^* (V_{n,f}^{re} - V_{n,f}^{re^*}) + \frac{\partial g}{\partial V^{im}}\Big|^* (V_{n,f}^{im} - V_{n,f}^{im^*}) \quad \forall n \in N, f \in F \tag{3.15}$$

$$I_{n,f}^{Dim} = h^* + \frac{\partial h}{\partial V^{re}}\Big|^* (V_{n,f}^{re} - V_{n,f}^{re^*}) + \frac{\partial h}{\partial V^{im}}\Big|^* (V_{n,f}^{im} - V_{n,f}^{im^*}) \quad \forall n \in N, f \in F \tag{3.16}$$

Therefore, the LP model that represents the steady-state operation of an unbalanced network is shown as

$$
\begin{array}{c}
\min \alpha \\
\text{subject to : } (3.6){-}(3.9), (3.15), \text{and } (3.16)
\end{array}
\tag{3.17}
$$

3.2.2 Power-Based Mathematical Formulation

An additional representation to determine the steady-state operation point of an unbalanced network using mathematical optimization is presented. This formulation is based on the active and reactive power flow through the circuits and the voltage magnitudes in the network. Analogue to Figs. 3.1 and 3.2 shows a single branch of an unbalanced network, depicting the power flows through the circuits. Thus, each vector of complex power represents the sum of the active and reactive powers, i.e., $\vec{S} = P + jQ$.

For the following formulation let the sets F, L, and N, represent phases, circuits, and nodes, respectively. Besides, $\vec{V}_{n,f}$ is the voltage vector at node n and phase f, with magnitude $V_{n,f}$ and angle $\theta_{n,f}$; while, $V_{n,f}^{qdr}$ is the squared voltage magnitude at node n and phase f. Furthermore, $I_{mn,f}$ and $I_{mn,f}^{qdr}$ are the current magnitude and the squared current magnitude through circuit mn and phase f, respectively; while $P_{mn,f}$ and $Q_{mn,f}$ are the active and reactive power flows arriving at node n, respectively.

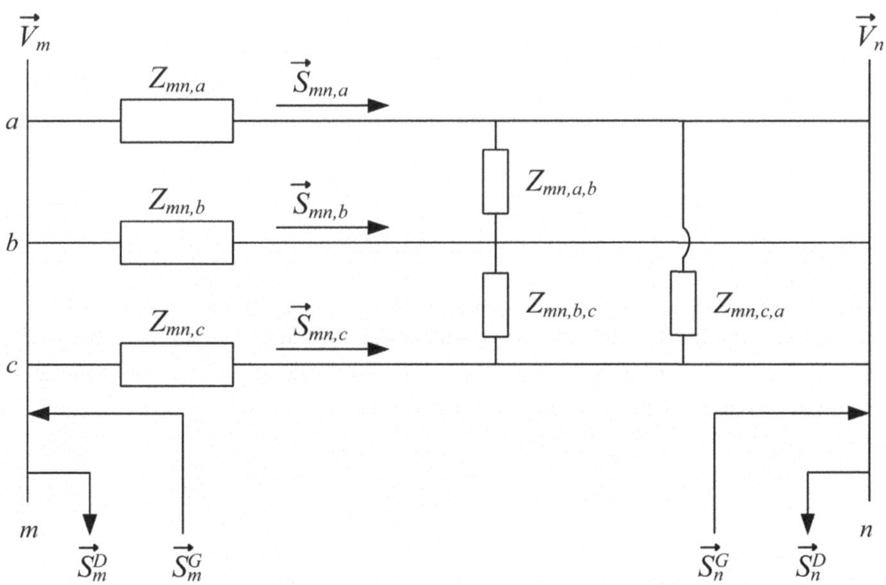

Fig. 3.2 Power-based representation of a single branch of an unbalanced electric distribution network

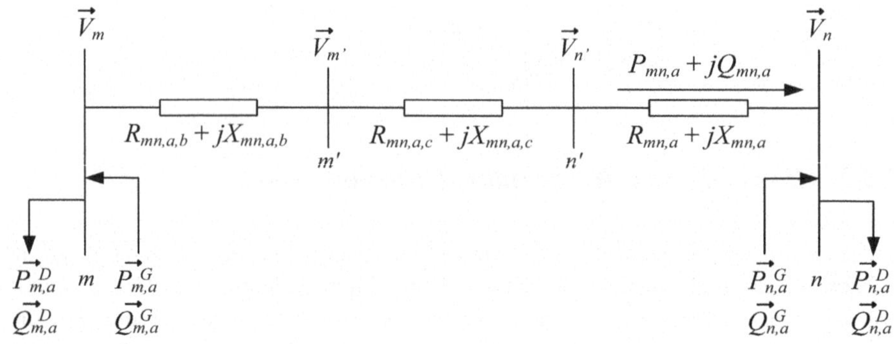

Fig. 3.3 Single branch equivalent for phase a

$S^{G/D}_{n,f}, P^{G/D}_{n,f}$ and $Q^{G/D}_{n,f}$ are the generated/demanded apparent, active, and reactive powers, respectively. Finally, $Z_{mn,f,h}, R_{mn,f,h}$, and $X_{mn,f,h}$ are the impedance, resistance, and reactance for circuit mn between phases f and h, respectively.

Hereby, $\vec{I}_{mn}$ in (3.2) must be expressed in terms of the active and reactive powers as shown in (3.18).

$$\vec{I}_{f,mn} = \left(\frac{\vec{S}_{f,mn}}{\vec{V}_{f,n}}\right)^* = \left(\frac{P_{f,mn}+jQ_{f,mn}}{\vec{V}_{f,n}}\right)^* \tag{3.18}$$

Analyzing only phase A from (3.2), Fig. 3.3 presents a simplified equivalent which divides the voltage drop considering two fictitious nodes m' and n'.

Therefore, the mathematical equations for each branch-segment are shown in (3.19).

$$\begin{aligned}\vec{V}_{m,a} - \vec{V}_{m',a} &= Z_{mn,a,b}\vec{I}_{mn,b}\\ \vec{V}_{m',a} - \vec{V}_{n',a} &= Z_{mn,a,c}\vec{I}_{mn,c}\\ \vec{V}_{n',a} - \vec{V}_{n',a} &= Z_{mn,a}\vec{I}_{mn,a}\end{aligned} \tag{3.19}$$

In order to reach a general expression for the voltage drop, each term of (3.19) is analysed separately. Initially, the first term can be written as shown in (3.20) where the term $\vec{V}^*_{m',a}/\vec{V}^*_{m',a}$ is added to enable some algebraic manipulations aiming to reach a approximated linear expression for the voltage drop.

$$\left(\vec{V}_{m,a} - \vec{V}_{m',a}\right) = \left(R_{mn,a,b} + jX_{mn,a,b}\right)\frac{P_{mn,b} - jQ_{mn,b}}{\vec{V}^*_{n,b}}\left(\frac{\vec{V}^*_{m',a}}{\vec{V}^*_{m',a}}\right) \tag{3.20}$$

Applying the simplification shown in (3.21), and replacing $(R_{mn,a,b}$ and $X_{mn,a,b})$ as presented in (3.22), constraint (3.23) is reached.

$$\frac{\vec{V}^*_{m',a}}{\vec{V}^*_{n,b}} \approx \frac{V_{m',a}}{V_{n,b}} \angle\left(\theta_{m,b} - \theta_{m,a}\right) \approx \frac{V_{m',a}}{V_{n,b}} \angle - 120^{\circ} \tag{3.21}$$

$$\left(\tilde{R}_{mn,a,b} + j\tilde{X}_{mn,a,b}\right) = \left(R_{mn,a,b} + jX_{mn,a,b}\right)\left(1\angle - 120^{\circ}\right) \tag{3.22}$$

$$\vec{V}_{m,a}\vec{V}^*_{m',a} - V^2_{m',a} = \frac{V_{m',a}}{V_{n,b}}\left(\tilde{R}_{mn,a,b} + j\tilde{X}_{mn,a,b}\right)\left(P_{mn,b} - jQ_{mn,b}\right) \tag{3.23}$$

By separating (3.23) into its real and imaginary parts, (3.24) and (3.25) are obtained.

$$V_{m,a}V_{m',a}\cos\left(\theta_{m,a} - \theta_{m',a}\right) = \frac{V_{m',a}}{V_{n,b}}\left(\tilde{R}_{mn,a,b}P_{mn,b} + \tilde{X}_{mn,a,b}Q_{mn,b}\right) + V^2_{m',a} \tag{3.24}$$

$$V_{m,a}V_{m',a}\sin\left(\theta_{m,a} - \theta_{m',a}\right) = \frac{V_{m',a}}{V_{n,b}}\left(\tilde{X}_{mn,a,b}P_{mn,b} - Q_{mn,b}\tilde{R}_{mn,a,b}\right) \tag{3.25}$$

Later, by adding the square power of (3.24) and (3.25), the expression for the voltage drop corresponding to the first segment of Fig. 3.3 is shown in (3.26).

$$V^2_{m,a} - V^2_{m',a} = 2\frac{V_{m',a}}{V_{n,b}}\left(\tilde{R}_{mn,a,b}P_{mn,b} + \tilde{X}_{mn,a,b}Q_{mn,b}\right) + \tilde{Z}^2_{mn.a.b}I^2_{mn,b} \tag{3.26}$$

Therefore, in order to avoid the nonlinearities shown in (3.26), assume $V_{m',a}/V_{n,a} \approx 1$, and the variables V^2 and I^2 are replaced by V^{sqr} and I^{sqr}, respectively. Hence,

$$V^{sqr}_{m,a} - V^{sqr}_{m',a} = 2\left(\tilde{R}_{mn,a,b}P_{mn,b} + \tilde{X}_{mn,a,b}Q_{mn,b}\right) + \tilde{Z}^2_{mn.a.b}I^{sqr}_{mn,b} \tag{3.27}$$

For the second term of (3.19), an analogue expression is obtained as shown in (3.27).

$$V^{sqr}_{m',a} - V^{sqr}_{n',a} = 2\left(\tilde{R}_{mn,a,c}P_{mn,c} + \tilde{X}_{mn,a,c}Q_{mn,c}\right) + \tilde{Z}^2_{mn.a.c}I^{sqr}_{mn,c} \tag{3.28}$$

Finally, for the last term of (3.19), the voltage drop is given by:

$$\left(\vec{V}_{n',a} - \vec{V}_{n,a}\right) = \left(R_{mn,a} + jX_{mn,a}\right)\frac{P_{mn,a} - jQ_{mn,a}}{\vec{V}^*_{n,a}} \tag{3.29}$$

For this segment it is considered that $R_{mn,a} + jX_{mn,a} = \tilde{R}_{mn,a} + j\tilde{X}_{mn,a}$, reaching the expression presented in (3.30).

$$V_{n',a}^{sqr} - V_{n,a}^{sqr} = 2\left(\tilde{R}_{mn,a}P_{mn,a} + \tilde{X}_{mn,a}Q_{mn,a}\right) + \tilde{Z}_{mn.a}^{2}I_{mn,a}^{sqr} \quad (3.30)$$

Therefore, the whole voltage drop in circuit mn is defined by (3.31), which considers coupling effects between the phases.

$$V_{m,f}^{sqr} - V_{n,f}^{sqr} = \sum_{h \in F}\left\{2 * \left(\tilde{R}_{mn,fh}P_{mn,f} + \tilde{X}_{mn,fh}Q_{mn,f}\right) + \tilde{Z}_{mn,fh}^{2}I_{mn,f}^{sqr}\right\} \quad \forall mn \in L, f \in F \quad (3.31)$$

In addition, the active and reactive power balances for each node are represented by (3.32) and (3.33), respectively. The calculation of the circuit current is given by (3.34).

$$\sum_{km \in L} P_{km,f} - \sum_{mn \in L}\left(P_{mn,f} + P_{mn,f}^{L}\right) + P_{m,f}^{G} = P_{m,f}^{D} \quad \forall m \in N, f \in F \quad (3.32)$$

$$\sum_{km \in L} Q_{km,f} - \sum_{mn \in L}\left(Q_{mn,f} + Q_{mn,f}^{L}\right) + Q_{m,f}^{G} = Q_{m,f}^{D} \quad \forall m \in N, f \in F \quad (3.33)$$

$$V_{n,f}^{sqr}I_{mn,f}^{sqr} = P_{mn,f}^{2} + Q_{mn,f}^{2} \quad \forall mn \in L, f \in F \quad (3.34)$$

To determine the power loss (P^L and Q^L), the complex power loss is initially expressed as

$$S_{mn,f}^{L} = \sum_{h \in \Omega_f} Z_{mn,f,h}\left(\frac{P_{mn,h} + jQ_{mn,h}}{\vec{V}_{m,h}}\right)^{*}\left(\frac{P_{mn,f} + jQ_{mn,f}}{\vec{V}_{m,f}}\right) \quad (3.35)$$

Equation (3.35) can be also written as

$$S_{mn,f}^{L} = \sum_{h \in \Omega_f} Z_{mn,f,h}\frac{\left(P_{mn,h} + jQ_{mn,h}\right)^{*}\left(P_{mn,f} + jQ_{mn,f}\right)}{\vec{V}_{m,f}\vec{V}_{m,h}\angle\left(\theta_{m,f} - \theta_{m,h}\right)} \quad (3.36)$$

Furthermore, replacing (3.22) in (3.36), it is obtained:

$$S_{mn,f}^{L} \approx \sum_{h \in \Omega_f} \tilde{Z}_{mn,f,h}\frac{\left(P_{mn,h} + jQ_{mn,h}\right)^{*}\left(P_{mn,f} + jQ_{mn,h}\right)}{\vec{V}_{n,f}\vec{V}_{n,h}} \quad (3.37)$$

Later, (3.37) and (3.38) are reached by separating the real and imaginary parts of (3.37). These constraints represent the power loss in an unbalanced network.

$$P^L_{mn,f} = \sum_{h\in\Omega_f} \tilde{R}_{mn,f,h} \frac{\left(P_{mn,f}P_{mn,h} + Q_{mn,f}Q_{mn,h}\right)}{V_{n,f}V_{n,h}} + \tilde{X}_{mn,f,h} \frac{\left(-Q_{mn,f}P_{mn,h} + P_{mn,f}Q_{mn,h}\right)}{V_{n,f}V_{n,h}} \quad \forall mn \in L, f \in F \tag{3.38}$$

$$Q^L_{mn,f} = \sum_{h\in\Omega_f} \tilde{X}_{mn,f,h} \frac{\left(P_{mn,f}P_{mn,h} + Q_{mn,f}Q_{mn,h}\right)}{V_{n,f}V_{n,h}} + \tilde{R}_{mn,f,h} \frac{\left(Q_{mn,f}P_{mn,h} - P_{mn,f}Q_{mn,h}\right)}{V_{n,f}V_{n,h}} \quad \forall mn \in L, f \in F \tag{3.39}$$

Finally, (3.40) represents the complete NLP model for the steady-state operation of an unbalanced network.

$$\begin{gathered} \min \alpha \\ \text{subject to } (3.31) - (3.34), (3.38), \text{ and } (3.39) \end{gathered} \tag{3.40}$$

The NLP model (3.40) contain nonlinearities [in constraints (3.34), (3.38), and (3.39)] that must be addressed to reach a LP model. Equation (3.34) has a product of two variables $\left(V^{sqr}_{n,f} \text{ and } I^{sqr}_{mn,f}\right)$ in the left-hand side and the square of two variables $\left(P_{mn,f} \text{ and } Q_{mn,f}\right)$ in the right-hand side. The left-hand side of (3.34) is linearized replacing the variable $V^{sqr}_{n,f}$ by an estimated value $\tilde{V}^2_{n,f}$. On the other hand, the right-hand side is approximated using a piecewise linearization technique (see Appendix 1). Hence, (3.41) shows the complete linear expression used to approximate (3.34).

$$\tilde{V}^2_{n,f} I^{sqr}_{mn,f} = f\left(P_{mn,f}, \overline{P_{mn,f}}, \Lambda\right) + f\left(Q_{mn,f}, \overline{Q_{mn,f}}, \Lambda\right) \quad \forall mn \in L, f \in F \tag{3.41}$$

Furthermore, as represented by (3.38) and (3.39), power loss can be approximated by any of the following options:

<u>Option–A</u>: Use actual, historic, or estimated values for the voltages and the power flows, in order to reach approximated values for P^L and Q^L in (3.32) and (3.33), disregarding (3.38) and (3.39), as shown in (3.42) and (3.43).

$$\sum_{km\in L} P_{km,f} - \sum_{mn\in L} \left(P_{mn,f} + \tilde{P}^L_{mn,f}\right) + P^G_{m,f} = P^D_{m,f} \quad \forall m \in N, f \in F \tag{3.42}$$

$$\sum_{km\in L} Q_{km,f} - \sum_{mn\in L} \left(Q_{mn,f} + \tilde{Q}^L_{mn,f}\right) + Q^G_{m,f} = Q^D_{m,f} \quad \forall m \in N, f \in F \tag{3.43}$$

In not fully observable distribution networks, a two-stage approach is recommended to estimate the power loss $\tilde{P}^L_{km,f}$ and $\tilde{Q}^L_{km,f}$. In the first stage, the LP is solved disregarding the power loss, i.e., $\tilde{P}^L_{km,f}$ and $\tilde{Q}^L_{km,f}$ are equal to zero. Later, the solution of stage one is used to initialize stage two and the LP model is once again solved.

If option A is chosen, the complete LP model will have the following form:

$$\begin{array}{c}\min \alpha \\ \text{subject to } (3.31), \text{and } (3.41)\text{–}(3.43)\end{array} \tag{3.44}$$

<u>Option–B</u>: Use Taylor's approximation around an estimated point for the power flows and voltages $\left(P^*_{mn,f}, P^*_{mn,h}, V^*_{m,f} \text{and } V^*_{m,h}\right)$. Let the functions $g(P^*_{mn,f}, P^*_{mn,h}, Q^*_{mn,f}, Q^*_{mn,h}, V^*_{m,f}, V^*_{m,h})$ and $h(P^*_{mn,f}, P^*_{mn,h}, Q^*_{mn,f}, Q^*_{mn,h}, V^*_{m,f}, V^*_{m,h})$ be equal to the right part of (3.38) and (3.39), respectively. Equations (3.45) and (3.46) show the Taylor's approximation used to determine the power loss.

$$\begin{aligned} P^L_{mn,f} = g^* &+ \frac{\partial g}{\partial P_{mn,f}}\bigg|^* (P_{mn,f} - P^*_{mn,f}) + \frac{\partial g}{\partial P_{mn,h}}\bigg|^* (P_{mn,h} - P^*_{mn,h}) \\ &+ \frac{\partial g}{\partial Q_{mn,f}}\bigg|^* (Q_{mn,f} - Q^*_{mn,f}) + \frac{\partial g}{\partial Q_{mn,h}}\bigg|^* (Q_{mn,h} - Q^*_{mn,h} \quad \forall mn \in L, f \in F \\ &+ \frac{\partial g}{\partial V_{m,f}}\bigg|^* (V_{m,f} - V^*_{m,f}) + \frac{\partial g}{\partial V_{m,h}}\bigg|^* (V_{m,h} - V^*_{m,h}) \end{aligned} \tag{3.45}$$

$$\begin{aligned} Q^L_{mn,f} = h^* &+ \frac{\partial h}{\partial P_{mn,f}}\bigg|^* (P_{mn,f} - P^*_{mn,f}) + \frac{\partial h}{\partial P_{mn,h}}\bigg|^* (P_{mn,h} - P^*_{mn,h}) \\ &+ \frac{\partial h}{\partial Q_{mn,f}}\bigg|^* (Q_{mn,f} - Q^*_{mn,f}) + \frac{\partial h}{\partial Q_{mn,h}}\bigg|^* (Q_{mn,h} - Q^*_{mn,h}) \quad \forall mn \in L, f \in F \\ &+ \frac{\partial g}{\partial V_{m,f}}\bigg|^* (V_{m,f} - V^*_{m,f}) + \frac{\partial g}{\partial V_{m,h}}\bigg|^* (V_{m,h} - V^*_{m,h}) \end{aligned} \tag{3.46}$$

Hence, the complete LP model is given by (3.47).

$$\begin{array}{c}\min \alpha \\ \text{subject to } (3.31)\text{–}(3.33), (3.41), (3.45), \text{and } (3.46)\end{array} \tag{3.47}$$

3.2.3 Performance and Accuracy

Two LP formulations to determine the steady-state operation point of an unbalanced network via mathematical optimization were presented. These representations are the base of the complete mathematical optimization framework for the distribution network operation. Different objectives related to technical and/or economic constraints can be pursued embedding these formulations in multi-period and multi-scenario optimization (see Appendix 2). Therefore, the quality of studies developed hereinafter will rely on their level of accuracy.

The performance and accuracy of both formulations is evaluated in the IEEE123-node test system [11]. All LP models were written in the mathematical language AMPL [12], and solved using CPLEX [3]. The case study had the following characteristics:

- The whole conventional demand of the distribution network was 1.62 MVA (40.61%), 1.05 MVA (26.36%), and 1.32 MVA (33.03%), connected to phases A, B, and C, respectively.
- All loads were connected in wye configuration.
- All demands were considered as constant power loads.

The LP formulations presented (3.17) and (3.44) are expressed as shown in (3.48) and (3.49), aiming to minimize the total active power generation; considering nominal voltage at the substation node. Moreover, as the accuracy of both unbalanced formulations depends on the precision of the assumed operation point (i.e., V^{re^*}, V^{im^*} for the current-based formulation, and P^L and Q^L for the power-based formulation), a two-stage approach was used for both formulations to obtain a better approximation for the operation point (see Appendix 3).

$$\begin{gathered} \min \sum_{f \in F} \left(V^{re}_{S,f} I^{Gre}_{S,f} + V^{im}_{S,f} I^{Gim}_{S,f} \right) \\ \text{subject to}: (3.6)-(3.9), (3.15), \text{and} (3.16) \end{gathered} \tag{3.48}$$

$$\begin{gathered} \min \sum_{f \in F} P^{G}_{S,f} \\ \text{subject to } (3.31) \text{ and } (3.41)-(3.43) \end{gathered} \tag{3.49}$$

A comparison of the two methods (Current-based and power-based power flow) was made analyzing the voltage magnitude profile obtained from each when compared to the one obtained from the solution of a conventional power flow. To solve this conventional power flow, the specialized software OpenDSS [13] was selected. For both formulations, Table 3.1 shows the maximum error percentage in the voltage magnitude, for each phase, compared against the OpenDSS; as well as the minimum voltage magnitude in the system.

Although both formulations show high accuracy when compared with OpenDSS results, it can be seen from Table 3.1, that the current-based LP formulation outperforms the power-based representation. This is an important fact to be taken

Table 3.1 Comparison of the LP formulations against OpenDSS

Phase	A		B		C	
LP formulation	Current-based	Power-based	Current-based	Power-based	Current-based	Power-based
Error (%)	0.01	0.02	0.01	0.05	0.11	0.30
Vmin (pu)	0.8996	0.8997	0.9599	0.9595	0.9241	0.9223

under consideration by the distribution network optimizer when choosing one LP formulation for an electric distribution network optimization algorithm. Nevertheless, the complexity due to troublesome constraints associated with operational limits and the inclusion of different technologies and devices should be thoroughly analyzed.

3.3 Operational Constraints

The LP formulations presented in (3.17) and (3.44) are expected to serve as the central engine of optimization analysis, mathematically formalizing decision–making processes regarding several objectives. Hereby, an essential component of any optimization strategy or algorithm related to electricity distribution is the service quality. Although several electrical quantities can be measured and limited to guarantee good quality in the service, voltage magnitude limits and thermal limits in conductors and transformers are the most commonly used in steady-state studies to ensure the proper operation of the electric distribution network. Therefore, the mathematical representation of these limits and their inclusion in the LP formulations are presented below.

3.3.1 Voltage Magnitude

Ensuring a good quality service in the electric distribution network, the voltage magnitude is limited within a range established by regulatory policies. This range is mathematically expressed as shown in (3.50), in terms of the minimum and maximum values of the voltage magnitude ($\underline{V}$ and $\bar{V}$, respectively). Therefore, the inclusion of (3.50) in the LP formulations is presented below.

$$\underline{V} \leq \left|\vec{V}\right| \leq \bar{V} \tag{3.50}$$

Current-Based Representation—Voltage Magnitude Limits: The current-based representation is written in terms of the currents and voltages real and imaginary parts. Hence, (3.50) is rewritten in (3.51), limiting the square power of the voltage magnitude. Nevertheless, (3.51) presents nonlinearities which have to be dealt with in order to include this limit in the LP model.

$$\underline{V}^2 \leq V_{n,f}^{re2} + V_{n,f}^{im2} \leq \bar{V}^2 \quad \forall n \in N, f \in F \tag{3.51}$$

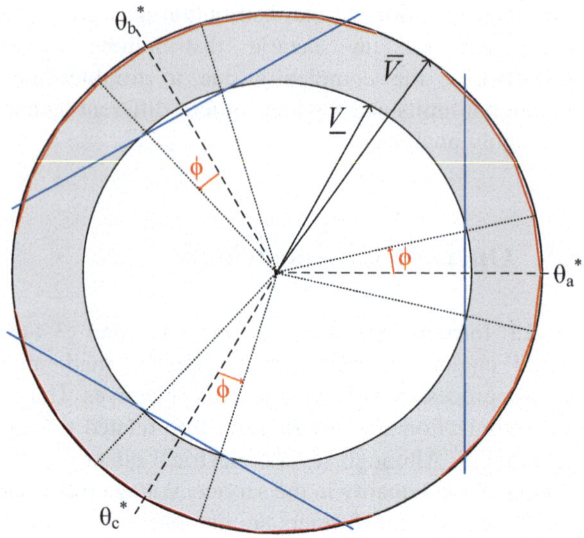

Fig. 3.4 Constraints for voltage limits in the current-based formulation

To reach a linear approximation for (3.51), a set of linear constraints limiting the feasible region for the voltage real and imaginary components is implemented. Figure 3.4 depicts the sets of lines used to approximate the upper and lower voltage limits (red lines and blue lines, respectively). In each phase, the upper limit is replaced by a set of $2 * \ell$ line segments; where ℓ lines are built clockwise starting from an estimated operation angle θ^*, and ℓ lines are built counterclockwise. For the lower limit, a single line is used for each phase also built around θ^*. Equation (3.52) and (3.53) are the mathematical expressions for the linear approximations of the upper and lower voltage limits, respectively.

$$V_{n,f}^{im} \leq \frac{\sin(\varphi_{n,f,i}^{+})-\sin(\varphi_{n,f,i}^{-})}{\cos(\varphi_{n,f,i}^{+})-\cos(\varphi_{n,f,i}^{-})}\left[V_{n,f}^{re} - \bar{V}\cos(\varphi_{n,f,i}^{-})\right] + \bar{V}\sin(\varphi_{n,f,i}^{-}) \quad \forall i \in -\ell\ldots\ell, n \in N, f \in F \tag{3.52}$$

$$V_{n,f}^{im} \leq \frac{\sin(\varphi_{n,f}^{o+})-\sin(\varphi_{n,f}^{o-})}{\cos(\varphi_{n,f}^{o+})-\cos(\varphi_{n,f}^{o-})}\left[V_{n,f}^{re} - \underline{V}\cos(\varphi_{n,f}^{o+})\right] + \underline{V}\sin(\varphi_{n,f}^{o+}) \quad \forall n \in N, f \in F \tag{3.53}$$

where $\varphi_{n,f,i}^{+} = \theta_{n,f}^{*} + \ell_i\phi$; $\varphi_{n,f,i}^{-} = \theta_{n,f}^{*} + (\ell_i - 1)\phi$; $\theta_{n,f}^{*}$ is the operation angle for bus n in phase f; ℓ_i is the i-element from the set of line segments; and ϕ is the angle of the arc corresponding to each line segment. Finally, $\varphi_{n,f,i}^{o+} = \theta_{n,f}^{*} + \phi$; and $\varphi_{n,f,i}^{o-} = \theta_{n,f}^{*} - \phi$.

Power-Based Representation—Voltage Magnitude Limits: On the other hand, to include the voltage magnitude limits in the power-based representation, (3.50) is rewritten in order to limit the square power of the voltage magnitude, as shown in (3.54).

$$\underline{V}^2 \le V_{n,f}^{sqr} \le \bar{V}^2 \quad \forall n \in N, f \in F \tag{3.54}$$

3.3.2 Circuit Currents

Thermal limits are often the main constraint in electric distribution networks. Hence, a proper operation of the electric distribution network must maintain the current in all circuits within the conductor thermal limitations; i.e., the magnitude of the current in all circuits must be held under their ampacities. Constraint (3.55) shows the general mathematical expression for the current limit, which is added to each LP formulation as

$$|\vec{I}| \le \bar{I} \tag{3.55}$$

Current-Based Representation—Circuit Current Limits: Similar to the voltage magnitude limit, (3.55) is squared and expressed in terms of the real and imaginary parts of the circuit currents, as shown in (3.56). Later, the nonlinearities are avoided through a piecewise linearization technique, reaching a linear expression for the circuit current limit in terms of the real and imaginary parts of the current (3.57).

$$I_{mn,f}^{re2} + I_{mn,f}^{im2} \le \bar{I}_{mn}^2 \quad \forall mn \in L, f \in F \tag{3.56}$$

$$f\left(I_{mn,f}^{re}, \bar{I}, \Lambda\right) + f\left(I_{mn,f}^{im}, \bar{I}, \Lambda\right) \le \bar{I}_{mn}^2 \quad \forall mn \in L, f \in F \tag{3.57}$$

Power-Based Representation—Circuit Current Limits: Analogous to (3.50), (3.58) must be added to the power-based LP formulation to limit the current in each circuit.

$$I_{mn,f}^{sqr} \le \bar{I}_{mn}^2 \quad \forall mn \in L, f \in F \tag{3.58}$$

3.3.3 Transformer Capacity

In steady-state studies of the electric distribution network where different voltage levels are taken into account, thermal limits regarding the distribution and substation transformers are a key aspect. Therefore, the inclusion of the transformer capacity in the mathematical optimization framework for the distribution network operation is presented below. Hence, the mathematical expression (3.59) limiting the transformer apparent power must be fulfilled for each transformer.

$$|\vec{S}| \leq \bar{S} \tag{3.59}$$

Hereby, let $TR \subseteq N$ be the set of nodes where a transformer is installed; while $P_{\wp}^{TR}$ and $Q_{\wp}^{TR}$ the active and reactive power, respectively, for transformer $\wp$. Thus, the general expression for each transformer capacity is presented in (3.60), and linearized in (3.61) through piecewise linearization technique.

$$P_{\wp}^{TR\,2} + Q_{\wp}^{TR\,2} \leq \overline{S_{\wp}^{TR}}^{2} \quad \forall \wp \in TR \tag{3.60}$$

$$f(P_{\wp}^{TR}, \overline{S_{\wp}^{TR}}, \Lambda) + f(Q_{\wp}^{TR}, \overline{S_{\wp}^{TR}}, \Lambda) \leq \overline{S_{\wp}^{TR}}^{2} \quad \forall \wp \in TR \tag{3.61}$$

Current-Based Representation—Transformer Capacity: Besides (3.61), for the current-based formulation, $P_{\wp}^{TR}$ and $Q_{\wp}^{TR}$ have to be calculated. Equations (3.62) and (3.63) calculate, for transformer $\wp$, the active and reactive powers in terms an operation point for the real and imaginary voltages $\left(V_{\wp,f}^{re/im*}\right)$ and the currents $\left(I_{\wp,f}^{re/im}\right)$ at the secondary windings.

$$P_{\wp}^{TR} = \sum_{f \in F} V_{\wp,f}^{re^*} I_{\wp,f}^{re} + V_{\wp,f}^{im^*} I_{\wp,f}^{im} \quad \forall \wp \in TR \tag{3.62}$$

$$Q_{\wp}^{TR} = \sum_{f \in F} -V_{\wp,f}^{re^*} I_{\wp,f}^{im} + V_{\wp,f}^{im^*} I_{\wp,f}^{re} \quad \forall \wp \in TR \tag{3.63}$$

Power-Based Representation—Transformer Capacity: Likewise, for the power-based formulation, (3.64) and (3.65) show the mathematical expression for $P_{\wp}^{TR}$ and $Q_{\wp}^{TR}$ in terms of the active an reactive power flowing through each phase of the transformer ($P_{\wp,f}$ and $Q_{\wp,f}$, respectively).

$$P_{\wp}^{TR} = \sum_{f \in F} P_{\wp,f} \quad \forall \wp \in TR \tag{3.64}$$

$$Q_{\wp}^{TR} = \sum_{f \in F} Q_{\wp,f} \quad \forall \wp \in TR \tag{3.65}$$

Finally, the complete LP formulations, considering operational limits are presented in (3.66) and (3.67), for the current-based and power-based representations, respectively.

$$\begin{gathered} \min \alpha \\ \text{subject to}: (3.6) - (3.9), (3.15), (3.16), (3.52), (3.53), (3.57), \text{and}(3.61) - (3.63) \end{gathered} \tag{3.66}$$

$$\begin{gathered} \min \alpha \\ \text{subject to}: (3.31), (3.41) - (3.43), (3.54), (3.58), (3.61), (3.64), \text{and}(3.65) \end{gathered} \tag{3.67}$$

3.4 Load Representation

Nowadays, electric distribution networks must deal with social, technical and environmental challenges in order to successfully satisfy present-day consumers. The world's electrical energy consumption is expected to have an annual growth rate of about 2.2% until 2040 [14], which will substantially impact the operation of future networks. Hence, network operators are continuously challenged as they are in charge of meeting customer demands and optimize energy sources, while guaranteeing a reliable service. As a result, improvements in the network load modelling are continuously demanded within the grid operator efforts for predicting system behavior.

Besides the rapid growth of the conventional demand, electric distribution networks face issues related to the progressive integration of new technologies. The rise of new loads (e.g., electric vehicles), which cannot be pigeonholed in traditional classifications, require a special attention. Thus, this section presents the load modelling and the inclusion of special loads within the mathematical optimization framework.

3.4.1 Type of Loads: Voltage Dependent Load Models

In electric distribution networks, loads are traditionally classified as residential, industrial or commercial. This rough classification was elaborated in order to group loads that share features related to usage patterns. These well-studied patterns are commonly affected by factors such as location, weather, cultural habits, and type of human works. However, since large concentrated loads called for detailed classifications and special representation in order to determine operational characteristics, studies have been performed aiming to achieve a more precise categorization for different types of loads [15]. Hereby, a new classification following the voltage dependency of the actual loads becomes more important in the representation as the network gets closer to individual loads. The demand of a distribution network is classified into loads that can be represented as

- constant power loads,
- constant impedance loads,
- constant current loads, or
- a combination of those.

Although load modelling in electric distribution networks is a well-studied topic, approached by several researches related to voltage and angular system stability, this issue has to be also taken into account on decision-making algorithms related to the steady-state operation of the grid. As discussed in [16–18], the effectiveness of several mathematical models in electric distribution networks is highly dependent on the accuracy of the load representation. Dependence on the voltage magnitude and frequency is considered in the load models; mathematically, this dependence can be represented by static and dynamic load models described by the traditional ZIP model. In this regards, two static models are commonly studied for the representation of the active and reactive demanded powers (P^D and Q^D, respectively): the polynomial load model and the exponential load model shown in (3.68) and (3.69), respectively [19].

$$\begin{aligned} P^D &= P^o\left(P^{Zo}\frac{V^2}{V^{o2}} + P^{Io}\frac{V}{V^o} + P^{Po}\right) \\ Q^D &= Q^o\left(Q^{Zo}\frac{V^2}{V^{o2}} + Q^{Io}\frac{V}{V^o} + Q^{Po}\right) \end{aligned} \tag{3.68}$$

$$\begin{aligned} P^D &= P^o\left(\frac{V}{V^o}\right)^{\alpha}\frac{1 + K_{pf}(f_r - f_r^o)}{f_r^o} \\ Q^D &= Q^o\left(\frac{V}{V^o}\right)^{\beta}\frac{1 + K_{qf}(f_r - f_r^o)}{f_r^o} \end{aligned} \tag{3.69}$$

where P^o, Q^o, and V^o, are the nominal active and reactive power and bus voltage, respectively. P^{Zo}, P^{Io}, and P^{Po}, are the active power percentage of the total load classified as constant impedance, constant current, and constant power, respectively. Likewise, Q^{Zo}, Q^{Io}, and Q^{Po}, are the reactive power percentage for constant impedance, constant current, and constant power, respectively. In the polynomial load model (3.68), the loads are treated as a combination of constant impedance constant current and/or constant power; hence the sum of these coefficients will represent the total load, as shown in (3.70). On the other hand, in the exponential load model (3.69) the load voltage dependency is generalized, and the demanded active and reactive powers vary according to the voltage exponents α and β. These voltage exponents depend on the type and composition of the load.

$$\begin{aligned} P^{Zo} + P^{Io} + P^{Po} &= 1 \\ Q^{Zo} + Q^{Io} + Q^{Po} &= 1 \end{aligned} \tag{3.70}$$

Moreover, f_r represents the frequency of the bus voltage and f_r^o represents the nominal frequency. The coefficients K_{pf} and K_{qf} are the frequency sensitivities for the active and reactive power loads, respectively. Nevertheless, for the exponential model the effects associated with frequency may be disregarded, as shown in (3.71). Hence, with an appropriate adjustment of the constants α and β the model can be restricted to the steady-state analysis case (i.e., dependence directly on the voltage magnitude). Appropriate values for these constants may be found in previous works, such as [20].

$$\begin{aligned} P^D &= P^o\left(\frac{V}{V^o}\right)^\alpha \\ Q^D &= Q^o\left(\frac{V}{V^o}\right)^\beta \end{aligned} \tag{3.71}$$

In some decision-making processes for electric distribution networks, the load voltage dependency is a key aspect of the suitable representation of the network operation, e.g., volt-var control. Hence, they must be included in the LP formulations presented in (3.17) and (3.44), where the demanded active and reactive powers were considered as constant values.

Current-Based Representation—Polynomial load model: In order to include the load voltage dependency in the LP problem presented in (3.17), the values of the demanded powers have to be replaced by (3.68) in (3.13) and (3.14). Therefore, the expressions representing the functions g and h are presented in (3.72) and (3.73).

$$\begin{aligned} g = {} & \frac{P^o_{n,f} V^{re}_{n,f}}{V^{re2}_{n,f} + V^{im2}_{n,f}} \left(\frac{P^{Zo}_{n,f}}{V^{o2}_n} \left(V^{re2}_{n,f} + V^{im2}_{n,f} \right) + \frac{P^{Io}_{n,f}}{V^o_n} \sqrt{\left(V^{re2}_{n,f} + V^{im2}_{n,f} \right)} + P^{Po}_{n,f} \right) \\ & + \frac{Q^o_{n,f} V^{im}_{n,f}}{V^{re2}_{n,f} + V^{im2}_{n,f}} \left(\frac{Q^{Zo}_{n,f}}{V^{o2}_n} \left(V^{re2}_{n,f} + V^{im2}_{n,f} \right) + \frac{Q^{Io}_{n,f}}{V^o_n} \sqrt{\left(V^{re2}_{n,f} + V^{im2}_{n,f} \right)} + Q^{Po}_{n,f} \right) \\ & \forall n \in N, f \in F \end{aligned} \tag{3.72}$$

$$
\begin{aligned}
h = & \frac{P^{o}_{n,f} V^{im}_{n,f}}{V^{re2}_{n,f} + V^{im2}_{n,f}} \left(\frac{P^{Zo}_{n,f}}{V^{o2}_{n}} \left(V^{re2}_{n,f} + V^{im2}_{n,f} \right) + \frac{P^{Io}_{n,f}}{V^{o}_{n}} \sqrt{\left(V^{re2}_{n,f} + V^{im2}_{n,f} \right)} + P^{Po}_{n,f} \right) \\
& - \frac{Q^{o}_{n,f} V^{re}_{n,f}}{V^{re2}_{n,f} + V^{im2}_{n,f}} \left(\frac{Q^{Zo}_{n,f}}{V^{o2}_{n}} \left(V^{re2}_{n,f} + V^{im2}_{n,f} \right) + \frac{Q^{Io}_{n,f}}{V^{o}_{n}} \sqrt{\left(V^{re2}_{n,f} + V^{im2}_{n,f} \right)} + Q^{Po}_{n,f} \right) \\
& \forall n \in N, f \in F
\end{aligned}
\tag{3.73}
$$

Current-Based Representation—Exponential load model: Likewise, to include the exponential load model in (3.17), (3.13) and (3.14) have to be rewritten, as shown in (3.74) and (3.75).

$$
g = P^{o}_{n,f} \frac{V^{re}_{n,f}}{V_{o}^{\alpha_{n,f}}} \left(V^{re^2}_{n,f} + V^{im^2}_{n,f} \right)^{\frac{\alpha_{n,f}}{2} - 1} + Q^{o}_{n,f,t} \frac{V^{im}_{n,f}}{V_{o}^{\beta_{n,f}}} \left(V^{re^2}_{n,f} + V^{im^2}_{n,f} \right)^{\frac{\beta_{n,f}}{2} - 1} \quad \forall n \in N, f \in F \tag{3.74}
$$

$$
h = P^{o}_{n,f} \frac{V^{im}_{n,f}}{V_{o}^{\alpha_{n,f}}} \left(V^{re^2}_{n,f} + V^{im^2}_{n,f} \right)^{\frac{\alpha_{n,f}}{2} - 1} - Q^{o}_{n,f,t} \frac{V^{re}_{n,f}}{V_{o}^{\beta_{n,f}}} \left(V^{re^2}_{n,f} + V^{im^2}_{n,f} \right)^{\frac{\beta_{n,f}}{2} - 1} \quad \forall n \in N, f \in F \tag{3.75}
$$

Power-Based Representation—Polynomial load model: For the power-based linear representation shown in (3.44), the expressions for the demanded power (3.68) are directly added to the formulation; notwithstanding, the nonlinearities of these equations must be dealt with, as shown in (3.76) and (3.77).

$$
P^{D}_{n,f} = P^{o}_{n,f} \left(P^{Zo}_{n} \frac{V^{sqr}_{n,f}}{V^{o2}_{n}} + P^{Io}_{n} \frac{V^{sqr}_{n,t}}{V^{*}_{n,t} V^{o}_{n}} + P^{Po}_{n} \right) \quad \forall n \in N, f \in F \tag{3.76}
$$

$$
Q^{D}_{n,f} = Q^{o}_{n,f} \left(Q^{Zo}_{n} \frac{V^{sqr}_{n,f}}{V^{o2}_{n}} + Q^{Io}_{n} \frac{V^{sqr}_{n,t}}{V^{*}_{n,t} V^{o}_{n}} + Q^{Po}_{n} \right) \quad \forall n \in N, f \in F \tag{3.77}
$$

Hence, the complete LP formulation, considering the polynomial model for the load voltage dependency, is given by (3.78).

$$
\begin{gathered}
\min \alpha \\
\text{subject to}: (3.31)-(3.33), (3.41), (3.45), (3.46), (3.74), \text{and} (3.75)
\end{gathered}
\tag{3.78}
$$

Power-Based Representation—Exponential load model: Finally, to include the exponential load model in the power-based LP formulation, expressions (3.71) are added. To avoid the nonlinearities associated to these expressions, (3.71) is rewritten as shown in (3.79) and (3.80).

$$P^D_{n,f} = P^o_{n,f}\left(\frac{V^{sqr}_{n,f}}{V^{o2}_n}\right)^{\frac{\alpha_{n,f}}{2}} \quad \forall n \in N, f \in F \tag{3.79}$$

$$Q^D_{n,f} = Q^o_{n,f}\left(\frac{V^{sqr}_{n,f}}{V^{o2}_n}\right)^{\frac{\beta_{n,f}}{2}} \quad \forall n \in N, f \in F \tag{3.80}$$

Later, Taylor's approximation is applied to linearize (3.79) and (3.80).

$$P^D_{n,f} = P^o_{n,f}\left(\frac{V^{sqr*}_{n,f}}{V^{o2}_n}\right)^{\frac{\alpha_{n,f}}{2}} + \frac{\alpha_{n,f}}{2}\frac{P^o_{n,f}}{\left(V^o_n\right)^{\frac{\alpha_{n,f}}{2}}}V^{sqr*\left(\frac{\alpha_{n,f}}{2}-1\right)}_{n,f}\left(V^{sqr}_{n,f} - V^{sqr*}_{n,f}\right) \quad \forall n \in N, f \in F \tag{3.81}$$

$$Q^D_{n,f} = Q^o_{n,f}\left(\frac{V^{sqr*}_{n,f}}{V^{o2}_n}\right)^{\frac{\beta_{n,f}}{2}} + \frac{\beta_{n,f}}{2}\frac{Q^o_{n,f}}{\left(V^o_n\right)^{\frac{\beta_{n,f}}{2}}}V^{sqr*\left(\frac{\beta_{n,f}}{2}-1\right)}_{n,f}\left(V^{sqr}_{n,f} - V^{sqr*}_{n,f}\right) \quad \forall n \in N, f \in F \tag{3.82}$$

Therefore, the complete LP formulation, considering the exponential load voltage dependency model, is given by (3.83).

$$\begin{gathered} \min \alpha \\ \text{subject to}: (3.31)-(3.33), (3.41), (3.45), (3.46), (3.81), \text{and } (3.82) \end{gathered} \tag{3.83}$$

3.4.2 *Special Loads: Plug-In Electric Vehicles*

A large number of EVs is expected to be integrated to the transport sector in the upcoming years, as an to environmental concerns related to the reduction of greenhouse gas emissions [21]. From the point of view of the customers, EVs represent an economical option in response to high fuel costs. On the other hand, for the electric distribution network, EVs represent an additional load which need to be attended, increasing the conventional demand in several ways, depending on the charging place [22]. Hence, EVs are new loads in networks which have to be taken into account in optimization studies for the grid.

EVs recharge their batteries from the distribution network, and an uncontrolled charging of large fleets can cause overloads, voltage limit violations, and excessive energy losses [23]. Hence, the EV charging coordination (EVCC) problem have to be tackled as part of the distribution network operation, and has received much attention in recent years [24, 25]. Furthermore, the ability of EVs to inject power into the grid (also known as vehicle-to-grid (V2G) technology), providing ancillary services to the network, also represent a highly studied subject [25, 26].

In order to include the EVCC within the electric distribution network optimization framework, let EV be the set of EVs plugged into the grid. P_e^{EV} is the power injected/drawn by EV e, and it is equal to the sum of the maximum charging and discharging powers $(\bar{P}_e^{EV+}$ and $\bar{P}_e^{EV-}$, respectively) multiplied by the binary variables y_e and z_e, which represent the charging or discharging state, as shown in (3.84). Moreover, (3.85) ensures only one action for the EV (e.g., charging, discharging, or idle). Due to the binary nature of these variables, a MILP model is obtained as the result of their inclusion.

$$P_e^{EV} = \bar{P}_e^{EV+} y_e - \bar{P}_e^{EV-} z_e \quad \forall e \in EV \tag{3.84}$$

$$y_e + z_e \leq 1 \quad \forall e \in EV \tag{3.85}$$

On the other hand, EVs storage capacity also needs to be taken into account, i.e., maximum energy limit $\left(\bar{E}_e^{EV}\right)$ and, for V2G applications, the maximum depth of discharge (DoD) must be always fulfilled. Hence, if an EV is charged/discharged constantly at P_e^{EV} during a time interval Δt, (3.86) ensures that the state of charge (SOC) is always maintained between the pre-established limits.

$$\min\left(E_e^{EVi}, \bar{E}_e^{EV} DoD\right) \leq E_e^{EVi} + \Delta t(\bar{P}_e^{EV+} y_e \eta_e^{EV+} - \bar{P}_e^{EV-} z_e \eta_e^{EV-}) \leq \bar{E}_e^{EV} \quad \forall e \in EV \tag{3.86}$$

where E_e^{EVi} is the initial SOC for EV e; while, η_e^{EV+} and η_e^{EV-} are the charging and discharging efficiencies, respectively.

The interaction of the EVs with the grid is integrated in the steady-state operation as follows:

Current-Based Representation—EV: Eqs. (3.87) and (3.88) represent the EV active and reactive powers in terms of the voltage operation point $\left(V_e^*\right)$ where the EV is plugged, and the EV current $\left(I_e^{EVre}\right)$. Considering that EVs will only exchange active power. Moreover, (3.89) and (3.90) are the extensions of (3.8) and (3.9) taking into account the EV current injection in each node.

$$P_e^{EV} = V_e^{re*} I_e^{EVre} + V_e^{im*} I_e^{EVim} \quad \forall e \in EV \tag{3.87}$$

$$0 = -V_e^{re*} I_e^{EVim} + V_e^{im*} I_e^{EVre} \quad \forall e \in EV \tag{3.88}$$

$$I^{Gre}_{m,f} + \sum_{km \in L} I^{re}_{km,f} - \sum_{mn \in L} I^{re}_{mn} - \left(\sum_{km \in L} B_{km,f} + \sum_{mn \in L} B_{mn,f} \right) \frac{V^{im}_{m,f}}{2} = I^{Dre}_{m,f} + \sum_{e \in EV} I^{EVre}_{e} \gamma_{e,m,f} \quad \forall m \in N, f \in F \tag{3.89}$$

$$I^{Gim}_{m,f} + \sum_{km \in L} I^{im}_{km,f} - \sum_{mn \in L} I^{im}_{mn} - \left(\sum_{km \in L} B_{km,f} + \sum_{mn \in L} B_{mn,f} \right) \frac{V^{re}_{m,f}}{2} = I^{Dim}_{m,f} + \sum_{e \in EV} I^{EVim}_{e} \gamma_{e,m,f} \quad \forall m \in N, f \in F \tag{3.90}$$

where $\gamma_{x,m,f}$ is a binary parameter that takes a value of 1 if the device x is connected at node m and phase f.

Power-Based Representation—EV: For this formulation, (3.91) represents the active power balance in each node, taking into account the EV active power injection/consumption.

$$\sum_{km \in L} P_{km,f} - \sum_{mn \in L} \left(P_{mn,f} + P^{L}_{mn,f} \right) + P^{G}_{m,f} = P^{D}_{m,f} + \sum_{e \in EV} P^{EV}_{e} \gamma_{e,m,f} \quad \forall m \in N, f \in F \tag{3.91}$$

Therefore, the complete MILP formulations for an unbalanced network, considering EV operation is presented in (3.92) and (3.93) for the current-based and power-based formulations, respectively.

$$\begin{gathered} \min \alpha \\ \text{subject to} : (3.6)-(3.9), (3.15), (3.16), \text{and}(3.84)-(3.90) \end{gathered} \tag{3.92}$$

$$\begin{gathered} \min \alpha \\ \text{subject to} : (3.31), (3.41)-(3.43), (3.84)-(3.86), \text{and}(3.91) \end{gathered} \tag{3.93}$$

3.5 Distributed Generation

Since the decade of 2000s, distributed generation has continuously grown among electric distribution networks, motivated by economic, environmental, technical, and market related features [27, 28]. Due to the flexibility of DG as a power source, distribution networks have been transformed from a passive network to an active network. Nowadays, DG plays an important role in the operation, structure and design of networks; therefore, several researches have been developed to model the integration of DG units in the network operation [29–31].

DG units are integrated into the electric distribution network in places that were not originally adapted to connect them can create several problems for distribution networks in terms of stability and power quality; particularly, when large amounts of DG units are connected to high impedance networks. In addition, integrating renewable sources of DG, such as wind or solar power, can mean new challenges to the network operation. Furthermore, according to the capacity of the DG units, the network can become an active one, attending loads without the need of the energy purchased from the main grid. Therefore, the inclusion of the DG in the study of the network operation is imperative [2].

On the other hand, DG can also offer several advantages to the electric distribution network, i.e., improving system reliability, reducing energy losses, reducing transmission and distribution line costs, and alleviating congestion in the grid. Moreover, the installation of small-scale DG units, close to loads, may delay or avoid investments in additional transmission or distribution infrastructure. In addition, certain types of DGs also have the ability to offer ancillary services, such as reactive power support, voltage control, and frequency control.

Typically, in mathematical representations for DG units, the models of synchronous generators (SiGs), induction generators (IGs), and doubly-fed induction generators (DFIGs) are disregarded. DG units are commonly modelled by a simple representation and coupling elements are not detailed. Hence, a simple mathematical representation is presented in (3.94)–(3.97) for the generation limits of DG units.

$$(P_n^{DG})^2 + (Q_n^{DG})^2 \leq (\bar{S}_n^{DG})^2 \quad \forall n \in DG \tag{3.94}$$

$$Q_n^{DG} \leq P_n^{DG} \tan(\arccos(pf_n^{DG})) \quad \forall n \in DG \tag{3.95}$$

$$\underline{Q_n^{DG}} \leq Q_n^{DG} \leq \overline{Q_n^{DG}} \quad \forall n \in DG \tag{3.96}$$

$$P_n^{DG} \geq 0 \quad \forall n \in DG \tag{3.97}$$

where $DG \subseteq N$ represents the set of nodes in which a DG unit is connected. P_n^{DG} and Q_n^{DG} are the active and reactive powers of DG unit n; while, pf_n^{DG} is the minimum power factor, $\underline{Q}_n^{DG}$ and $\bar{Q}_n^{DG}$ are the minimum and maximum reactive power limits, and, S_n^{DG} is the maximum apparent power. Therefore, (3.94) shows the nonlinear representation for the apparent power limit, and it is approximated in (3.98) via a piecewise linearization technique. Constraints (3.95) and (3.96) limit the reactive power in terms of the power factor and the maximum and minimum reactive power limits, respectively. Finally, (3.97) ensures non-negativity for the active power of the DG unit.

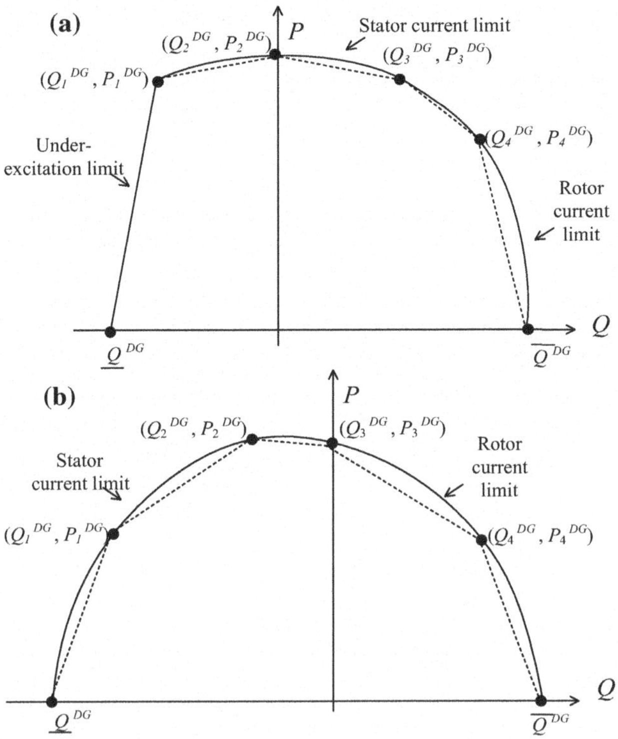

Fig. 3.5 Capability curves: **a** DFIG and **b** SiG [29]

$$f\left(P_n^{DG}, \overline{P_n^{DG}}, \Lambda\right) + f\left(Q_n^{DG}, \overline{Q_n^{DG}}, \Lambda\right) \leq (\bar{S}_n^{DG})^2 \quad \forall n \in DG \tag{3.98}$$

On the other hand, an improved and realistic model for DG units, considering the capability curves (see Fig. 3.5) for SiGs and DFIGs is presented in [29]. These types of generators are widely used in DG applications, e.g., wind turbines, biomass-based CHP generation systems, and small hydroelectric plants.

Figure 3.5 defines the points $(Q_{1,n}^{DG}, P_{1,n}^{DG}), (Q_{2,n}^{DG}, P_{2,n}^{DG}), (Q_{3,n}^{DG}, P_{3,n}^{DG})$, and $(Q_{4,n}^{DG}, P_{4,n}^{DG})$, which are used to obtain linear expressions for the DG operation constraints, as

$$P_n^{DG} \leq \frac{P_{1,n}^{DG}}{Q_{1,n}^{DG} - \underline{Q}_n^{DG}} (Q_n^{DG} - \underline{Q}_n^{DG}) \quad \forall n \in DG \tag{3.99}$$

$$P_n^{DG} \leq \frac{P_{2,n}^{DG} - P_{1,n}^{DG}}{Q_{2,n}^{DG} - Q_{1,n}^{DG}} (Q_n^{DG} - Q_{2,n}^{DG}) + P_{2,n}^{DG} \quad \forall n \in DG \tag{3.100}$$

Table 3.2 Linearization points for the linearization of distributed generators capability curves [29]

	SiG	DFIG
$(Q_{1,n}^{DG}, P_{1,n}^{DG})$	the intersection between the under-excitation and armature current limits	half of the arc of the armature current limit between points $(\underline{Q}_n^{DG}, 0)$ and $(Q_{2,n}^{DG}, P_{2,n}^{DG})$
$(Q_{2,n}^{DG}, P_{2,n}^{DG})$	the intersection between the armature current limit and the P axis	the intersection between the armature current and field current limits,
$(Q_{3,n}^{DG}, P_{3,n}^{DG})$	half of the arc of the armature current limit between points $(Q_{2,n}^{DG}, P_{2,n}^{DG})$ and $(Q_{4,n}^{DG}, P_{4,n}^{DG})$	the intersection between the field current limit and the P axis
$(Q_{4,n}^{DG}, P_{4,n}^{DG})$	the intersection between the armature current and field current limits	the half of the arc of the field current limit between points $(Q_{3,n}^{DG}, P_{3,n}^{DG})$ and $(\bar{Q}_n^{DG}, 0)$

$$P_n^{DG} \leq \frac{P_{3,n}^{DG} - P_{2,n}^{DG}}{Q_{3,n}^{DG} - Q_{2,n}^{DG}} (Q_n^{DG} - Q_{3,n}^{DG}) + P_{3,n}^{DG} \quad \forall n \in DG \tag{3.101}$$

$$P_n^{DG} \leq \frac{P_{4,n}^{DG}}{Q_{4,n}^{DG} - \bar{Q}_n^{DG}} (Q_n^{DG} - \bar{Q}_n^{DG}) \quad \forall n \in DG \tag{3.102}$$

Table 3.2 shows how these points were obtained for each type of generator.

Moreover, the power injection of the DG units connected in the system is integrated in the electric distribution network steady-state formulations as follows:

Current-Based Representation—DG Units: Assuming three-phase DG units, (3.103) and (3.104) represent the DG unit active and reactive powers in terms of the voltage operation point and their currents $\left(I_{n,f}^{DG}\right)$. Finally, the current injection due to DG units is included in the current balance for each node, as:

$$P_n^{DG}/3 = V_{n,f}^{re*} I_{n,f}^{DGre} + V_{n,f}^{im*} I_{n,f}^{DGim} \quad \forall n \in DG, \ f \in F \tag{3.103}$$

$$Q_n^{DG}/3 = -V_{n,f}^{re*} I_{n,f}^{DGim} + V_{n,f}^{im*} I_{n,f}^{DGre} \quad \forall n \in DG, \ f \in F \tag{3.104}$$

$$I^{Gre}_{m,f} + I^{DGre}_{m,f} + \sum_{km \in L} I^{re}_{km,f} - \sum_{mn \in L} I^{re}_{mn} - \left(\sum_{km \in L} B_{km,f} + \sum_{mn \in L} B_{mn,f} \right) \frac{V^{im}_{m,f}}{2} = I^{Dre}_{m,f} \quad \forall m \in N, f \in F \tag{3.105}$$

$$I^{Gim}_{m,f} + I^{DGim}_{m,f} + \sum_{km \in L} I^{im}_{km,f} - \sum_{mn \in L} I^{im}_{mn} - \left(\sum_{km \in L} B_{km,f} + \sum_{mn \in L} B_{mn,f} \right) \frac{V^{re}_{m,f}}{2} = I^{Dim}_{m,f} \quad \forall m \in N, f \in F \tag{3.106}$$

Power-Based Representation—DG Units: For this formulation, active and reactive powers are included in the power balance as:

$$\sum_{km \in L} P_{km,f} - \sum_{mn \in L} \left(P_{mn,f} + P^{L}_{mn,f} \right) + P^{G}_{m,f} + P^{DG}_{m}/3 = P^{D}_{m,f} \quad \forall m \in N, f \in F \tag{3.107}$$

$$\sum_{km \in L} Q_{km,f} - \sum_{mn \in L} \left(Q_{mn,f} + Q^{L}_{mn,f} \right) + Q^{G}_{m,f} + Q^{DG}_{m}/3 = Q^{D}_{m,f} \quad \forall m \in N, f \in F \tag{3.108}$$

Therefore, the complete LP formulations for an unbalanced network, considering DG units is presented by:

$$\begin{gathered} \min \alpha \\ \text{subject to}: (3.6)-(3.9), (3.15), (3.16), (3.95)-(3.98), \text{ and } (3.103)-(3.106) \end{gathered} \tag{3.109}$$

$$\begin{gathered} \min \alpha \\ \text{subject to}: (3.31), (3.41)-(3.43), (3.95)-(3.98), (3.107), \text{and } (3.108) \end{gathered} \tag{3.110}$$

for the current-based and power-based formulations, respectively.

3.5.1 Renewable DG

Different from the dispatchable DG, in which the electric distribution network operator controls the active and reactive powers injection for each DG unit, renewable generation depends on availability of renewable resources (e.g., wind speed and solar irradiance). Nowadays, renewable DG has taken an important role in the decentralization of energy production [32]. Due to difficulties related to DG forecasting when it operates from renewable energy sources, additional considerations must be taken into account to properly include this type of DG into a distribution network optimization framework. Moreover, with the increase in penetration of these technologies, issues related to voltage profiles, energy losses, restoration actions, and network reinforcements have to be addressed.

In this regard, (3.111) represents an additional constraint for renewable DG units, and models an active power curtailment. This technique is used in order to avoid undesired levels of power injection from renewable sources which can lead to voltage rises and high energy losses [33, 34].

$$\hat{P}_n^{DG} = P_n^{DG} + \tilde{P}_n^{DG} \quad \forall n \in DG \tag{3.111}$$

where $\hat{P}_n^{DG}$ and $\tilde{P}_n^{DG}$ are the maximum available power and the power curtailment for DG unit n. Under this optimization scheme, $\hat{P}_n^{DG}$ will depend on the availability related to the renewable energy source (e.g., wind speed and solar irradiance); hence, multi-scenario approaches are mainly used to tackle this problems.

3.6 Energy Storage Devices

Energy storage systems have been foregrounded as an answer to conciliate time-difference between excessive generation and peak demand. In recent years, energy storage devices prices have declined, which in turn, raised the usage of these technologies in the electric distribution networks. For electric distribution, battery-based energy storage systems (BESS) are the most common type of storage. This is because other storage technologies such as super capacitors and flywheels are characterized by their high energy cost and are primarily applied on high power, short duration applications. Hence, due to the growth in the utilization of BESS and their constant interaction with renewable DG and EVs, their inclusion in the distribution network steady-state operation must be addressed [18, 35, 36].

3.6.1 BESS Operation

The BESS power drawn or injected from/to the grid must be taken into account in the steady-state operation of the system. Thus, let SD be the set of BESSs plugged into the grid. P_u^{SD} is the power injected/drawn by BESS u, and it is equal to the sum of two non-negative variables that represent the ESS charging and discharging powers $\left(P_u^{SD+}\right.$ and P_u^{SD-}, respectively), as shown in (3.112). Moreover, (3.113) and (3.114) limit the variables P_u^{SD+} and P_u^{SD-}, in terms of the BESS maximum charging/discharging power $\left(\bar{P}_u^{SD}\right)$ and the binary variables w_u and x_u.

Equation (3.115) ensures only one action for the BESS (e.g., charging, discharging, or idle). Due to the integer nature of these variables, once they are included in the distribution network optimization framework, one obtains a MILP problem.

$$P_u^{SD} = P_u^{SD+} - P_u^{SD-} \quad \forall u \in SD \tag{3.112}$$

$$0 \leq P_u^{SD+} \leq \bar{P}_u^{SD} w_u \quad \forall u \in SD \tag{3.113}$$

$$0 \leq P_u^{SD-} \leq \bar{P}_u^{SD} x_u \quad \forall u \in SD \tag{3.114}$$

$$w_u + x_u \leq 1 \quad \forall u \in SD \tag{3.115}$$

On the other hand, BESS have physical constraints regarding their storage capacity, i.e., the maximum energy limit $\left(\bar{E}_u^{SD}\right)$ and the maximum *DoD* must be always fulfilled. Hence, if the P_u^{SD} is maintained during a time interval Δt, (3.116) keeps the BESS energy level between the pre-established limits.

$$\bar{E}_u^{SD} DoD \leq E_u^{SDi} + \Delta t (P_u^{SD+} \eta_u^+ - P_u^{SD-} \eta_u^-) \leq \bar{E}_u^{SD} \quad \forall u \in SD \tag{3.116}$$

where E_u^{SDi} is the initial SOC for BESS u; while, η_u^+ and η_u^- are the charging and discharging efficiencies, respectively.

The interaction of the BESSs with the grid is integrated in the steady-state operation as follows:

Current-Based Representation—BESS: Initially, in (3.117) and (3.118) the active and reactive powers injected or consumed by the BESS are expressed in terms of the voltage operation point $\left(V_u^*\right)$ where the BESS is connected and the BESS current $\left(I_u^{SDre}\right)$. Considering that BESSs will only inject/drawn active power. Moreover, (3.119) and (3.120) are the extensions of (3.8) and (3.9) taking into account the BESS current injection in each node.

$$P_u^{SD} = V_u^{re*} I_u^{SDre} + V_u^{im*} I_u^{SDim} \quad \forall u \in SD \tag{3.117}$$

$$0 = -V_u^{re*} I_u^{SDim} + V_u^{im*} I_u^{SDre} \quad \forall u \in SD \tag{3.118}$$

$$\begin{aligned} & I_{m,f}^{Gre} + \sum_{km \in L} I_{km,f}^{re} - \sum_{mn \in L} I_{mn}^{re} - \left(\sum_{km \in L} B_{km,f} + \sum_{mn \in L} B_{mn,f} \right) \frac{V_{m,f}^{im}}{2} \\ & = I_{m,f}^{Dre} + \sum_{u \in SD} I_u^{SDre} \gamma_{u,m,f} \end{aligned} \quad \forall m \in N, f \in F \tag{3.119}$$

$$I^{Gim}_{m,f} + \sum_{km \in L} I^{im}_{km,f} - \sum_{mn \in L} I^{im}_{mn} - \left(\sum_{km \in L} B_{km,f} + \sum_{mn \in L} B_{mn,f} \right) \frac{V^{re}_{m,f}}{2} = I^{Dim}_{m,f} + \sum_{u \in SD} I^{SDim}_{u} \gamma_{u,m,f} \quad \forall m \in N, f \in F \tag{3.120}$$

where $\gamma_{x,m,f}$ is a binary parameter that takes a value of 1 if the device x is connected at node m and phase f.

Power-Based Representation—BESS: For this formulation, (3.121) represents the active power balance in each node, taking into account the BESS active power injection/consumption.

$$\sum_{km \in L} P_{km,f} - \sum_{mn \in L} \left(P_{mn,f} + P^{L}_{mn,f} \right) + P^{G}_{m,f} = P^{D}_{m,f} + \sum_{u \in SD} P^{SD}_{u} \gamma_{u,m,f} \quad \forall m \in N, f \in F \tag{3.121}$$

Therefore, the complete MILP formulations for an unbalanced network, considering BESS is presented in (3.122) and (3.123) for the current-based and power-based formulations, respectively.

$$\begin{gathered} \min \alpha \\ \text{subject to : } (3.6)–(3.9), (3.15), (3.16), \text{and } (3.112)–(3.121) \end{gathered} \tag{3.122}$$

$$\begin{gathered} \min \alpha \\ \text{subject to : } (3.31), (3.41)–(3.43), (3.112)–(3.116), \text{and } (3.121) \end{gathered} \tag{3.123}$$

3.7 Voltage and Reactive Power Control Devices

Voltage optimization and reactive power control have been widely used in power systems as tools to improve energy efficiency and quality [16, 17]. In electric distribution networks, the management of voltage magnitudes variations together with the reactive power flows is known as volt-var control (VVC). The main objective of the VVC is to determine control actions for the devices related to voltage management and reactive power flow management. The classical devices controlled within a VVC scheme are on-load tap changers (OLTCs), voltage regulators (VRs), and switched capacitor banks (SCBs). Hence, the optimization of the VVC will pursue a proper distribution network operation, while maximizing or minimizing an objective imposed by the distribution network operator, e.g., power loss reduction, minimization of voltage deviation, or maximization of energy efficiency.

In this framework the mathematical modeling for the optimization of the VVC problem is presented. A solution for the VVC problem will provide the number of enabled/disabled modules in every SCB, and the tap position for the OLTCs and VRs. Hence, the mathematical representation for the switchable SCBs, OLTCs, and VRs, are shown for both LP formulations. Due to the integer nature of the variables that model the operation of the VVC devices, the obtain formulations corresponds to a MILP problem.

3.7.1 Capacitor Banks

The inclusion of switchable SCBs in the electric distribution network operation will represent an injection of reactive power that will depend on the number of SCB modules enabled. Thus, let $CB \subseteq N$ be the set of nodes where a three-phase SCB is installed. B_n is an integer variable that represents the number of modules enabled from the SCB connected at node n; $\bar{B}_n$ is the maximum number of SCB modules; Q_n^{cb} is the reactive power delivered; and Q_n^{esp} is the reactive power capacity of each module. Equation (3.124) represents the reactive power injected by the modules of the switchable SCBs, while the maximum number of operating modules for each BC is modeled by (3.125).

$$Q_n^{cb} = B_n Q_n^{esp} \quad \forall n \in CB \tag{3.124}$$

$$0 \leq B_n \leq \bar{B}_n \quad \forall n \in CB \tag{3.125}$$

Furthermore, in a multi-period optimization where the SCB operations permitted along the entire time period must be limited, (3.126) must be taken into account.

$$\sum_{t \in T} \left| B_{n,t} - B_{n,t-1} \right| \leq \Delta^{cb} \quad \forall n \in CB \tag{3.126}$$

where T is the set of time intervals; $B_{n,t}$ is the number of modules enabled from the SCB connected at node n in time t; and Δ^{cb} is the maximum number of operations allowable over the time period.

The reactive power injection Q_n^{cb} is included in the steady-state operation of the distribution network as follows:

Current-Based Representation—Capacitor Banks: The active and reactive power of the switchable SCBs are represented by (3.127) and (3.128), considering that the value for the active power injection of every SCB will always be equal to zero.

Moreover, the current balances in each node are updated to include the injection due to the SCB reactive power, as shown in (3.129) and (3.130).

$$0 = V^{re*}_{n,f} I^{cbre}_{n,f} + V^{im*}_{n,f} I^{cbim}_{n,f} \quad \forall n \in CB, f \in F \tag{3.127}$$

$$\frac{Q^{cb}_n}{3} = -V^{re*}_{n,f} I^{cbim}_{n,f} + V^{im*}_{n,f} I^{cbre}_{n,f} \quad \forall n \in CB, f \in F \tag{3.128}$$

$$I^{Gre}_{m,f} + \sum_{km \in L} I^{re}_{km,f} - \sum_{mn \in L} I^{re}_{mn} - \left(\sum_{km \in L} B_{km,f} + \sum_{mn \in L} B_{mn,f} \right) \frac{V^{im}_{m,f}}{2} = I^{Dre}_{m,f} - I^{cbre}_{n,f} \quad \forall n \in N, f \in F \tag{3.129}$$

$$I^{Gim}_{m,f} + \sum_{km \in L} I^{im}_{km,f} - \sum_{mn \in L} I^{im}_{mn} - \left(\sum_{km \in L} B_{km,f} + \sum_{mn \in L} B_{mn,f} \right) \frac{V^{re}_{m,f}}{2} = I^{Dim}_{m,f} - I^{cbim}_{n,f} \quad \forall n \in N, f \in F \tag{3.130}$$

where $I^{cbre}_{n,f}$ and $I^{cbre}_{n,f}$ are the real and imaginary parts of the current injected in phase *f*, by the SCB connected and node *n*.

Power-Based Representation—Capacitor Banks: For this formulation, the Q^{cb} is included in the reactive power balance as shown in (3.131).

$$\sum_{km \in L} Q_{km,f} - \sum_{mn \in L} \left(Q_{mn,f} + Q^{L}_{mn,f} \right) + Q^{G}_{m,f} = Q^{D}_{m,f} - Q^{cb}_{m} \quad \forall m \in N, f \in F \tag{3.131}$$

3.7.2 On-Load Tap Changers and Voltage Regulators

OLTCs and VRs are the devices in charge of controlling the voltage magnitudes in the electric distribution network, within a VVC environment. These devices adjust their input voltage through tap changing, and their operation can be represented under the same mathematical formulation. Thus, let $RT \subseteq L$ be the set of circuits where a VR is installed. $tp_{mn,f}$ is the integer variable that defines the tap position for the VR installed in circuit *mn*, in phase *f*; while, $Tp_{mn,f}$ is the maximum number of taps; and $\%R_{mn}$ is the regulation percentage.

Independent of the steady-state formulation adopted, (3.132) represents the minimum and maximum limits of the tap position. Analogue to CB, the number of tap changes permitted along the entire time period must be limited in a multi-period optimization; hence, under such scenario (3.133) must be taken into account.

$$-Tp_{mn} \leq tp_{mn,f} \leq Tp_{mn} \quad \forall mn \in RT, f \in F \tag{3.132}$$

$$\sum_{t \in T} \left| tp_{mn,f,t} - tp_{mn,f,t-1} \right| \leq \Delta^{vr} \quad \forall mn \in RT, f \in F \tag{3.133}$$

Current-Based Representation—OLTCs and VRs: In this formulation, (3.134) and (3.135) represent the real and imaginary regulated voltage; while, (3.136) and (3.137) represent the real and imaginary regulated current on each VR.

$$V^{re}_{n,f} = (1 + \%R_{mn} tp_{mn,f}/Tp_{mn}) V^{re}_{m,f} \quad \forall mn \in RT, f \in F \tag{3.134}$$

$$V^{im}_{n,f} = (1 + \%R_{mn} tp_{mn,f}/Tp_{mn}) V^{im}_{m,f} \quad \forall mn \in RT, f \in F \tag{3.135}$$

$$I^{re}_{km,f} = (1 + \%R_{mn} tp_{mn,f}/Tp_{mn}) I^{re}_{mn,f} \quad \forall mn \in RT, f \in F \tag{3.136}$$

$$I^{im}_{km,f} = (1 + \%R_{mn} tp_{mn,f}/Tp_{mn}) I^{im}_{mn,f} \quad \forall mn \in RT, f \in F \tag{3.137}$$

Equations (3.134)–(3.137) represent the operation of VRs and OLTCs in terms of the real and imaginary voltages and currents of the electric distribution network. Nevertheless, the nonlinearities presented have to be addressed, i.e., the product of the decision variables $tp_{mn,f}$ and, $V_{m,f}$ or $I_{mn,f}$ on the real and imaginary components. In this regard, the integer number of steps is represented as a set of binary variables $bt_{mn,f}$ and the products $tp_{mn,f} V_{m,f}$, $tp_{mn,f} I_{mn,f}$ are substituted by auxiliary variables $V^{c}_{mn,f,k}$ and $I^{c}_{mn,f,k}$, respectively.

A linear extension for (3.134)–(3.137) is presented in (3.138)–(3.151), where (3.138) and (3.139) represent the calculation of the regulated voltage, and, (3.140) and (3.141) the calculation of the regulated current. Constraint (3.142) associates the set of binary variables with the tap integer variable. Equations (3.143) and (3.144), and (3.145) and (3.146), define the auxiliary variables $V^{c}_{mn,f,k}$ and $I^{c}_{mn,f,k}$ respectively while (3.147) and (3.148), and (3.149) and (3.150) describe their limits. Finally, (3.151) represents the sequencing of the binary variable $bt_{mn,f}$.

$$V^{re}_{n,f} = (1 - \%R_{mn}) V^{re}_{m,f} + \sum_{k=1}^{2Tp_{mn}} \frac{\%R_{mn}}{Tp_{mn}} V^{c(re)}_{mn,f,k} \quad \forall mn \in RT, f \in F \tag{3.138}$$

$$V^{im}_{n,f} = (1 - \%R_{mn}) V^{im}_{m,f} + \sum_{k=1}^{2Tp_{mn}} \frac{\%R_{mn}}{Tp_{mn}} V^{c(im)}_{mn,f,k} \quad \forall mn \in RT, f \in F \tag{3.139}$$

$$I_{km,f}^{re} = (1 - \%R_{mn})I_{mn,f}^{re} + \sum_{k=1}^{2Tp_{mn}} \frac{\%R_{mn}}{Tp_{mn}} I_{mn,f,k}^{c(re)} \quad \forall mn \in RT, f \in F \tag{3.140}$$

$$I_{km,f}^{im} = (1 - \%R_{mn})I_{mn,f}^{im} + \sum_{k=1}^{2Tp_{mn}} \frac{\%R_{mn}}{Tp_{mn}} I_{mn,f,k}^{c(im)} \quad \forall mn \in RT, f \in F \tag{3.141}$$

$$\sum_{k=1}^{2Tp_{mn}} bt_{mn,f,k} - Tp_{mn} = tp_{mn,f} \quad \forall mn \in RT, f \in F \tag{3.142}$$

$$\left| V_{m,f}^{re} - V_{mn,f,k}^{c(re)} \right| \leq \bar{V}(1 - bt_{mn,f,k}) \quad \forall mn \in RT, f \in F, k = 1 \ldots 2Tp_{mn} \tag{3.143}$$

$$\left| V_{m,f}^{im} - V_{mn,f,k}^{c(im)} \right| \leq \bar{V}(1 - bt_{mn,f,k}) \quad \forall mn \in RT, f \in F, k = 1 \ldots 2Tp_{mn} \tag{3.144}$$

$$\left| V_{mn,f,k}^{c(re)} \right| \leq \bar{V} bt_{mn,f,k} \quad \forall mn \in RT, f \in F, k = 1 \ldots 2Tp_{mn} \tag{3.145}$$

$$\left| V_{mn,f,k}^{c(im)} \right| \leq \bar{V} bt_{mn,f,k} \quad \forall mn \in RT, f \in F, k = 1 \ldots 2Tp_{mn} \tag{3.146}$$

$$\left| I_{mn,f}^{re} - I_{mn,f,k}^{c(re)} \right| \leq \bar{I}_{mn}(1 - bt_{mn,f,k}) \quad \forall mn \in RT, f \in F, k = 1 \ldots 2Tp_{mn} \tag{3.147}$$

$$\left| I_{mn,f}^{im} - I_{mn,f,k}^{c(im)} \right| \leq \bar{I}_{mn}(1 - bt_{mn,f,k}) \quad \forall mn \in RT, f \in F, k = 1 \ldots 2Tp_{mn} \tag{3.148}$$

$$\left| I_{mn,f,k}^{c(re)} \right| \leq \bar{I}_{mn} bt_{mn,f,k} \quad \forall mn \in RT, f \in F, k = 1 \ldots 2Tp_{mn} \tag{3.149}$$

$$\left| I_{mn,f,k}^{c(im)} \right| \leq \bar{I}_{mn} bt_{mn,f,k} \quad \forall mn \in RT, f \in F, k = 1 \ldots 2Tp_{mn} \tag{3.150}$$

$$bt_{mn,f,k} \leq bt_{mn,f,k-1} \quad \forall mn \in RT, f \in F, k = 1 \ldots 2Tp_{mn} \tag{3.151}$$

Power-Based Representation—OLTCs and VRs: For the power-based formulation, $V_{m,f}^{sqr}$ is altered by the square of the regulation ratio, which is expressed in terms of the regulation percentage, the tap integer value, and the maximum tap, as shown in (3.152).

$$V_{n,f}^{sqr} = \left(1 + \%R_{mn} \frac{tp_{mn,f}}{Tp_{mn}}\right)^2 V_{m,f}^{sqr} \quad \forall mn \in RT, f \in F \tag{3.152}$$

In order to cope with the nonlinearities observed in (3.152), $tp^2_{mn,f}$ is represented as a set of binary variables $bt_{mn,f}$, and the product $tp^2_{n,f} V^{sqr}_{m,f}$ is represented using the auxiliary variables $V^c_{mn,f}$, as shown in set (3.153)–(3.157).

$$V^{sqr}_{n,f} = \sum_{k=1}^{2Tp_{mn}} \left[\frac{\%R_{mn}}{Tp_{mn}} \left(\frac{(2k-1)\%R_{mn}}{Tp_{mn}} + 2(1 - \%R_{mn}) \right) V^c_{mn,f,k} \right] + V^{sqr}_{m,f}(1 - \%R_{mn})^2 \quad \forall mn \in RT, f \in F \tag{3.153}$$

$$\underline{V}^2(1 - bt_{mn,f,k}) \leq V^{sqr}_{m,f} - V^c_{mn,f,k} \quad \forall mn \in RT, f \in F, k = 1 \ldots 2Tp_{mn} \tag{3.154}$$

$$V^{sqr}_{m,f} - V^c_{mn,f,k} \leq \bar{V}^2(1 - bt_{mn,f,k}) \quad \forall mn \in RT, f \in F, k = 1 \ldots 2Tp_{mn} \tag{3.155}$$

$$\underline{V}^2 bt_{mn,f,k} \leq V^c_{mn,f,k} \leq \bar{V}^2 bt_{mn,f,k} \quad \forall mn \in RT, f \in F, k = 1 \ldots 2Tp_{mn} \tag{3.156}$$

$$bt_{mn,f,k} \leq bt_{mn,f,k-1} \quad \forall mn \in RT, f \in F, k = 2 \ldots 2Tp_{mn} \tag{3.157}$$

Therefore, the complete MILP formulations for VVC optimization considering operational limits, are presented in (3.158) and (3.159), for the current-based and power-based representations, respectively.

$$\begin{gathered} \min \alpha \\ \text{subject to}: (3.6)-(3.9), (3.15), (3.16), (3.52), (3.53), (3.57), (3.61)-(3.63), (3.124), (3.125), \\ (3.127)-(3.130), (3.132), \text{and } (3.138)-(3.151) \end{gathered} \tag{3.158}$$

$$\begin{gathered} \min \alpha \\ \text{subject to}: (3.31), (3.41)-(3.43), (3.54), (3.58), (3.61), (3.64), (3.65), (3.124), (3.125), \\ (3.131), (3.132), \text{and } (3.153)-(3.157) \end{gathered} \tag{3.159}$$

3.8 Mathematical Framework Application in Control Approaches

Two control applications are presented in this section to assess the presented mathematical optimization framework. Initially, the EVCC problem is tackled using the current-based formulation, as shown in [25]. Later, a voltage control using the power-based LP formulation is shown, solving a VVC scheme to reduce voltage deviation and guarantee proper operation of the electric distribution network.

3.8.1 Electric Vehicle Charging Coordination Problem

The EV charging coordination problem consists in determining the optimal schedule for charging the EV batteries aiming an economical operation of the electric distribution network, while maintaining a suitable and efficient system operation. Hereby, a multi-period MILP formulation was proposed in [25] to solve the optimal charging coordination of EVs in unbalanced distribution networks considering V2G technology and DG. The steady-state operation of the grid is represented using the current-based formulation. The MILP formulation is embedded in a step-by-step control method that considers randomness in EV arrival, departure, and initial SOC.

The multi-period approach studies a specific time period which is divided into several time intervals. The control method finds an optimal schedule for the energy exchange between EV batteries and the grid. This method solves the proposed MILP model at the beginning of each time interval, constructing a step-by-step solution over the entire time period. The solution presents a charging schedule for each EV, which is generated between arrival and departure, ideally dispatching a fully charged battery.

The objective function of the EVCC problem, presented in (3.160), seeks to minimize the cost of the energy provided by the substation and the DG units as well as to reduce energy curtailment in EVs (if an EV cannot be completely charged, the unserved energy is considered as an energy curtailment).

$$\min \sum_{f \in F} \sum_{t \in T} \alpha^{G}_{S,t} \Delta_t \left(V^{re}_{S,f,t} I^{Gre}_{S,f,t} + V^{im}_{S,f,t} I^{Gim}_{S,f,t} \right) + \sum_{n \in N} \sum_{t \in T} \alpha^{DG}_{n,t} \Delta_t P^{DG}_{n,t} + \sum_{e \in EV} \beta E^{SH}_{e} \tag{3.160}$$

where $\alpha^{G}_{S,t}$ and $\alpha^{DG}_{n,t}$ are the energy costs at the substation and for each DG unit in time interval t, respectively. E^{SH}_{e} is the energy curtailment for EV e, while β is the EV curtailment cost (typically a high value to avoid curtailment).

Furthermore, the steady-state operation of the distribution network was modeled using (3.6), (3.7), (3.15), (3.16), (3.161), and (3.162). Constraints (3.52), (3.53), and (3.57), were used represent the operational limits. The DG units were modeled using (3.95), (3.96), (3.97), (3.103), and (3.104); an additional limit for the active power was also employed. Finally, the operation of the EVs was represented by (3.84)-(3.88), and (3.163).

$$\begin{aligned} & I^{Gre}_{m,f} + I^{DGre}_{m,f} + \sum_{km \in L} I^{re}_{km,f} - \sum_{mn \in L} I^{re}_{mn} - \left(\sum_{km \in L} B_{km,f} + \sum_{mn \in L} B_{mn,f} \right) \frac{V^{im}_{m,f}}{2} \\ & = I^{Dre}_{m,f} + \sum_{e \in EV} I^{EVre}_{e} \gamma_{e,m,f} \\ & \forall m \in N, f \in F \end{aligned} \tag{3.161}$$

$$
\begin{aligned}
&I^{Gim}_{m,f} + I^{DGim}_{m,f} + \sum_{km\in L} I^{im}_{km,f} - \sum_{mn\in L} I^{im}_{mn} - \left(\sum_{km\in L} B_{km,f} + \sum_{mn\in L} B_{mn,f}\right)\frac{V^{re}_{m,f}}{2} \\
&= I^{Dim}_{m,f} + \sum_{e\in EV} I^{EVim}_{e}\gamma_{e,m,f} \\
&\forall m \in N, f \in F
\end{aligned}
\tag{3.162}
$$

$$
\bar{E}^{EV}_{e} = E^{EVi}_{e} + \sum_{t\in T} \Delta t\left(\bar{P}^{EV+}_{e} y_{e,t}\eta^{EV+}_{e} - \bar{P}^{EV-}_{e} z_{e,t}\eta^{EV-}_{e}\right) + E^{SH}_{e} \quad \forall e \in EV
\tag{3.163}
$$

The proposed model was tested in the IEEE 123-node test system [11], and the following considerations were taken into account:

- Phases A, B, and C of the electric distribution network were charged with 1.42 MVA (40.7%), 0.915 MVA (26.2%), and 1.155 MVA (33.1%), respectively.
- The time period was set from 18:00 to 08:00 h, divided into half-hour time intervals.
- Two types of EV batteries were considered: 50 kWh Tesla EVs and 20 kWh Nissan Leafs. The charging maximum power was 10 and 4 kW, and for EV-V2Gs the discharging maximum power was 5 and 2 kW, respectively.
- Hourly energy cost and load variation were considered.
- The arrival and departure time intervals were generated based on the two chi-squared probability functions with 8 and 4 degrees of freedom.
- The initial SOC of the EVs was generated using the normal-based probability function with mean value and the standard deviation of 15 and 10, respectively.
- The minimum voltage limit was set at 0.90 pu.
- The maximum current was 500 A for all feeders.
- 400 EVs were plugged into the grid.
- 40% of EVs were considered to have V2G technology.

The model was implemented in the mathematical programming language AMPL [12] and solved with the commercial solver CPLEX [3]. Initially, the dumb charge case is presented. Here, the EV recharge was done without any charging coordination, i.e., the EV batteries started an uninterrupted charging process as soon as they were plugged into the electric distribution network. Later, several control scenarios were analyzed. Figure 3.6 shows the energy exchange between the EVs

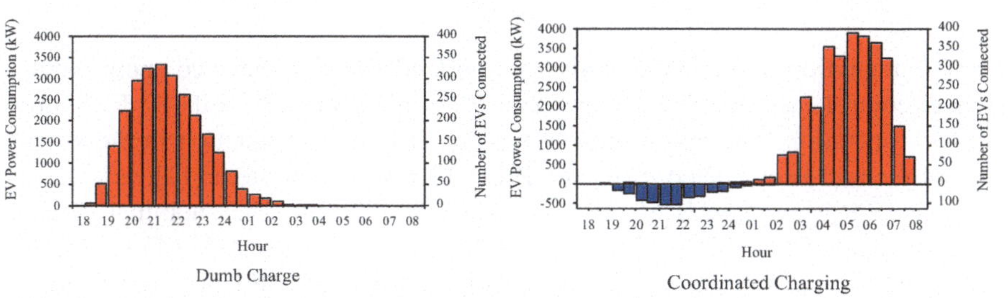

Fig. 3.6 EV active power exchange [25]

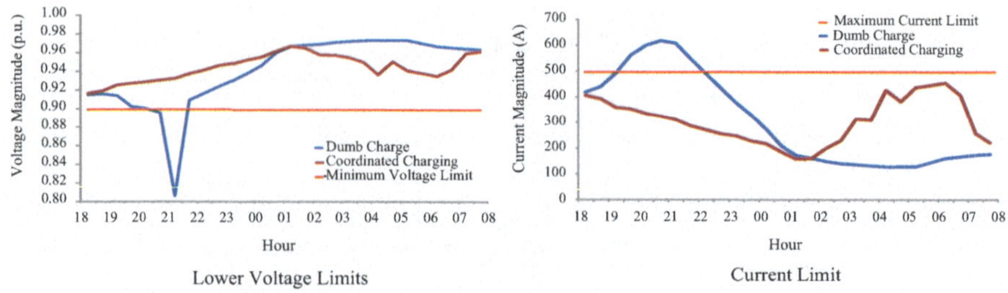

Fig. 3.7 Operational limits [25]

and the grid for the dumb charge and a coordinated charge scenario, in which all EVs were considered 'Tesla EV'. For the coordinated charge case, the objective function was reduced by 25% when compared to the dumb charge case, and no curtailment was presented.

The power related to the charging and discharging of EVs is shown in red and blue, respectively. Following convention, the EV charging power and the EV-V2G discharging power are given in positive and negative values, respectively. It can be seen that without any coordination (Dumb Charge), the EV batteries were continuously charged upon arrival; hence, the peak load for this case was between 21:00 and 22:00. In the coordinated charge case, this peak was shifted to the low-cost time intervals. This represented a reduction of almost 1 MW in the total active power demand.

Figure 3.7 presents the voltage and thermal limits for these cases. For the dumb charge, voltage and current limit violations were presented. These breaches were avoided when the EV charging control was enabled. Hence, it is stated that the EVCC in the distribution network is beneficial not only for peak load reduction, but also for maintaining the proper operation of the grid.

Therefore, the step-by-step methodology based on the current-based MILP formulation was proved efficient to find an optimal charging schedule for EVs in unbalanced network considering V2G technology.

3.8.2 Voltage Control Problem

High voltage drops along a radial distribution feeder lead to elevated energy losses. Hence, control methods for voltage optimization are crucial in daily distribution network operation. The application of the power-based formulation in the voltage control method is evaluated using the IEEE 13-node test system [11].

Figure 3.8 shows the IEEE 13-node test system, which has nominal conventional demand of 1.31 MVA (34.2%), 1.16 MVA (30.3%), and 1.36 MVA (35.5%), connected to phases A, B, and C, respectively. For this test, all loads are considered to be connected in wye-configuration and classified as constant power. Moreover,

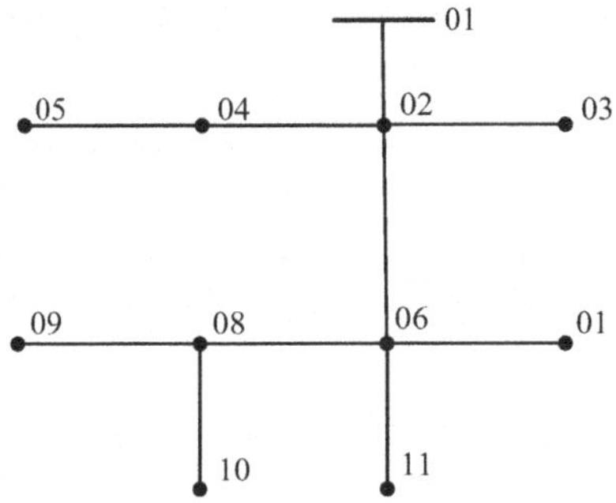

Fig. 3.8 IEEE 13-node test system

Table 3.3 Voltage profile of the IEEE 13-node test system, under nominal demand (pu)

Bus\Phase	A	B	C
1	1.0000	1.0000	1.0000
2	0.9555	0.9695	0.9347
3	0.9523	0.9675	0.9319
4	–	0.9591	0.9378
5	–	0.9557	0.9389
6	0.9221	0.9723	0.8585
7	0.9138	0.9734	0.8545
8	0.9200	–	0.8546
9	–	–	0.8508
10	0.9138	–	–
11	0.9221	0.9723	0.8585

the nominal voltage is 4.16 kV, and the voltage magnitude at the substation is fixed at 1.0 pu. Working under nominal demand, the test system shown in Fig. 3.8, presents the voltage profile shown in Table 3.3.

In order to improve the voltage profile of the electric distribution network, the minimum and maximum voltage limits are set at 0.93 and 1.05 pu, respectively. An OLTC and two switchable SCB are added to the grid in order to fulfill voltage limit requirements. It is desired to use these devices to keep the voltage between the established limits, while minimizing the voltage deviation at each bus for 4 different load levels, 100, 70, 50, and 30% of the nominal demand (i.e., it is expected to maintain the voltage magnitude of all buses as close as possible to the nominal value).

Hence, the control actions for each device should be determined in order to fulfill the operational constraints related to the voltage limit, while minimizing voltage deviation. In this matter, consider:

- The OLTC is installed at the substation and controls the voltage magnitude at node 1.
- The OLTC can vary the input voltage magnitude in a 5% regulation ratio, distributed in 8 tap positions (± 4).

- The switchable SCBs are installed at node 7 and 9.
- Each switchable SCB has 6 × 500 kvar, modules.

To determine the control actions for the OLTC and the BCs, the power-based LP formulation is chosen. Initially the voltage deviation (ψ) is expressed as the absolute value of the difference between nominal voltage (V_{nom}^2) and the square value of the bus voltage $\left(V_{n,f}^{sqr}\right)$, as shown in (3.164). Thus, (3.164) is linearized in (3.165) and (3.166) taking advantage that the deviation is minimized in the objective function.

$$\left|V_{nom}^2 - V_{n,f}^{sqr}\right| = \psi_{n,f} \quad \forall n \in N, f \in F \tag{3.164}$$

$$V_{nom}^2 - V_{n,f}^{sqr} \leq \psi_{n,f} \quad \forall n \in N, f \in F \tag{3.165}$$

$$-\left(V_{nom}^2 - V_{n,f}^{sqr}\right) \leq \psi_{n,f} \quad \forall n \in N, f \in F \tag{3.166}$$

The LP formulation presented in Sect. 3.7 is used, and the complete MILP is shown as

$$\begin{gathered} \min \sum_{n \in N} \sum_{f \in F} \psi_{n,f} \\ \text{subject to : } (3.31), (3.41) - (3.43), (3.54), (3.58), (3.61), (3.64), (3.65), (3.124), (3.125), \\ (3.131), (3.132), (3.153) - (3.157), (3.165), \text{and } (3.166) \end{gathered} \tag{3.167}$$

The mathematical formulation was written in the mathematical language AMPL [12], and solved using CPLEX [3]. The model was solved for the four load levels, finding the best configuration for each case. Table 3.4 presents for each case the tap position for the OLTC, the number of enabled modules for the SCBs, and the total deviation. Besides, it can be seen that the voltage limits were fulfilled for each case in which the devices were taken into account.

Table 3.4 Summary of the results for the Voltage Control Problem

Case	Loading (%)	OLTC-Tap	SCB-7 modules active/ total	SCB-9 modules active/ total	Voltage deviation (pu)	Voltage limits
w/o devices	100	–	–	–	4.0177	Breached
I	100	+3	6/6	6/6	1.2943	Fulfilled
II	70	+2	6/6	6/6	0.7122	Fulfilled
III	50	+1	6/6	5/6	0.4937	Fulfilled
IV	30	+1	1/6	3/6	0.2783	Fulfilled

Table 3.5 Voltage profile of the IEEE 13-node test system, after voltage control implementation (pu)

Phase	A		B		C	
Bus\Case	w/o devices	I	w/o devices	I	w/o devices	I
1	1.0000	1.0375	1.0000	1.0375	1.0000	1.0375
2	0.9555	0.9980	0.9695	1.0067	0.9347	1.0017
3	0.9523	0.9949	0.9675	1.0048	0.9319	0.9991
4	–	–	0.9591	0.9967	0.9378	1.0046
5	–	–	0.9557	0.9934	0.9389	1.0056
6	0.9221	0.9693	0.9723	1.0076	0.8585	0.9594
7	0.9138	0.9622	0.9734	1.0092	0.8545	0.9566
8	0.9200	0.9672	–	–	0.8546	0.9601
9	–	–	–	–	0.8508	0.9608
10	0.9138	0.9613	–	–	–	–
11	0.9221	0.9693	0.9723	1.0076	0.8585	0.9594

To better illustrate the solution found through mathematical optimization, the voltage profile for the solution determined by the MILP formulation under nominal demand, is shown in Table 3.5. It can be seen that the MILP formulation found a solution which guarantee the compliance of the voltage limits.

Voltage limit violations were highlighted in red. The efficiency of the power-based MILP formulation was proven as the solutions for all cases improved the voltage profile, keeping all voltage magnitudes in between the limits, while minimizing the voltage deviation.

3.9 Comparative Overview and Discussion

The formulations presented, efficiently model the steady-state operation of unbalanced networks; constituting a mathematical framework that can be used by planners and operators as a tool inside optimization methods and algorithms, aiming to optimize specific goals. Although both formulations target the same objective, the planner/operator can choose the one that better accommodates and fits the problem that he is aiming to tackle. In order to make this decision, the following considerations must be addressed and well-thought:

1. Although both formulations show high accuracy determining the steady-state operation point, the current-based model slightly outperforms the power-based, as shown in Sect. 3.2.3.

2. Due to its representation in terms of the real and imaginary parts of voltages and currents, modeling operational limits within the current-based formulation requires a high number of constraints. This fact conveys to higher computational burden, which can slow down the solution process especially in large-scale test systems.
3. In optimization methods or algorithms applied over non-stressed test systems, where operational limits are not a concern to the optimizer (e.g., demand response or market-based optimizations), the current-based formulation may highlight as a better option.
4. As mentioned, both formulations can handle optimizations algorithms taking into account smart grid devices plugged into the network. Nevertheless, the mathematical representation of the considered smart grid devices can sway the formulation choice; e.g., in a volt-var approach, the power-based formulation is recommended, as the volt-var devices influence directly over the voltage magnitude and reactive power flows.
5. Finally, the load behavior is also an important feature to take into account as it impacts directly in the operation point estimation. Both LP formulations rely on the accuracy of the estimated operation point. In this regard, in little observable distribution networks, estimating voltages is an easier task to the planner/operator than estimating power flows along the grid. Hence, the current-based formulation will suit better to this application.

It is important to remark that every optimization problem targeted in electric distribution network will bring specific considerations that have to be analyzed to make the best choice.

Appendix 1: Piecewise Linearization Technique

The piecewise linearization is a technique in which a nonlinear function is approximated using a set of piecewise linear functions [37]. Widely used in engineering, this technique is often employed to cope with quadratic nonlinearities, helping to reach LP models. Typically, a function f is defined in order to calculate the square value of a variable σ, represented as $\sigma^+ + \sigma^-$ and limited by the interval $[0, \bar{\sigma}]$. This type of function has a general structure, as

$$f(\sigma, \bar{\sigma}, \Lambda) = \sum_{\lambda=1}^{\Lambda} \phi_{\sigma,\lambda} \Delta_{\sigma,\lambda} \quad (3.168)$$

$$\sigma = \sigma^+ - \sigma^- \quad (3.169)$$

$$\sigma^{+} + \sigma^{-} = \sum_{\lambda=1}^{\Lambda} \Delta_{\sigma,\lambda} \tag{3.170}$$

$$0 \leq \Delta_{\sigma,\lambda} \leq \bar{\sigma}/\Lambda \quad \forall \lambda \in \Lambda \tag{3.171}$$

$$\phi_{\sigma,\lambda} = (2\lambda - 1)\bar{\sigma}/\Lambda \quad \forall \lambda \in \Lambda \tag{3.172}$$

The parameter $\phi_{\sigma,\lambda}$ is calculated to compute the contribution of $\Delta_{\sigma,\lambda}$ in each step of the discretization. The parameter $\bar{\sigma}$ represents the maximum value of σ, while Λ is the number of discretizations used in the linearization.

It is important to remark that this approach is limited to maximizing strictly concave functions or minimizing convex functions. If the application of this technique under different conditions is desired, the inclusion of binary variables and additional constraints is mandatory.

Appendix 2: Multi-period and Multi-scenario Extension

Typically, optimization analyses in electric distribution network operation are done along a time window in which several control actions have be defined and they may be dependent among them; this is known as multi-period optimization. For example, the day-ahead operation planning is typically divided in one-hour time windows, and the decisions from one hour may or may not affect the decisions regarding the next time intervals. Thus, mathematical formulations for the distribution network operation should be able to handle multi-period optimization analyses. In this regard, the LP formulations presented can be easily adapted to handle several time intervals. Hence, a new index associated to the time interval is added to the variables that represent the distribution network operation (e.g., voltages, currents, and power flows).

Furthermore, adaptability to multi-scenario optimizations is also required in an optimization framework for electric distribution network to model the uncertainty in the grid. The multi-scenario optimization is a method usually employed to solve stochastic programming problems in which some of the variables or parameters are of uncertain nature (e.g., EV behavior, renewable DG availability, and demand variations). The uncertainties are represented through a set of scenarios and each one with an associated probability, i.e., a multi-scenario model will provide an optimal solution on average, considering all the scenarios simultaneously. Analogue to the multi-period case, a new index is added to the uncertain variables associated to each scenario. Thereby, the objective function of the problem is calculated as the expected value due to the inclusion of the probabilities related to each scenario.

Appendix 3: Estimated Steady-State Operation Point

As mentioned in Sect. 3.2.3, the accuracy of the presented three-phase formulations relies on the precision of the estimated operation point. High quality estimations will minimize the error corresponding to some approximations in voltage magnitudes and some linearization techniques (e.g., Taylor's linearization). In order to obtain a suitable estimated operation point, the followings techniques might be employed:

1. A two-stage approach, in which a first stage solves the LP model using a flat start (e.g., assuming nominal voltages and disregarding power). Later, the solution of the first stage is used to initialize the second stage in which the LP model is once again solved from the already calculated operating point.
2. Using historical data, where historical data is used in order to determine the estimated values. Typically, the operator's knowledge and experience are crucial to select previous operating points which have occurred under similar loading and generation scenarios.
3. Using the previous time interval operating point is another technique for the estimation of the operating point. This approach is commonly used on small time interval optimization approaches in which abrupt changes in the demand are not expected (e.g., EVCC problems).

It is important to remark that the estimation of the operation point is an important issue to be taken into account when applying the presented formulations. Furthermore, the technique chosen to determine the estimated operation point will depend on the information available and the characteristics of the problem that is being tackled.

References

1. J.A. Momoh., *Electric Power System Applications of Optimization*, 2nd edn. (CRC Press, Boca Raton, 2009)
2. L. Aleixo et al., A general framework for active distribution network planning, in *CIGRE Symposium 2013*, no. April 2013 (2013), pp. 1–8
3. "IBM ILOG CPLEX V12.1 User's Manual for CPLEX." CPLEX Division, Incline Village, NV (2009)
4. MOSEK ApS, *The MOSEK optimization tools manual—version 6.0*. Copenhagen, 2009
5. G. O. Inc., Gurobi Optimizer reference manual, www.Gurobi.com, vol. 6 (2014), p. 572
6. R.H. Byrd, J. Nocedal, R.A. Waltz, Knitro: an integrated package for nonlinear optimization, in *Large-Scale Nonlinear Optimization*, ed. by G. Di Pillo, M. Roma (Springer US, Boston, MA, 2006), pp. 35–59
7. P. Bonami, J. Lee, BONMIN Users' Manual, (2007)
8. D. Shirmohammadi, H.W. Hong, A. Semlyen, G.X. Luo, A compensation-based power flow method for weakly meshed distribution and transmission networks. IEEE Trans. Power Syst. **3**(2), 753–762 (1988)

9. R.G. Cespedes, New method for the analysis of distribution networks. IEEE Trans. Power Deliver. **5**(1), 391–396 (1990)
10. J.F. Franco, M.J. Rider, R. Romero, A mixed-integer linear programming model for the electric vehicle charging coordination problem in unbalanced electrical distribution systems. IEEE Trans. Smart Grid **6**(5), 2200–2210 (2015)
11. IEEE/PES, Distribution Test Feeders, 123-bus Feeder.
12. R. Fourer, D.M. Gay, B.W. Kernighan, *AMPL: a modeling language for mathematical programming*, 2nd edn. (Brooks/Cole-Thomson Learning, Pacific Grove, CA, 2003)
13. EPRI, Open Distribution System Simulator, (2012)
14. International energy outlook 2013. (2013)
15. P.L. Dandeno et al., System load dynamics-simulation effects and determination of load constants. IEEE Trans. Power Appar. Syst. **PAS-92** (1973)
16. A. Padilha-Feltrin, D. Quijano, J.R. Mantovani, Volt-VAR multiobjective optimization to peak-load relief and energy efficiency in distribution networks. IEEE Trans. Power Deliver. **30**(2), 618–626 (2015)
17. H. Ahmadi, J.R. Martí, H.W. Dommel, A framework for volt-VAR optimization in distribution systems. IEEE Trans. Smart Grid **6**(3), 1473–1483 (2015)
18. C. Sabillon, O. Melgar Dominguez, J. Franco, M. Lavorato, M.J. Rider, Volt-VAR control and energy storage device operation to improve the electric vehicle charging coordination in unbalanced distribution networks. IEEE Trans. Sustain. Energy (2017)
19. L.M. Korunovic, S. Sterpu, S. Djokic, K. Yamashita, S. M. Villanueva, J.V. Milanovic, Processing of load parameters based on Existing Load Models, in *2012 3rd IEEE PES Innovative Smart Grid Technologies Europe (ISGT Europe)* (2012), pp. 1–6
20. Task force on load representation for dynamic performance IEEE, Bibliography on load models for power flow and dynamic performance simulation. IEEE Trans. Power Syst. **10** (1), 523–538 (1995)
21. N. Anglani, F. Fattori, G. Muliere, Electric vehicles penetration and grid impact for local energy models, in *2012 IEEE International Energy Conference and Exhibition (ENERGYCON)* (2012), pp. 1009–1014
22. K. Clement-Nyns, E. Haesen, J. Driesen, The impact of charging plug-in hybrid electric vehicles on a residential distribution grid. IEEE Trans. Power Syst. **25**(1), 371–380 (2010)
23. D. Wu, D.C. Aliprantis, K. Gkritza, Electric energy and power consumption by light-duty plug-in electric vehicles. IEEE Trans. Power Syst. **26**(2), 738–746 (2011)
24. A. O'Connell, A. Keane, D. Flynn, Rolling multi-period optimization to control electric vehicle charging in distribution networks, in *2014 IEEE PES General Meeting|Conference Exposition* (2014), p. 1
25. C. Sabillon Antunez, J.F. Franco, M.J. Rider, R. Romero, A new methodology for the optimal charging coordination of electric vehicles considering vehicle-to-grid technology. IEEE Trans. Sustain. Energy **7**(2), 596–607 (2016)
26. A.T. Al-Awami, E. Sortomme, Coordinating vehicle-to-grid services with energy trading. IEEE Trans. Smart Grid **3**(1), 453–462 (2012)
27. T. Ackermann, G. Andersson, L. Söder, Distributed generation: a definition. Electr. Power Syst. Res. **57**(3), 195–204 (2001)
28. K. Qian, C. Zhou, Y. Yuan, X. Shi, M. Allan, Analysis of the environmental benefits of Distributed Generation, in *IEEE Power and Energy Society 2008 General Meeting: Conversion and Delivery of Electrical Energy in the 21st Century, PES* (2008)
29. A.C. Rueda-Medina, J.F. Franco, M.J. Rider, A. Padilha-Feltrin, R. Romero, A mixed-integer linear programming approach for optimal type, size and allocation of distributed generation in radial distribution systems. Electr. Power Syst. Res. **97**, 133–143 (2013)
30. J.A.P. Lopes, N. Hatziargyriou, J. Mutale, P. Djapic, N. Jenkins, Integrating distributed generation into electric power systems: a review of drivers, challenges and opportunities. Electr. Power Syst. Res. **77**(9), 1189–1203 (2007)
31. M. Bollen, F. Hassan, *Integration of Distributed Generation in the Power System* (2011)

32. Y.M. Atwa, E.F. El-Saadany, MMa. Salama, R. Seethapathy, Optimal renewable resources mix for distribution system energy loss minimization. IEEE Trans. Power Syst. **25**(1), 360–370 (2010)
33. S. Weckx, C. Gonzalez, J. Driesen, Combined central and local active and reactive power control of PV inverters. IEEE Trans. Sustain. Energy **5**(3), 776–784 (2014)
34. R. Tonkoski, L.A.C. Lopes, T.H.M. El-Fouly, Coordinated active power curtailment of grid connected PV inverters for overvoltage prevention. IEEE Trans. Sustain. Energy **2**(2), 139–147 (2011)
35. M. Ross, R. Hidalgo, C. Abbey, G. Joos, Analysis of energy storage sizing and technologies, in *EPEC 2010—IEEE Electrical Power and Energy Conference: "Sustainable Energy for an Intelligent Grid"* (2010)
36. M.S. ElNozahy, T.K. Abdel-Galil, M.M.A. Salama, Probabilistic ESS sizing and scheduling for improved integration of PHEVs and PV systems in residential distribution systems. Electr. Power Syst. Res. **125**, 55–66 (2015)
37. J.F. Franco, M.J. Rider, M. Lavorato, R. Romero, A mixed-integer LP model for the reconfiguration of radial electric distribution systems considering distributed generation. Electr. Power Syst. Res. **97**, 51–60 (2013)

Chapter 4
Multi-stage Primary-Secondary Planning Considering Wholesale-Retail Markets

Mehrdad Setayesh Nazar, Alireza Heidari and Mahmood Reza Haghifam

Abstract This chapter presents an approach for Integrated Distributed Generation and primary-secondary network Expansion Planning (IDGNEP) in the presence of wholesale and retail markets. The presented method uses a unified model to explore the impacts of retail market participants on the IDGNEP procedure. While the theory and practice of IDGNEP have advanced over the years, the Non-Utility Retail Market Participants (NURMPs) and Customers' Active MicroGrids (CAMGs) introduce some other resources which can also be included in distribution network planning exercises. An electric distribution network may interchange energy with wholesale/retail market participants and downward CAMGs. When the volume of the energy interchanged between the network and NURMPs/CAMGs is comparable with the volume of electricity delivered to the end users, the IDGNEP results may considerably be different from the condition that no energy is interchanged. The presented model of IDGNEP is a Mixed Integer Non Linear Programming (MINLP) problem and the introduced algorithm decomposes the IDGNEP problem into multi sub-problems to achieve an optimal expansion planning of a network, in which the investment and operational costs are minimized, while the reliability of the network is maximized. Demand Side Management (DSM) programs, Distribution Automation (DA) investment alternatives and NURMP and CAMGs contribution scenarios which may significantly change the network's resources are considered in IDGNEP formulation. The algorithm was successfully tested for an urban distribution network.

M. S. Nazar (✉)
Faculty of Electrical Engineering, Shahid Beheshti University, Tehran, Iran
e-mail: m_setayesh@sbu.ac.ir

A. Heidari
School of Electrical Engineering and Telecommunication,
University of New South Wales, Sydney, Australia
e-mail: Alireza.heidari@unsw.edu.au

M. R. Haghifam
School of Electrical and Computer Engineering, Tarbiat Modares University,
Tehran, Iran
e-mail: haghifam@modares.ac.ir

F. Shahnia et al. (eds.), *Electric Distribution Network Planning*,
Power Systems, https://doi.org/10.1007/978-981-10-7056-3_4

Keywords Distribution network expansion planning · Genetic algorithm (GA) Optimization · Active microgrids

4.1 Introduction

Distribution network electric loads are usually supplied through a common electric distribution network and each end user is coupled with the main utility grid (denoted as 'main grid') through the point of common coupling as shown in Fig. 4.1 [1]. However, many of the end users may have Distributed Generation (DG) facilities that supply energy to their energy-intensive industries and they may behave as dispatchable loads by reducing their electricity withdrawal from the distribution network and increasing the utilization of their electricity generation systems. In addition, the main grid may transact electricity with upward wholesale electricity market and its downward CAMGs [1].

Based on the end users' electrical load group characteristics, land ownership and operational constraints, the main grid can be segmented into different operational zones. In addition, for an open access electric distribution network, CAMGs and end users of different zones can transact energy with each other and they may form various power exchange patterns. However, any retail market electricity transactions between end users is analysed and approved by the Distribution System Operator (DSO) in advance and then the transactions can be performed. Energy interchanging between a network and wholesale and retail markets may change the

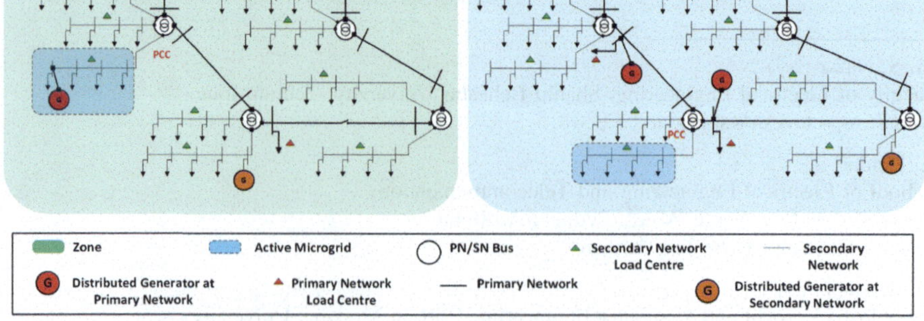

Fig. 4.1 Schematic diagram of the large primary/secondary electric distribution network

network's resources, costs and reliability. Thus, the optimal resource planning and operation of an energy interchanging network may be different from the ordinary ones [2–5].

The DG and primary-secondary Network Expansion Planning (DGNEP) problem consists of determining of the location, the capacity, and the time of installation of generation devices and network devices, depend on load growth conditions, reliability criteria, DSM programs, DA investment alternatives and NURMPs/CAMGs contribution scenarios.

The DGNEP is logical in light of demands and electric distribution network optimal operation. However, the main operation decisions are critical due to the two-way DSO and NURMPs/CAMGs interactions what will happen now to what will happen later based on the networks constraints.

This book chapter is on the Integrated DGNEP (IDGNEP) algorithm that considers the wholesale market price uncertainties, and NURMPs/CAMGs bid/offer price scenarios. Further, it uses the optimal reconfiguration procedure to investigate the adequacy of network resources under normal and contingent conditions.

4.2 Problem Modelling and Formulation

The wholesale market is assumed as a mandatory power pool. The DSO submits purchase bids to the wholesale market as a wholesale market participant. In this two-sided auction model, the independent system operator runs an auction market, settles the market and returns the settlements to the wholesale market participants.

The retail market is assumed as a voluntary power market, in which the NURMPs can contract bilaterally with the each other. The residual load is supplied by the DSO. The DSO may use the UDGs or purchase energy from the NURMPs/CAMGs and/or wholesale market to supply the residual load [4, 6, 7].

A NURMP can be classified as Non-Utility Distributed Generation (NUDG) and Dispatchable/Non-Dispatchable Load (DL/NDL) as shown in Fig. 4.2.

The NUDG and DL/NDL can be categorized as [4]:

(1) A NUDG that might be dispatchable by paying an appropriate capacity and energy fee is known as Dispatchable NUDG (DNUDG). The NUDG can participate in the retail market and the generation cost of this type of NURMP can be formulated as (4.1):

$$C_{DNUDG} = \left(\alpha.Cap_{DNUDG} + \sum_{i=1}^{Np} \beta_i.P_{gi}^{DNUDG}.\tau_i \right).\sigma \tag{4.1}$$

(2) A NUDG that might be non-dispatchable technically or economically is known as Non-Dispatchable NUDG (NDNUDG). The NDNUDG can participate in the retail market and the power generation cost of this type of DG can be formulated as (4.2):

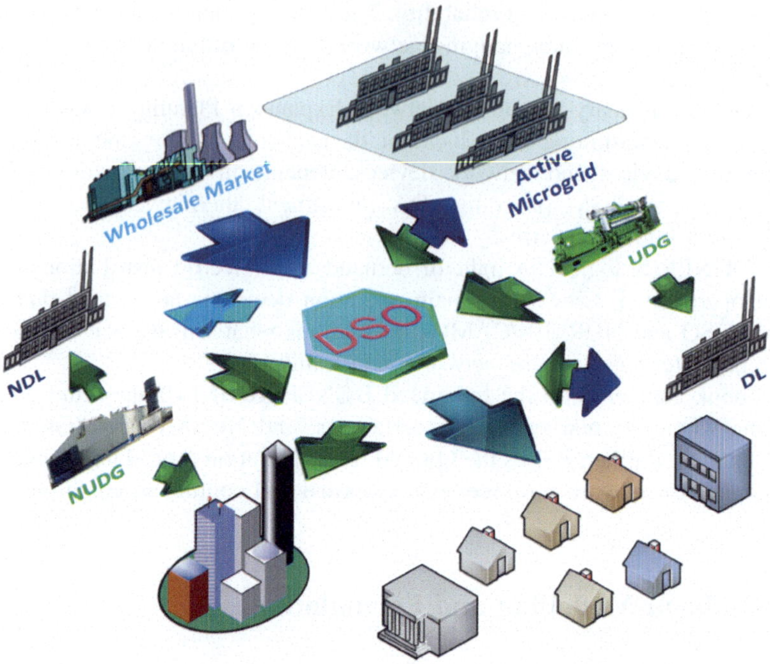

Fig. 4.2 Schematic diagram of the considered wholesale/retail market and active microgrid

$$C_{NDNUDG} = \left(\sum_{i=1}^{Np} \eta_i . P_{gi}^{NDNUDG} . \tau_i \right) . \sigma \tag{4.2}$$

(3) A NURMP that reduces its electric demand can be represented as DL or NDL based on the technical and economic parameters. The DL/NDL can participate in the retail market and the costs of load reduction in DL and NDL are formulated as (4.3) and (4.4), respectively:

$$C_{DL} = \left(\gamma . Cap_{DL} + \sum_{i=1}^{Np} \chi_i . P_{di}^{DL} . \tau_i \right) . \sigma \tag{4.3}$$

$$C_{NDL} = \left(\sum_{i=1}^{Np} \varsigma_i . P_{di}^{NDL} . \tau_i \right) . \sigma \tag{4.4}$$

A customer's active microgrid can be classified as Customer's Dispatchable MicroGrid (CDMG) and Customer's Non-Dispatchable MicroGrid (CNDMG).

The CDMG and CNDMG can be categorized as

(1) A CAMG that might be dispatchable is known as CDMG. The CDMG sells its power to the DSO at its point of common coupling and it can't participate in the retail market. The generation/load reduction cost of this type of CAMGs can be formulated as (4.5):

$$C_{CDMG} = \left(\sum_{i=1}^{Np} \varpi_i . P_{gi}^{CDMG} . \tau_i \right) . \sigma \tag{4.5}$$

(2) A CAMG that might be non-dispatchable is known as CNDMG. The CNDMG may have intermittent power generation units and it can't sell its power to the DSO or participate in the retail market. This type of CAMG is modelled as an embedded generation of a load point.

The DSO makes optimal decisions throughout IDGNEP horizon with incomplete information and it determines the optimal values of problem decision variables that consist of the location, the capacity, and the time of installation the UDGs and network devices. The DSO uses an estimated data of NURMPs/CDMGs location, type and capacity to determine optimal generation schedules of its generation units, electricity transactions with wholesale market and NURMPs/CDMGs, and contingency-based load shedding alternatives.

The IDGNEP problem is subject to the three sources of uncertainty: wholesale electricity market prices, NURMPs/CDMGs contribution scenarios, and network contingencies.

Thus, the uncertainty can be modelled as a multi-stage decision making problem based on a scenario driven algorithm. An iterative four stage optimization procedure is presented that minimizes the total network costs based on the wholesale market price scenarios at the first stage. At the second stage, the algorithm determines the optimal device allocations and capacity selections. At the third stage, the algorithm finds the optimal NURMPs/CDMGs contribution scenarios to maintain network reliability at the third stage. Finally, at the fourth stage, the optimal restoration problem investigates the adequacy of network resources under contingent conditions.

4.2.1 First Stage Problem Formulation

At the first stage, the DSO determines the number of wholesale market scenarios for each bi-annual periods and it estimates the electric loads, upward wholesale market prices, and power exchanges with upward network and NURMPs/CDMGs contribution scenarios. Based on the wholesale market price scenarios, the first stage minimizes the cost allocation problem that consists of whole investment and operational costs and customer interruption cost for the bi-annually periods of planning years. The objective function of the first stage problem can be written as (4.6):

$$
\begin{aligned}
Min\ C_1 = & \sum_{i=1}^{Nyear}\sum_{j=1}^{Nzone}[\sum_{\Omega} C_{ij\Omega}.\varphi_{ij\Omega} \\
& + W_{WM\,Purchased\_energy} * \sum_{k=1}^{WM\_Sc} \varphi_{WM\,ijk}^{Purchase\_Scenario}.MCP_{ijk}.E_{ijk}^{WM} \\
& + W_{NURMP\,Purchased\_energy} * \sum_{k=1}^{NURMP\_Sc} \varphi_{NURMP\,ijk}^{Purchase\_Scenario}.price_{ijk}^{NURMP}.E_{ijk}^{NURMP} \\
& + W_{CDMG\,Purchased\_energy_k} * \sum_{k=1}^{CDMG\_Sc} \varphi_{CDMG\,ijk}^{Purchase\_Scenario}.price_{ijk}^{CDMG}.E_{ijk}^{CDMG} \\
& + W_{CIC} * \sum_{k=1}^{N\_Critical\_Outage} CIC_{ijk}] \\
\Omega = & \{Sub, Feed, RPS, UDG, DSM, DA\}
\end{aligned}
\tag{4.6}
$$

The first stage objective function can be decomposed into three groups: (1) substation, feeder, RPS, UDG, DSM and DA investment plus aggregated operation costs, (2) the costs of purchased energy from upward network and NURMP and CDMG as the second group that is calculated in the third stage problem, (3) customer interruption cost as the third group. The first term (investment term) of (4.6) is described in the second stage problem and the customer interruption cost term is determined in the fourth stage problem. The Decision Variable Set (DVS) of the first stage problem can be written as (4.7):

$$
\begin{aligned}
DVS^1 &= [\varphi_\Omega, \varphi_{WM}^{Purchase\_Scenario}, \varphi_{NURMP}^{Purchase\_Scenario}, \varphi_{CDMG}^{Purchase\_Scenario}] \\
\Omega &= \{Sub, Feed, RPS, UDG, DSM, DA\}
\end{aligned}
\tag{4.7}
$$

The technical constraints are categorized into: device loading constraints, the entire load centres to be served constraints and dc load flow constraints.

The dc load flow constraints for network can be represented as (4.8):

$$f_1(x, u, z) = 0 \tag{4.8}$$

where x, u, z are problem variables, controls and network topology, respectively. Technical constraints can be compactly represented as (4.9):

$$g_1(x, u, z) \leq 0 \tag{4.9}$$

The first stage results can be used in the second stage to determine where and when an investment is needed. For every first stage scenario, the second stage problem optimizes the characteristics of network devices and their technical parameters for each period of the planning years.

4.2.2 Second Stage Problem Formulation

At the second stage, the DSO estimates the network's electric loads, upward wholesale market prices, and power exchanges with upward network and NURMPs/CDMGs for quarter of year periods. At this stage, the DSO optimizes the network device selection and allocation parameters. The second stage objective function can be presented as (4.10):

$$
\begin{aligned}
& Min\ C_2^a = \sum_{i=1}^{Nyear} \sum_{j=1}^{Nzone} [\sum_{\Gamma} C_{ij\Gamma} \cdot \psi_{ij\Gamma} + \sum_{\Delta} C_{ij\Delta} \cdot \psi_{ij\Delta}] \\
& \Gamma = \{Nsub, N\_Fr, N\_UDG, N\_DSM, N\_DA, N\_RPS\} \\
& \Delta = \{N\_NURMP, N\_CDMG\} \\
& a \in \text{First stage problem state space}
\end{aligned} \tag{4.10}
$$

The second stage objective function can be decomposed into two groups: (1) substation, feeder, UDG, DSM, DA and RPS allocation and capacity selection, (2) the NURMP and CDMG contribution scenarios as the second group that is calculated in the third stage problem.

The DVS of the second stage problem can be presented as (4.11):

$$
\begin{aligned}
& DVS^2 = [\psi_{\Gamma}, \psi_{\Delta}] \\
& \Gamma = \{Nsub, N\_Fr, N\_UDG, N\_DSM, N\_DA, N\_RPS\} \\
& \Delta = \{N\_NURMP, N\_CDMG\}
\end{aligned} \tag{4.11}
$$

The constraints can be categorized into: voltage constraints, device loading constraints, power balance at the network nodes under normal and contingent conditions, the entire load centres to be served constraints, uniqueness parameter selection constraints, operation of radial network under normal and contingent conditions and reliability constraints.

Technical and uniqueness parameter selection constraints can be presented as

$$g_2(x, u, z) \le 0 \tag{4.12}$$

Load flow constraints are:

$$f_2(x, u, z) = 0 \tag{4.13}$$

4.2.3 Third Stage Problem Formulation

Based on the second stage results, the contribution scenarios of NURMP/CDMGs is investigated for cost minimization. The third stage problem explores the monthly NURMP bid/offer state space and finds the best NURMPs/CDMGs contribution scenarios [2]. The DSO supplies the residual load of the retail market, thus, the third stage objective function can be stated as (4.14):

$$\begin{aligned} Min\ C_3^b = & \sum_{i=1}^{Nyear} \sum_{j=1}^{Nzone} \sum_{k=1}^{Np} [\sum_{\Lambda} C_{ijk\Lambda} . \phi_{ijk\Lambda} \\ & + W_{Purchased\_energy_k} * (MCP_{ijk} + Trans.\_service\_price_{ijk}).E_{ijk}^{WM}] \\ & \Lambda = \{N\_UDG, N\_NDNUDG, N\_DNUDG, N\_DL, N\_NDL, N\_CDMG\} \\ & b \in \text{ Second stage problem state space} \end{aligned} \quad (4.14)$$

The third stage objective function consists of UDG, NURMP and CDMG contribution scenarios. The DSO uses the scenario driven information to describe NDNUDG, DNUDG, DL, NDL and CDMG contribution scenarios. By selection of the best NURMP/CDMG contribution scenarios, the DSO optimizes the decision variables of (4.14).

The DVS of the third stage problem can be presented as (4.15):

$$\begin{aligned} & DVS^3 = [\psi_\Lambda] \\ & \Lambda = \{N\_UDG, N\_NDNUDG, N\_DNUDG, N\_DL, N\_NDL, N\_CDMG\} \end{aligned} \quad (4.15)$$

The third stage technical selection constraints can be presented as

$$g_3(x, u, z) \le 0 \quad (4.16)$$

Load flow constraints are:

$$f_3(x, u, z) = 0 \quad (4.17)$$

If the NURMPs/CDMGs contribution scenarios are fixed, the feasibility of network restoration can be investigated. This problem is a slave problem of the third stage problem that optimizes the network resource coordination in contingent conditions.

4.2.4 Fourth Stage Problem Formulation

The fourth stage problem tries to find the monthly optimal resource coordination in contingent conditions. The control variables of the network can be categorized as [2, 8]:

1. Discrete control variables of the network resources such as switching of tie switches and capacitors, and
2. Continuous control variables of the network resources such as UDG, DNUDG, CDMG and DL.

The fourth stage objective function is introduced as

$$\begin{aligned} &\text{Min } C_4^c = \sum_{i=1}^{Nyear} \sum_{j=1}^{Nzone} \sum_{k=1}^{Noutage} [CIC_{ijk} + \sum_{\Xi} \Delta C_{OP(\Xi)\,ijk}] \\ &\Xi = \{UDG, DNUDG, CDMG, DL\} \\ &c \in \text{Third stage problem state space} \end{aligned} \tag{4.18}$$

The fourth stage procedure investigates the adequacy of network resources for restoration of the most important loads. It tries to switch the tie switches and capacitors and optimize the network resource coordination. For new topology, the optimal coordination of resource coordination problem can be solved by custom optimal power flow. The optimization constraints are voltage drop, line loading and load flow constraints. The DVS of the fifth stage can be written as (4.19):

$$DVS^4 = [Cb^1, TS^1, \ldots, Cb^{N\_RPS}, TS^{N\_TS}] \tag{4.19}$$

$$Cb_i = 1 \quad \text{if the } i\text{ th element shunt is used, else it equals 0.} \tag{4.20}$$

$$TS_i = 1 \quad \text{if the } i\text{ th tie switch is used, else it equals 0.} \tag{4.21}$$

Technical and radial operation constraints can be compactly represented as

$$y_5^{\prime n}(x, u, z) \leq 0 \, \forall n \in \{0, 1, \ldots, Noutage\} \tag{4.22}$$

Load flow constraints are formulated as

$$f_5^{\prime n}(x, u, z) = 0 \, \forall n \in \{0, 1, \ldots, Noutage\} \tag{4.23}$$

4.3 Solution Algorithm

The mentioned model of IDGNEP is a scenario driven MINLP problem. The IDGNEP sub-problems are nonlinear and non-convex, if the first stage problem parameters are fixed. The fourth stage problem uses discrete control variables such as capacitor and line switching. It also generates new state spaces for solving the optimal restoration. Every effective switching will generate a new state space based on the load variations. The fourth-stage problem has a great state space and its solution algorithm must be able to effectively search this space. For optimization procedure, a Genetic Algorithm (GA) with variable fitness functions is used. The rates of the operators are adapted in a deterministic, reinforcement-based manner. The behaviour of each operator (that is, the specific way it operates) is modified by changing its parameter values. Figure 4.3 depicts the flowchart of the introduced multi-stage optimization algorithm. At first, the first stage problem is optimized for each bi-annual period of the planning years for wholesale market price scenarios. Then, the second stage problem optimality is investigated. At the third stage, the NURMP/CDMG contribution scenarios are optimized. Finally, at the fourth stage a feasible and optimal restoration problem solution is investigated.

In order to map the possible solutions of the problem, a binary basis codification is employed. The first, second, third and fourth stage GA Population (GAP) of decision variables can be presented as

$$GAP^i = DVS^i \quad \forall i = \{1, 2, 3, 4\} \tag{4.24}$$

To improve the performance and speed of the specified GA, a list of suitable candidates is selected for the first generation of the chromosomes. This population could be generated using engineering experience rules. In order to map the possible solutions of problem, a binary basis codification is employed.

Two operators, namely crossover and mutation, are applied to the first generation and as a result new chromosomes are generated. Figure 4.4 depicts the crossover process for the first stage problem. The crossover operator is employed for each of the decision variable set. Figure 4.5 depicts the mutation process for the first stage problem.

For implementation of operational constraints in the optimization process, a penalty factor representation is used. The final optimization fitness function of the multi-stage problem can be written as

$$\text{Max}\, \mathbb{Z} = \boldsymbol{M} - \boldsymbol{C} - \boldsymbol{W}.\boldsymbol{g}(\boldsymbol{u}, \boldsymbol{x}, z) - \boldsymbol{W}'.\boldsymbol{f}(\boldsymbol{u}, \boldsymbol{x}, z) \tag{4.25}$$

where $\mathbb{Z}$ and $\boldsymbol{M}$ are objective function and high number vectors, respectively. $\boldsymbol{W}$ and $\boldsymbol{W'}$ are weight factor vectors that can be increased linearly through iterations from zero to a very high number.

The Weighted Reliability Index (WRI) is used for stopping criteria, defined as

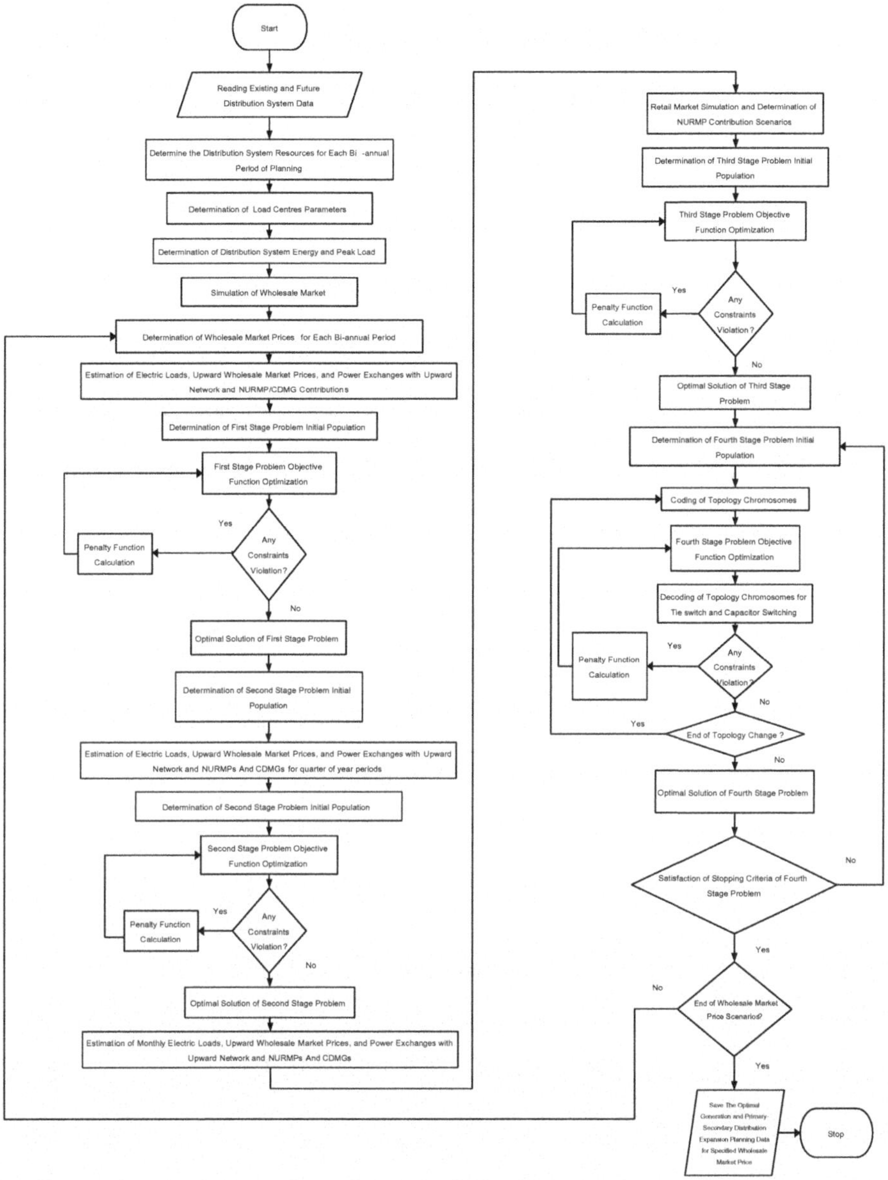

Fig. 4.3 Flowchart of introduced algorithm for solving of IDGNEP

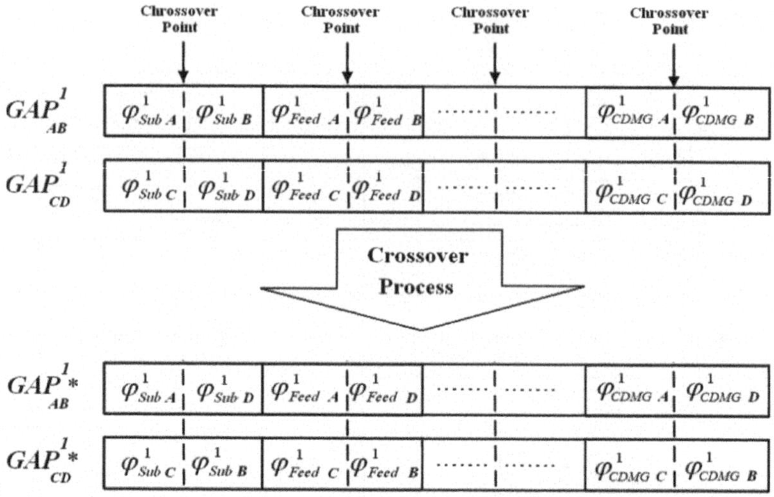

Fig. 4.4 Crossover process of the first stage problem chromosomes

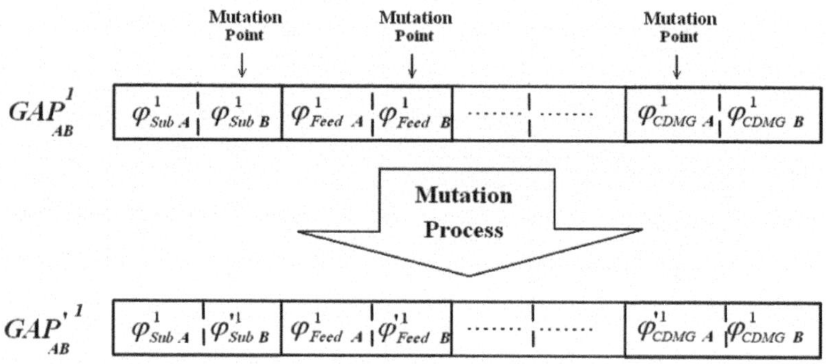

Fig. 5 Mutation process of the first stage problem chromosomes

$$WRI = wf_1' * SAIDI + wf_2' * SAIFI \tag{4.26}$$

where

$$SAIFI = \text{Total number of customer interruptions/total number of customers served.} \tag{4.27}$$

$$SAIDI = \text{Sum of customers' interruption duration/total number of customers.} \tag{4.28}$$

4.4 Numerical Results

The discussed algorithm was applied to an urban electric distribution network. This network was a part of a city network with 45,000 customers at the horizon year. The selected region will have about 7000 residential and commercial customers at the horizon year. Its primary network voltage is 20 kV and its secondary network voltage is 400 V. The time horizon was chosen 7 years into the future. The average repair time of the permanent faults is assumed as 2 h and the weight factors are 1.0.

The stopping criterion was selected as *WRI* < 8 for all the buses based on the utility reliability standard with $wf_1' = wf_2' = 1/2$ or the number of iterations > 2000.

By wholesale/retail market simulation, the most important NURMP contribution scenarios and the residual load of the network were determined for different periods and the NURMPs and CAMGs groups were recognized. The DSO investigates the optimum NURMP/CAMGs contribution scenarios for supplying of residual load at the normal and contingent conditions. By wholesale market simulation, different wholesale market price categories were calculated, which their values are shown in Monetary Units (MUs) in Table 4.1. Table 4.2 shows the total costs of NURMPs/CAMGs contribution. Table 4.3 shows the final load forecasting results considering NURMPs/CAMGs contribution.

The second stage problem determined the optimum allocation of network substations and feeder routing. Table 4.4 shows the final transformer capacity selection results that were determined in the second stage. This problem consists of using of the existing devices, which their life cycle is not expired, but the devices with greater capacity are needed. For example, a pole mounted transformer, which supplies a load, but the capacity of the existing transformer is not enough for supplying the load, must be changed by another one with greater capacity. The older transformer can be used in other substations to supply an appropriate load.

Table 4.5 depicts the final optimized IDGNEP costs. The transformer and feeder investment costs takes on a value 1921 billion MUs, which is decomposed in 159.1, 138.0, 129.2, 135.6, 107.8, 571.6 and 679.4 billion MUs for the first, second, third, fourth, fifth, sixth and seventh year of planning, respectively. Further, the DA and DSM and RPS investment costs takes on a value 4.792 billion MUs, which is decomposed in 1.0859, 1.7294, 0.5185, 0.5594, 0.2425, 0.2983 and 0.3585 billion MUs for the first, second, third, fourth, fifth, sixth and seventh year of planning, respectively. Finally, the NURMP and CAMG contribution costs takes on a value 8798 billion MUs, which is decomposed in 775.267, 899.48, 1063, 1261.2, 1504.6, 1576.8 and 1718 billion MUs for the first, second, third, fourth, fifth, sixth and seventh year of planning, respectively. The UDGs contribution factor is about 0.25, 0.28, 0.32, 0.35, 0.38, 0.41 and 0.43 for the first, second, third, fourth, fifth, sixth and seventh year of planning, respectively.

The fourth stage problem investigates the adequacy of network resources for restoration of the most important loads. The algorithm tries to switch the tie switches and capacitors and find a new set of network resources. The final and

Table 4.1 Estimated upward wholesale market price and network energy consumption

Year	First year	Second year	Third year	Fourth year	Fifth year	Sixth year	Seventh year
Estimated first scenario wholesale market price (MU/MWh)	7.50E + 01	7.80E + 01	8.25E + 01	8.53E + 01	8.81E + 01	9.24E + 01	9.63E + 01
Estimated first scenario energy consumption (kWh)	7.83E + 07	8.22E + 07	8.63E + 07	9.07E + 07	9.52E + 07	1.00E + 08	1.05E + 08
Estimated second scenario wholesale market price (MU/MWh)	9.23E + 01	9.59E + 01	1.01E + 02	1.05E + 02	1.08E + 02	1.14E + 02	1.18E + 02
Estimated second scenario energy consumption (kWh)	7.46E + 07	7.91E + 07	8.31E + 07	8.72E + 07	9.16E + 07	9.62E + 07	1.01E + 08

Table 4.2 Total costs of NURMPs/CAMGs contribution

Type of NURMP	First year cost (Million MUs)	Second year cost (Million MUs)	Third year cost (Million MUs)	Fourth year cost (Million MUs)	Fifth year cost (Million MUs)	Sixth year cost (Million MUs)	Seventh year cost (Million MUs)
DNUDG	225,520.84	261,668.25	310,848.66	384,383.45	476,444.07	502,622.19	547,858.19
NDNUDGG	31,083.62	36,070.26	42,561.34	50,540.28	62,851.62	65,716.39	69,093.95
DL	204,208.7472	238,021.6392	284,517.7848	327,717.3744	378,447.804	393,343.5096	429,672.7188
NDL	54,581.72	62,166.38	70,874.21	80,873.68	92,346.59	97,118.3	105,858.94
CDMG	226,408.1	263,589.4	311,137.7	368,826.7	438,987.6	460,783.1	503,111.6
CNDMG	33,464.28	37,967.61	43,095.52	48,942.75	55,592.55	57,299.47	62,456.42

Table 4.3 The final load forecasting results considering NURMPs/CAMGs contribution

Load centre	Load in first year (kVA)	Load in second year (kVA)	Load in third year (kVA)	Load in fourth year (kVA)	Load in fifth year (kVA)	Load in sixth year (kVA)	Load in seventh year (kVA)
1	130.8	126.9	145.1	148.3	162.0	167.6	177.7
2	305.2	326.5	345.2	375.4	394.7	410.9	437.1
3	135.6	134.7	141.7	155.6	157.5	176.4	171.5
4	253.2	266.3	284.8	306.6	321.2	332.0	360.1
5	190.1	215.0	218.0	225.8	240.4	259.8	273.7
6	317.7	325.6	356.2	367.1	387.3	421.5	438.0
7	294.7	319.0	343.4	354.1	370.6	392.4	419.3
8	53.6	57.0	60.2	75.7	66.9	72.1	83.7
9	221.3	237.6	241.8	263.9	275.7	295.9	310.5
10	82.6	86.0	99.1	100.4	100.2	105.6	116.3
11	278.1	297.2	312.4	329.4	356.1	367.3	384.4
12	217.3	239.1	252.0	251.6	276.6	289.5	307.4
13	312.6	330.2	358.9	375.1	399.5	410.1	434.8
14	158.2	162.1	180.8	188.5	199.0	212.0	226.4
15	343.2	357.2	387.6	399.4	429.9	455.6	473.9
16	784.5	835.1	884.3	935.0	997.2	1046.3	1109.5
17	569.7	606.0	646.2	675.6	721.2	762.8	797.8
18	618.9	695.8	748.4	786.8	825.7	879.6	918.0
19	61.4	74.1	71.9	72.6	85.9	78.5	92.6
20	288.4	303.7	324.9	342.8	348.5	376.6	388.7
21	40.8	54.1	52.8	44.5	51.8	51.2	62.7
22	366.7	383.4	406.5	434.6	461.8	479.9	509.6
23	40.3	45.9	43.2	44.9	43.0	52.3	48.4

(continued)

Table 4.3 (continued)

Load centre	Load in first year (kVA)	Load in second year (kVA)	Load in third year (kVA)	Load in fourth year (kVA)	Load in fifth year (kVA)	Load in sixth year (kVA)	Load in seventh year (kVA)
24	271.0	274.1	303.8	315.1	329.8	357.4	368.3
25	38.2	36.5	31.8	40.3	34.1	41.6	50.9
26	212.4	232.7	239.6	258.0	275.5	280.5	297.8
27	73.4	76.3	92.2	83.0	88.4	101.1	103.0
28	167.1	182.9	184.9	202.3	205.2	225.4	225.6
29	493.4	529.4	558.5	580.1	628.0	656.5	690.0
30	318.0	336.3	356.6	367.0	391.7	420.4	433.0
31	825.6	870.9	928.5	968.4	1032.3	1083.3	1154.4
32	886.1	942.0	995.1	1055.6	1112.3	1166.9	1234.2
33	371.4	394.5	414.3	437.3	472.8	492.6	520.1
34	213.3	218.5	228.5	249.8	255.7	275.6	290.9
35	131.9	140.8	153.5	162.1	169.5	176.9	190.8
36	145.5	155.8	164.2	174.4	174.7	188.8	196.5
37	165.0	181.4	193.4	202.1	212.8	222.0	236.5
38	496.6	522.2	559.9	580.8	619.5	650.1	686.3
39	320.4	329.8	353.0	375.0	396.9	411.7	435.3
40	489.1	519.5	544.3	575.1	604.5	638.9	670.3
41	616.0	646.5	678.9	724.2	764.8	803.0	854.5
42	162.7	183.8	193.4	192.2	214.5	213.0	233.9
43	496.5	520.9	556.7	588.2	615.1	657.8	688.3
44	306.1	336.3	343.4	373.2	384.9	409.9	438.6
45	214.5	225.4	238.5	255.5	255.6	271.9	293.1
46	168.3	172.5	186.6	195.7	210.6	211.4	236.2

(continued)

Table 4.3 (continued)

Load centre	Load in first year (kVA)	Load in second year (kVA)	Load in third year (kVA)	Load in fourth year (kVA)	Load in fifth year (kVA)	Load in sixth year (kVA)	Load in seventh year (kVA)
47	81.7	80.0	78.0	82.8	95.5	95.5	104.8
48	83.7	78.9	90.7	82.4	88.1	99.6	110.1
49	152.7	156.0	160.6	174.1	186.4	197.3	200.7
50	494.7	530.0	552.0	592.0	623.8	649.4	690.5
51	877.6	931.6	986.9	1052.1	1104.5	1179.0	1237.9
52	64.7	72.4	68.1	72.3	72.1	77.0	83.1
53	162.3	181.5	182.2	203.4	202.3	224.1	237.4
54	141.9	152.3	165.5	172.4	186.3	196.2	201.4
55	133.3	142.6	145.1	166.9	162.8	184.4	183.9
56	489.9	526.5	552.7	581.0	618.0	650.8	691.6
57	214.3	225.8	232.2	244.1	259.0	277.2	295.4
58	170.5	180.0	192.1	200.8	201.8	217.6	229.4
59	64.7	71.4	70.2	81.0	84.1	77.6	82.0
60	133.5	142.0	155.6	164.3	172.0	179.2	193.0
61	81.3	79.3	86.3	86.0	97.5	100.4	105.2
62	496.8	532.4	560.4	582.2	624.3	652.0	690.8
63	217.5	223.9	229.6	253.9	263.1	277.5	288.0
64	152.1	159.8	170.5	176.5	179.2	199.1	197.7
65	42.7	43.7	54.7	48.3	56.9	56.9	60.2
66	167.5	178.2	179.5	192.4	202.0	222.0	229.8
67	882.9	938.9	992.2	1046.4	1105.7	1171.5	1241.8
68	137.8	150.7	149.1	165.0	161.7	171.0	181.1
69	80.8	74.1	86.7	94.1	101.1	95.7	109.2

(continued)

Table 4.3 (continued)

Load centre	Load in first year (kVA)	Load in second year (kVA)	Load in third year (kVA)	Load in fourth year (kVA)	Load in fifth year (kVA)	Load in sixth year (kVA)	Load in seventh year (kVA)
70	170.5	176.3	182.2	197.6	215.0	223.6	233.6
71	39.1	47.4	50.7	52.9	48.7	54.7	54.3
72	151.1	148.3	164.8	168.9	181.4	185.3	207.4
73	128.7	149.8	153.3	163.0	171.4	178.2	183.0
74	494.7	531.3	555.3	592.0	618.8	654.3	694.3
75	214.1	221.8	228.4	247.7	266.8	280.8	292.9
76	889.5	941.9	993.0	1048.3	1111.9	1173.2	1232.9
77	494.0	520.1	554.7	587.0	620.9	661.0	685.2
78	149.9	156.3	157.2	168.8	178.3	196.5	206.1
79	82.3	76.8	84.1	91.8	97.9	94.1	108.0
80	162.8	175.5	185.6	202.7	213.9	213.3	229.6

Table 4.4 The final transformer capacity selection results

Load centre	First year capacity (kVA)	Second year capacity (kVA)	Third year capacity (kVA)	Fourth year capacity (kVA)	Fifth year capacity (kVA)	Sixth year capacity (kVA)	Seventh year capacity (kVA)
1	200	200	200	200	200	200	250
2	400	400	500	500	500	500	630
3	200	200	200	200	200	250	250
4	315	315	400	400	400	400	500
5	250	315	315	315	315	315	400
6	400	400	500	500	500	500	630
7	400	400	500	500	500	500	500
8	100	100	100	100	100	100	100
9	315	315	315	315	400	400	400
10	100	200	200	200	200	200	200
11	400	400	400	400	500	500	500
12	315	315	315	315	400	400	400
13	400	400	500	500	500	500	630
14	200	200	250	250	250	250	315
15	500	500	500	500	630	630	630
16	1000	1000	1250	1250	1250	1250	1600
17	800	800	800	800	1000	1000	1000
18	800	1000	1000	1000	1000	1250	1250
19	100	100	100	100	200	100	200
20	400	400	400	500	500	500	500
21	50	100	100	100	100	100	100
22	500	500	500	630	630	630	630
23	50	100	100	100	100	100	100

(continued)

Table 4.4 (continued)

Load centre	First year capacity (kVA)	Second year capacity (kVA)	Third year capacity (kVA)	Fourth year capacity (kVA)	Fifth year capacity (kVA)	Sixth year capacity (kVA)	Seventh year capacity (kVA)
24	400	400	400	400	400	500	500
25	50	50	50	50	50	50	100
26	250	315	315	315	400	400	400
27	100	100	200	100	200	200	200
28	200	250	250	250	250	315	315
29	630	630	800	800	800	800	1000
30	400	400	500	500	500	500	630
31	1000	1250	1250	1250	1250	1600	1600
32	1250	1250	1250	1250	1600	1600	1600
33	500	500	500	630	630	630	630
34	315	315	315	315	315	400	400
35	200	200	200	200	200	250	250
36	200	200	200	250	250	250	250
37	200	250	250	250	315	315	315
38	630	630	800	800	800	800	1000
39	400	400	500	500	500	500	630
40	630	630	800	800	800	800	800
41	800	800	800	1000	1000	1000	1250
42	200	250	250	250	315	315	315
43	630	630	800	800	800	800	1000
44	400	400	500	500	500	500	630
45	315	315	315	315	315	400	400
46	200	250	250	250	250	250	315

(continued)

Table 4.4 (continued)

Load centre	First year capacity (kVA)	Second year capacity (kVA)	Third year capacity (kVA)	Fourth year capacity (kVA)	Fifth year capacity (kVA)	Sixth year capacity (kVA)	Seventh year capacity (kVA)
47	100	100	100	100	200	200	200
48	100	100	200	100	200	200	200
49	200	200	200	250	250	250	250
50	630	630	800	800	800	800	1000
51	1250	1250	1250	1250	1600	1600	1600
52	100	100	100	100	100	100	100
53	200	250	250	250	250	315	315
54	200	200	200	250	250	250	250
55	200	200	200	200	200	250	250
56	630	630	800	800	800	800	1000
57	315	315	315	315	315	400	400
58	250	250	250	250	250	315	315
59	100	100	100	100	100	100	100
60	200	200	200	200	250	250	250
61	100	100	200	200	200	200	200
62	630	630	800	800	800	800	1000
63	315	315	315	315	315	400	400
64	200	200	250	250	250	250	250
65	100	100	100	100	100	100	100
66	200	250	250	250	250	315	315
67	1250	1250	1250	1250	1600	1600	1600
68	200	200	200	200	200	250	250
69	100	100	200	200	200	200	200

(continued)

Table 4.4 (continued)

Load centre	First year capacity (kVA)	Second year capacity (kVA)	Third year capacity (kVA)	Fourth year capacity (kVA)	Fifth year capacity (kVA)	Sixth year capacity (kVA)	Seventh year capacity (kVA)
70	250	250	250	250	315	315	315
71	50	100	100	100	100	100	100
72	200	200	200	200	250	250	250
73	200	200	200	200	250	250	250
74	630	630	800	800	800	800	1000
75	315	315	315	315	315	400	400
76	1250	1250	1250	1250	1600	1600	1600
77	630	630	800	800	800	800	1000
78	200	200	200	200	250	250	250
79	100	100	100	200	200	200	200
80	200	250	250	250	315	315	315

Table 4.5 Final IDGNEP results

Costs	First year cost	Second year cost	Third year cost	Fourth year cost	Fifth year cost	Sixth year cost	Seventh year cost
Total transformer and feeder installation Costs (1000 MUs)	1.60E + 08	1.40E + 08	1.30E + 08	1.40E + 08	1.10E + 08	5.70E + 08	6.80E + 08
Total DA and DSM and RPS installation costs (1000 MUs)	1.1E + 06	1.7E + 06	5.2E + 05	5.6E + 05	2.4E + 05	3.0E + 05	3.6E + 05
Total DNUDGs and NDNUDGs contribution costs (1000 MUs)	2.57E + 08	2.98E + 08	3.53E + 08	4.35E + 08	5.39E + 08	5.68E + 08	6.17E + 08
Total DLs and NDLs contribution costs (1000 MUs)	2.59E + 08	3.00E + 08	3.56E + 08	4.09E + 08	4.71E + 08	4.90E + 08	5.36E + 08
Total CAMGs contribution costs (1000 MUs)	2.61E + 08	3.03E + 08	3.54E + 08	4.18E + 08	4.95E + 08	5.18E + 08	5.66E + 08
Total UDG energy generation (kWh)	4.31E + 07	5.08E + 07	6.39E + 07	7.21E + 07	8.52E + 07	9.58E + 07	1.10E + 08
Total energy purchased from wholesale market (kWh)	1.03E + 08	9.99E + 07	9.78E + 07	9.07E + 07	8.74E + 07	8.41E + 07	8.71E + 07

optimum topology of the network has 2463 independent single short and long time failures. Based on the fourth stage procedure, the stopping criterion was selected as WRI. The optimized network topology is shown in Fig. 4.6 for the seventh year of planning years. The optimal tie switch allocations are also illustrated. Under normal conditions, the sectionalizers are normally opened and can be used for optimal restoration. The corresponding WRI index is shown in Fig. 4.7 for the seventh year of planning years. As Fig. 4.7 shows, the maximum WRI index takes on a value 8.72 at bus 5 for the seventh year of planning years.

Appendix

The notations used throughout this chapter are listed below:

E^{WM}	Total energy purchased from the wholesale market to supply residual load
E^{NURMP}	Total energy purchased from the NURMPs to supply residual load
E^{CAMG}	Total energy purchased from the CAMGs to supply residual load
Cap_{DNUDG}	Generation capacity prepared by DNUDG
Cap_{DL}	Load reduction capacity prepared by DL
C_{Ω}	Present worth of cost allocation of Ω set
C_{Γ}	Present worth of device installation cost of Γ set
C_{Δ}	Present worth of NURMP/CDMG contribution cost of Δ set
C_{Λ}	Present worth of contribution cost of Λ set
$C_{OP(\Xi)}$	Present worth of operation costs of Ξ set
N_Fr	Number of primary/secondary feeder type candidates for installation
MCP	Marginal clearing price of the wholesale market
$Nyear$	Number of planning years
$N\_Critical\_Outage$	Number of network critical outages
Np	Number of load curve periods
$Nzone$	Number of distribution network zones
N_RPS	Number of RPS installation candidates
N_DSM	Number of DSM installation candidates
N_DA	Number of DA installation candidates
N_UDG	Number of UDG installation alternatives
N_NURMP	Number of NURMP installation scenario at the second stage problem
N_CDMG	Number of CDMG contribution alternatives
N_NDNUDG	Number of NDNUDG contribution alternatives
N_DNUDG	Number of DNUDG contribution alternatives
N_DL	Number of DL contribution alternatives

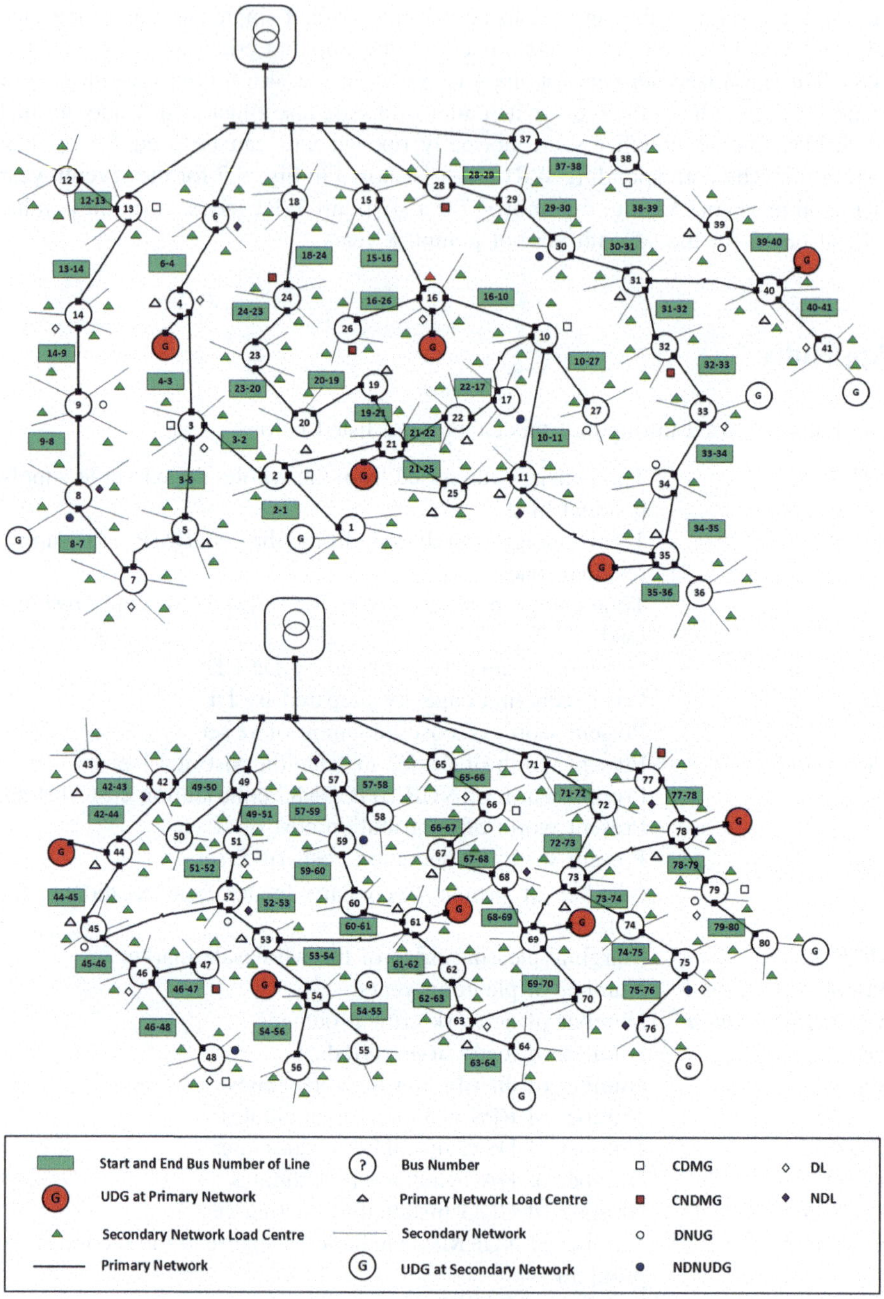

Fig. 4.6 The optimized topology for the seventh year of planning years

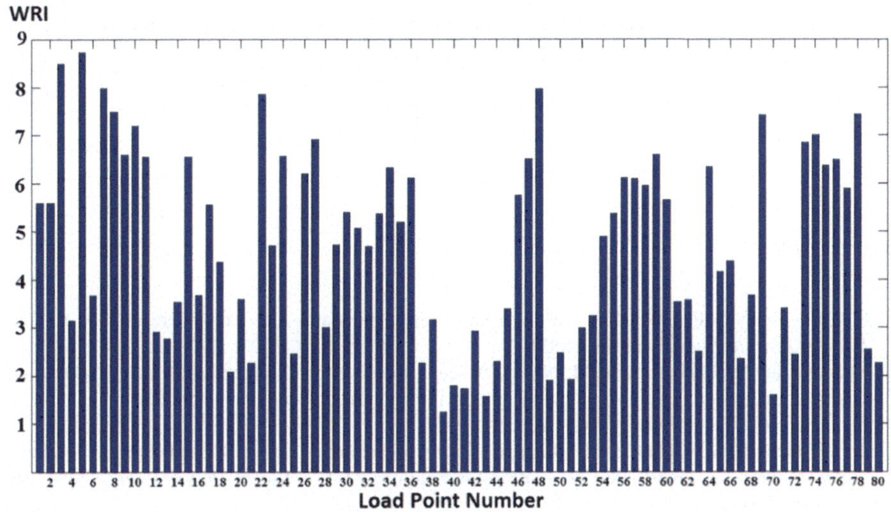

Fig. 4.7 The WRI at the seventh year of planning years

N_NDL	Number of NDL contribution alternatives
$Nsub$	Number of primary/secondary substations candidates
$price^{NURMP}$	Price of energy purchased from the NURMPs to supply residual load
$price^{CDMG}$	Price of energy purchased from the CDMGs to supply residual load
$Trans\_Service\_price$	Transmission service price of upward transmission network which delivers energy to the DSO
W	Weighting Factor
WM_Sc	Number of wholesale market price scenarios
$NURMP\_Sc$	Number of NURMP contribution scenarios
$CDMG\_Sc$	Number of CDMG contribution scenarios
α	Generation capacity fee of DNUDG
β	Energy generation fee of DNUDG
χ	Load reduction energy fee of DL
$\varphi_{WM}^{Purchase\_Scenario}$	Decision variable of purchasing energy from the wholesale market
$\varphi_{NURMP}^{Purchase\_Scenario}$	Decision variable of purchasing energy from NURMP
$\varphi_{CDMG}^{Purchase\_Scenario}$	Decision variable of purchasing energy from CMDG
γ	Load reduction option fee of DL
η	Load reduction energy fee of NDL
σ	Present worth factor
ς	Load reduction fee of DL
τ	Time duration of NURMP contribution
ϖ	CDMG power generation contribution

ψ	Decision variable for device installation or UDG/DNUUDG/CDMG/DL contribution
ϕ	Decision variable for UDG/DNUUDG/CDMG/DL contribution coordination problem

References

1. S. Chowdhury, S.P. Chowdhury, P. Crossley, Microgrids and active distribution networks, in *IET Renewable Energy Series* (2009)
2. M.S. Nazar, M.R. Haghifam, M. Nazar, A scenario driven multiobjective Primary-Secondary Distribution System Expansion Planning algorithm in the presence of wholesale–retail market. Int. J. Electr. Power Energy Syst. **40**, 29–45 (2012)
3. C.L.T. Borges, V.F. Martins, Multistage expansion planning for active distribution networks under demand and distributed generation uncertainties. Int. J. Electr. Power Energy Syst. **36**, 107–116 (2012)
4. M.S Nazar, M.R. Haghifam, Multiobjective electric distribution system expansion planning using hybrid energy hub concept. Electr. Power Syst. Res. **79**, 899–911, (2009)
5. P.S. Georgilakis, N.D. Hatziargyriou, A review of power distribution planning in the modern power systems era: Models, methods and future research. Electr. Power Syst. Res. **121**, 89–100 (2015)
6. A. Satchwell, R. Hledik, Analytical frameworks to incorporate demand response in long-term resource planning. Utilities Policy **28**, 73–81 (2014)
7. S. Sekizaki, I. Nishizaki, T. Hayashida, Electricity retail market model with flexible price settings and elastic price-based demand responses by consumers in distribution network. Int. J. Electr. Power Energy Syst. **81**, 371–386 (2016)
8. M.M. Aman, G.B. Jasmon, A.H.A. Bakar, H. Mokhlis, M. Karimi, Optimum shunt capacitor placement in distribution system: a review and comparative study. Renew. Sustain. Energy Rev. **30**, 429–439 (2014)

Chapter 5
Multi-agent Based Planning Considering the Behavior of Individual End-Users

Jan Kays

Abstract The volatile feed-in of distributed generation based on renewable energy sources as well as new and intelligent loads and storages require an appropriate consideration in the distribution grid planning process. With the conventional planning method being dependent on extreme scenarios, the consideration is very limited. Therefore, a new planning tool based on the concept of a multi-agent system is presented. In this system, every network user is represented by an agent, allowing not only the consideration of the volatile feed-in characteristics of renewable energy sources but also of the dependencies between the network users and their environment. Every network user is modeled as an agent of its own, guaranteeing the preservation of its individual character. Within this chapter, a system overview is given and the agent design process demonstrated on the example of the household load agent and the storage agent, including negotiations. This multi-agent system generates time series for all relevant system variables, defining detailed input parameters in the distribution grid planning process. The probabilities of occurrence of loading situations can be derived from the time series. For the first time, this allows for a detailed determination of the conditions in the up to now rarely measured medium and low voltage grids. As a consequence, new assumptions for the planning process are derivable, permitting a demand- and future-oriented grid planning and avoiding over-dimensioning of the grids.

Keywords Distribution grid planning · Multi agent system · Time series
Storage systems · Distributed energy resources

J. Kays (✉)
Amprion GmbH, Dortmund, Germany
e-mail: jan.kays@udo.edu

J. Kays
TU Dortmund University, Dortmund, Germany

F. Shahnia et al. (eds.), *Electric Distribution Network Planning*,
Power Systems, https://doi.org/10.1007/978-981-10-7056-3_5

5.1 Introduction

The requirements concerning the distribution grid planning have become more challenging in the last years. In contrast to transmission systems that have been planned precisely and computer-aided for over four decades already, the electric distribution network planning was simpler, due to several reasons. With unidirectional power flows from the higher voltage layers to the customers in the past, the needed grid structure has been rather simple. The distribution system operators (DSO) are faced to a large number of network areas, with many customers and assets. Neither the monitoring of every device, nor the installation of complicated protection systems had been justifiable from an economic point of view. Additionally, the collection of a vast amount of measurements was just impossible, when electronic data processing was not as powerful as today. For this reason, most of the low-voltage grids and with them most customers' power consumption are not measured [1].

Besides these reasons, the low deployment of distributed generation (DG) units and the inflexible demand have led to a deterministic planning practice, which is known as fit and forget [2]. Based on deliberated planning guidelines and the planners' experiences, the grid performance is evaluated on the basis of extreme scenarios for the combination of maximum load/minimum feed-in and maximum feed-in/minimum load. Without integrating the probabilities of occurrence for certain loading situations, this conventional planning [3] has its difficulties in finding an efficient solution to face the present challenges in the distribution grids.

With the evolution from passive to active distribution grids, new methods of planning and optimizing the distribution grids are mandatory [4]. Many different approaches have been developed in the past to facilitate the distribution grid planning process. An overview to different development trends is given in [3] as well as in [5–7], showing that optimization approaches often deal with an optimized placement and sizing of DG units. Considering the planning of the grid itself, approaches use inter alia Monte Carlo Simulations as in [8] or Particle Swarm Optimization (PSO) methods, like in [9], to determine necessary grid reinforcement measures. Others like [2, 10, 11] use probabilistic approaches to face uncertainties of the loading situations. The uncertainty of occurrence of future supply tasks is considered in [12], enabling the DSO to assort the priority of grid measures. Although the utilization of time series supports the evolution to active distribution networks [4] as well as the integration [13] and financial analysis of Smart Grid applications into the planning, most of the optimization approaches are based on input data that are derived from the conventional extreme scenarios.

Nevertheless, some approaches, as [14, 15], utilize at least measurements of installed smart meters to derive loading situations. Others like [16] analyze the positioning of storage systems on the basis of profiles. A battery management system for demand response purposes has been developed in [17]. Intelligent loads and their resulting impact on the system are analyzed in [18] or [19], as well as the optimized loading of plug-in electric vehicles is outlined in [20]. In [21], an agent-based model is combined with a PSO algorithm to determine necessary grid measures on the basis of time series. An agent-based system for the analysis of Smart Grid applications in a test feeder is outlined in [22]. A test bed for simulations of energy markets with high penetration of photovoltaics and loads with price sensitivity is published in [23]. The interaction of electric vehicles with DSOs is outlined in [24], implementing a multi agent system. An agent-based system for smart grid and market for the analysis of negotiations and synergies on a daily basis has been presented in [25]. Another agent-based system is presented in [26], combing electric vehicles, photovoltaic units and domestic loads to generate load profiles, as well as socio-demographic information. A simulation framework for market-based control in emergency situations in the distribution grids focuses on the operational aspects in [27]. The reviewed approaches either use extreme scenarios for optimization problems or time series that usually do not include functional dependencies between the network users that are existent in Smart Grid applications. Therefore, the impact of Smart Grid and Smart Market applications is difficult to port into the mid- and long term planning of the grids.

As a consequence, a new simulation environment on the basis of a multi-agent system has been developed and is presented in this chapter as key element of a new planning process, where probabilities of occurrence become more important. Within this simulation environment, the functional and chronological dependencies of the network users can be combined, resulting in detailed time series. These time series, containing implicitly the consequences of the dependencies during the simulation, constitute new input data for the following planning process. Especially the probability of occurrence and duration of loading situations are derivable and support the decision process for reinforcement measures. Including all relevant information, the generated time series can replace the conventional extreme scenarios to allow for adapted distribution grid planning.

The implementation and application of the developed simulation environment is demonstrated in a test grid with a variation of scenarios of a given supply task in this chapter. To start with, the applicability of multi-agent systems in the distribution grid planning is outlined. Then, the developed multi-agent simulation environment with a brief description of the most important aspects is introduced. An exemplary agent design for household load as well as storage systems is performed afterwards. The simulation environment is applied in a test grid. A conclusion and outlook closes the chapter.

5.2 The Application of Multi-agent Systems in the Distribution Grid Planning

When simulating electrical networks, the challenge occurs in the individual behavior of the network users, influencing each other. In the grids, many similar network users, like loads or DG units and storages, are existent, having complex interdependencies. With network users reacting to different input parameters, like market prices or the behaviors of others, the system gets emergent characteristics. Agent-based systems offer a method to meet these challenges by disassembling the complex problem to small parts that describe elements of the system.

A multi-agent system (MAS) is defined as a system that is built up of two or more (intelligent) agents, without having a superior system target [28]. These agents are either small entities of hardware or software in a simulation environment. In contrast to particle swarm optimization (PSO), MAS use various agents with individual characteristics and egoistic objectives. Therefore, MAS allow for a very realistic modeling of complex situations, whose overall structure is based on interaction of individuals. The internal behavior of agents is for the others in the multi-agent system a black box. They are only dependent on some input data x and an internal parameter set u to produce an output vector y. Consequently, the functional description of the agent behavior is defined as in (5.1) [29].

$$y = f(x, u) \tag{5.1}$$

Power systems present a good area of application of MAS, because they usually extend over large areas and many diverse and individually behaving participants and elements take part in them. Especially the ownership structure of distributed generation units with different operation philosophies and decisions implies an individual modeling of each unit. The concept of local data access and management in agents avoids the processing of large data. The usefulness of application in electric distribution networks depends on the challenges that need to be faced. With focus on these networks, the implementation of MAS makes sense, if [28]:

- interactions between entities are essential, like in control systems or power plants;
- a large group of entities interacts and the modeling of the complete explicit system behavior is impossible;
- sufficient local data for analysis and decision is available;
- new functionality needs to be integrated into an existing system;
- new functionality is to be integrated over time.

Therefore, MAS are already applied in electric distribution networks to simulate or coordinate different tasks. In [21], a smart grid test feeder is coordinated with an agent system. Additionally, MAS are used to control Smart Grids and μ-grids in [30, 31] as well as the voltage within a distribution grid in [32]. The integration of storage systems for in-house applications is realized with a MAS in [33]. The analysis of price effects on the grid is performed in [34] and time series, though neglecting interdependencies, are generated in [22]. Additional MAS applications in Smart Grids are outlined in [35].

Detailed input parameters are often a weak point within approaches that improve the distribution grid planning. Furthermore, the consideration of interdependencies between network users and external parameters is usually not integrated due to the complexity in large systems. Therefore, the concept of agent-based systems is an adequate solution to solve the aforementioned shortcomings in planning simulation systems. The decomposition of the complex electric distribution networks using a multi-agent system modeling allows for an individual modeling of all network users within a complex environment considering interdependencies between all network users.

Here, agents represent single network users, modeling their own objectives and realistic behaviors. Based on available environmental parameters and internal desires, every agent derives the most probable behavior of the represented network user. Existing interdependencies amongst the network users and dependencies on environmental parameters can be considered while designing the agents' behaviors. Besides the exchange of relevant output parameters of the agents, the more complex interdependencies include the necessity for collaboration as well as negotiation between the agents. As a consequence, the system is able to simulate the manifold interplay between the network users, the grid and the electricity market.

Additionally, the modular design concept of multi-agent systems facilitates the flexible analysis of different future supply tasks by easily adding or removing network users in the simulation environment.

5.3 Simulation Environment

The fact that there are many different types of distribution network users complicates the detailed and realistic modeling of their behavior for the analysis of occurring loading situations in the grid. The probabilistic network planning allows for taking into account the probability of loading situations. There, the application and appraisal of measured time series enable the derivation of frequency distribution functions. But especially in the low and medium voltage level, measurements are often unavailable, and thus the necessary input for the probabilistic modeling is not satisfying. Additionally, neither the input data for the distribution grid planning process nor the available planning methods are able to consider interdependencies between the network users. With the conventional distribution grids developing to smart grids with intelligent interacting users, this aspect becomes more important in the future but cannot be covered by existing methods.

The decomposition of complex problems is one of the main benefits of agent-based systems that are described previously. Therefore, the setup of an agent-based simulation environment is a feasible solution for facing the complex modeling problem of distribution grids, including the interdependencies between the network users. The modular characteristics of agent-based systems enables the individual and close to reality modeling of all existing network users, taking into account their real behavior, desires and objectives. The simulation of the agents' behaviors results in time series for all relevant information in the distribution grid. These time series can be analyzed subsequently, and recommendations for grid extensions are derivable.

5.3.1 Structures of the Simulation Environment

Aiming for getting realistic time series of asset loadings and nodal voltages in the distribution grid, the simulation environment needs an adequate representation of the network users within the analyzed grid. Therefore, every participant is modeled with an individual agent, which preserves characteristic properties of the represented user. These agents are modeled straightforwardly, fragmenting the complex modeling problem into small entities.

An overview of the agents with their tasks, their dependencies as well as their provided parameters to others is given in Table 5.1. Focusing on the simulation

Table 5.1 Overview of the modeled agents

Agent type	Task	Dependent on	Provides
Grid agent (1)	Performs power flow calculation, determines asset loading, initiation of voltage control concepts	Nodal power value vectors $\boldsymbol{P}_{n,load}(t)$ and $\boldsymbol{P}_{n,gen}(t)$ from (2)	Nodal voltage $\boldsymbol{U}(t)$, asset loading (lines, transformers),
Node agent (2)	Collection of nodal data (power consumption/feed-in), supports nodal negotiations	Power consumption $\boldsymbol{P}_{load}(t)$ Power feed-in $\boldsymbol{P}_{gen}(t)$ of connected users (6/7/8/9)	Nodal power balance, aggregated consumption $P_{n,load}(t)$ and feed-in $P_{n,gen}(t)$
Time agent (3)	Keep simulation synchronous	Acknowledgement of grid agent (1)	Simulation time step t
Weather agent (4)	Provides weather data	Time (3)	Weather data (temperature, solar radiation, wind speed, wind direction)
Market agent (5)	Represents external market behavior, provides the market price $p(t)$	Time (3)	Market price $p(t)$
Load agent (6)	Representation of residential loads, either with load profiles or probabilistic modeling	Time (3), weather (4)	Power consumption $P_{load}(t)$
Electric vehicle agent (7)	representation of electric vehicles, including driving behavior and different charging objective functions	time (3), nodal power balance $P_{n,load}(t)$ and $P_{n,gen}(t)$ (2)	power consumption $P_{load,EV}(t)$
DG unit agent (8)	Representation of the DG units, in the low voltage grid mainly PV units;	Time (3), weather (4)	Power feed-in $P_{gen}(t)$
Storage agent (9)	Represents small battery storage systems with different objective functions	Nodal power balance $P_{n,load}(t)$ and $P_{n,gen}(t)$ (2), market price $p(t)$ (5)	Power consumption or feed-in $P_{storage}(t)$

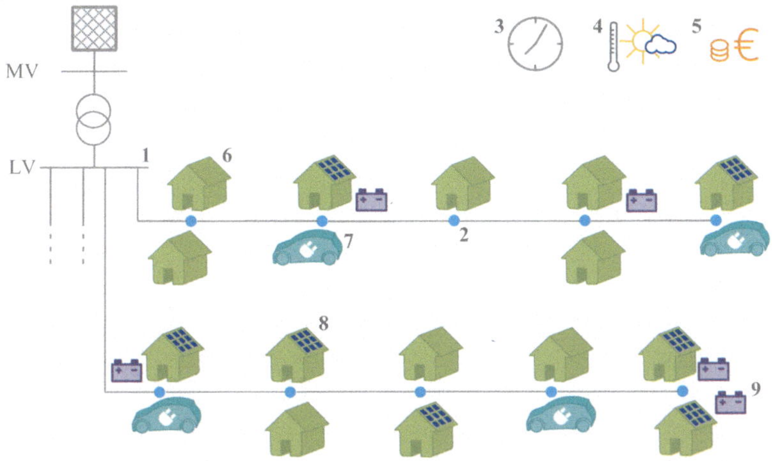

Fig. 5.1 Low voltage network user and influencing external variables

within low voltage (LV) networks, the relevant network users and external data sources for the simulation are also depicted in Fig. 5.1. However, the systematic representation of the elements is too extensive to be outlined completely in the following. For details of the elements, the consideration of [36] is recommended, the load agent, the DG agent as well as the storage agent are briefly outlined in the following subsection.

The resulting general system concept is depicted in Fig. 5.2 and explained in the following. Dependent on global information about the time, weather and market price, the elements, which are connected to a node, determine their behavior. If there are requests for negotiations on the nodal level, the agents initiate a negotiation and determine a solution. Negotiations are necessary, if conflicting situations occur while the agents pursue their objectives. In the following, the complex power flow calculation in the grid agent determines the current loading situation in the grid. If DG units or storage systems have implemented voltage control algorithm or power flow control mechanisms, their agents react on the grid agent's calculation results. So, negotiations and control algorithms are also feasible on a grid level. Principally possible, the resulting power balance in the analyzed grid can influence the market price for electricity in the market agent. This influence causes a new determination of the nodal elements' behavior, etc. Therefore, depending on the implementation, the feedback to the market agent can lead to a negotiation of the market price between the market agent on a global level and the consumers and producers on a nodal level.

The chosen agent-based design of the distribution system allows for a maximum of flexibility in the implementation of negotiations, control algorithms and dependencies of the network users within and between the different identified levels. The utilization of the system in the planning process enables the determination of relevant loading situations in dependency of the assumed interactions between the

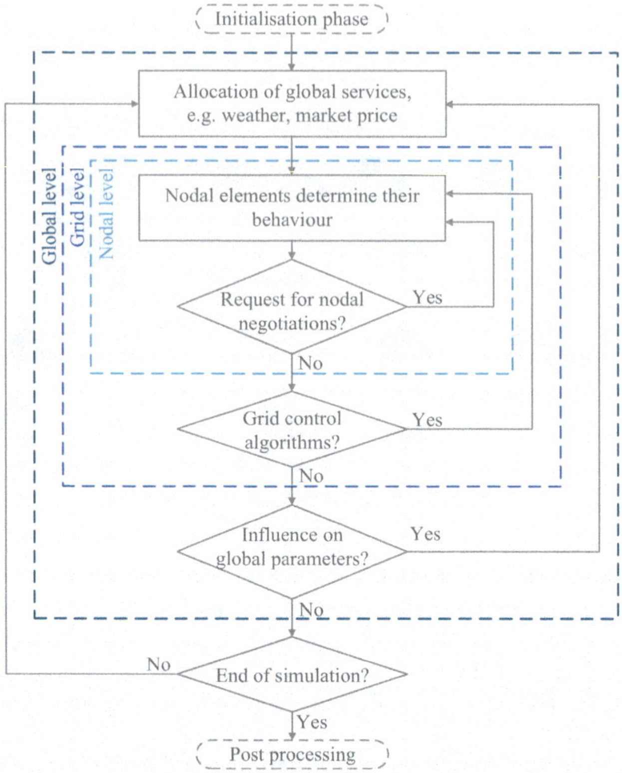

Fig. 5.2 General system concept

network users. Additionally, the simulation results can be aggregated and applied in the analysis of higher voltage levels. As a consequence of the modular construction of the system and its elements, any level of detail can be analyzed in the post processing of the simulation. The significance of the results is only influenced by the modeling approach of the elements. Consequently, for example, if the relevant data is available, the residential loads can be further decomposed to single domestic loads like washing machines, which can still negotiate the price for electric energy with the market agent on the global level.

5.4 Network Users as Agents

This subsection describes the exemplary modeling of relevant agents in the simulation environment. The way of representing the network users allows for keeping their individual characteristics. At first, the necessary information for the agent is analyzed, followed by the description of the defined objectives and its implementation in a logical behavior. A complete and detailed description of all agents in the simulation environment is given in [36] as well as in [37], where an additional agent model of heat pumps is outlined.

5.4.1 Household Load Agent

5.4.1.1 Defining Relevant Input and Output Parameters

Within the handled data, some parameters will not change during the simulation and are therefore static, while dynamic parameters will change. The static parameters are stored in the setup database of the MAS. An overview of the relevant static parameters for household agents is given in Table 5.2. The probability density function (PDF) data is required for the probabilistic calculation of the demanded power of a household. Following the analysis of given smart meter data in [38], generalized extreme value (GEV) distributions are used for the calculation. The necessary parameters to describe GEV distribution functions are the shape ξ, the location μ and the scale σ. For every PDF of every quarter-hour interval of working days, Saturdays and Sundays they are stored in tables $\boldsymbol{\xi}$, $\boldsymbol{\mu}$ and $\boldsymbol{\sigma}$ in the database.

During the simulation, some dynamic input variables are necessary for the household agent to adapt its behavior to the current situation. Besides the time t from the time agent, this is the market price for energy $p(t)$, provided by the market agent and relevant in case of sensitivity of the household behavior to the market price.

Based on the static and changing input parameters, the agent determines its output parameters, which are the active power $P_{Load,res}(\mathrm{t})$ and the reactive power $Q_{Load,res}(\mathrm{t})$. These values may differ to the internal variables $P_{Load}(\mathrm{t})$ and $Q_{Load}(\mathrm{t})$, due to the consideration of objectives that are described in the following. Afterwards, the agent sends these values to subscribed agents and stores them in a result database for further analysis purposes. Embedded in the simulation environment, the household agent communicates exclusively with the node agent of its node and the market agent, if it is sensitive to changing market prices.

The calculation of the demanded power can easily be adapted to other distribution functions, load profiles or specific characteristics of individual loads in the network within the agent. Therefore, the necessary diversity of loads in an area can be met.

Table 5.2 Static input parameters for household loads

Parameter	Variable	Unit
Annual energy consumption	E_a	kWh
Power factor	$\cos(\varphi)$	–
Connected node	*Node name*	–
Sensitivity to market price	*s_mp*	–
PDF data tables	$\boldsymbol{\xi}$, $\boldsymbol{\mu}$, $\boldsymbol{\sigma}$	–

5.4.1.2 Objectives and Behavior of the Household Agent

When receiving a new time step message from the node agent, the household agent determines its power consumption. This is done on a probabilistic basis. Let $\psi(\mathrm{t})$ be a random number in $\mathbb{R}^+$ with the distribution function in (5.2). Then, the resulting active $P_{Load}(\mathrm{t})$ and reactive power consumption $Q_{Load}(\mathrm{t})$ is calculated with (5.4) and (5.5).

$$F(\psi(\mathrm{t})) = \int_{-\infty}^{\psi(\mathrm{t})} f(x)dx \tag{5.2}$$

with

$$f(x) = G(x|\xi(t), \mu(t), \sigma(t))$$

$$f(x) = \exp\left\{-\left[1 + \xi(t)\left(\frac{x - \mu(t)}{\sigma(t)}\right)\right]^{-\frac{1}{\xi(t)}}\right\} \tag{5.3}$$

$$P_{Load}(t) = \psi \cdot \frac{E_a}{1000\ \mathrm{kWh}} \tag{5.4}$$

$$Q_{Load}(t) = P_{Load}(t) \cdot \tan(\varphi) \tag{5.5}$$

Due to the limited power consumption of households, $P_{Load}(\mathrm{t})$ is limited to a maximum of 20 kW, if the random generated value is higher. Otherwise, unrealistic maximum loadings in the grid could occur and adulterate the simulation results due to the characteristic of the used PDF.

The household agent represents the domestic load behavior. Therefore it objectives for satisfying at least the probabilistically calculated power demand $P_{Load}(\mathrm{t})$ in any case. The market price sensitivity s_{mp}, which is the indication for the participation in demand side management, can be either true or false. It is set in the database for every load agent either manually or randomly. If s_{mp} is true, the agent will try to reproduce the probable household customer behavior in minimizing the household's expenditures for electrical energy in every time step. Then, the resulting $P_{Load,res}(\mathrm{t})$ becomes a function of the calculated power demand $P_{Load}(\mathrm{t})$ and the market price $p(t)$. Considering these objectives, the household agent implements a resulting behavior, whose overview is depicted in Fig. 5.3.

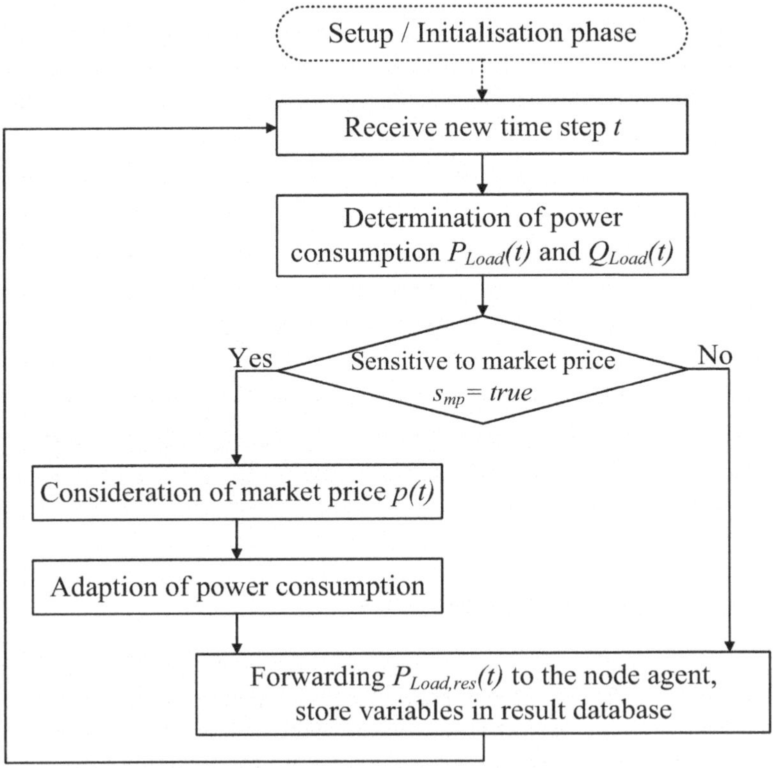

Fig. 5.3 Behavior of the household agent

5.4.2 Storage Agent

The designed storage agent is a representation of storages from a system operator's point of view. Therefore, the implemented battery model is rather simple, without the consideration of a specific technology (as the DSO does not know either), lifetime preservation algorithms or costs for the storage system operator. However, more detailed models can be also integrated within the simulation system.

5.4.2.1 Defining Relevant Input and Output Parameters

Some parameters, which the storage agent needs to know, will never change during the simulation. As they are internal and static, they are stored in a database that is read-out during the initialization phase. These parameters are listed in Table 5.3. Additionally, the relevant internal variables changing in every simulation time step are given in Table 5.4. The required external parameters are, depending on the pursued charging strategy, the nodal power consumption $P_{Load}(t)$ and power

Table 5.3 Static parameters for storage agents

	Variable	Unit
Rated peak power	$P_{r,BSS}$	kW
Storage capacity	E_{BSS}	kWh
Maximum charging power	$P_{charge,max}$	kW
Maximum discharging power	$P_{discharge,max}$	kW
Efficiency	η_{BSS}	–
Maximum depth of discharge	DoD_{max}	–
Connected node	Node name	–
Length of time interval	τ	h
Power factor	$\cos(\varphi)$	

feed-in $P_{PV}(t)$ as well as the market price $p(t)$ (see also Table 5.1). The resulting consumption or feed-in $P_{BSS,res}(t)$ and $Q_{BSS,res}(t)$ are the output variables of the storage agent. The static input parameters, stored in the setup database, are only retrieved in the initialization phase.

5.4.2.2 Objectives of the Storage Agents

Representing the real situation, the storage system operator might pursue different operation strategies. Three different main objective functions for the storage system operation can be identified: the network oriented behavior to reduce the grid loading, the self-consumption maximization of self-produced renewable energy of attached DG units and the market driven behavior to maximize the owner's revenues.

When applying the network-oriented behavior, the storage system is used to reduce the peaks of local distributed generation, namely PV units. It implies balancing load and generation locally. The high PV feed-in in noon hours in combination with low load consumption stresses the grid and raises the voltage level. Therefore, the storage system charges, if a given threshold $P_{th,PV}$ for the PV feed-in $P_{PV}(t)$ is exceeded:

$$P_{PV}(t) > P_{th,PV} \rightarrow P_{BSS}(t) > 0 \tag{5.6}$$

When the sum of local loads rises above the feed-in sum, the storage system will discharge the batteries to supply the loads:

$$|P_{Load}(t)| > |P_{PV}(t)| \rightarrow P_{BSS}(t) < 0 \tag{5.7}$$

The second possible strategy aims for maximizing the self-consumption of a DG unit operator. Looking at one node, this means charging when generation is dominant at a period and discharging when the load is higher. If the feed-in of the

DG unit is independent from the local load, the storage system can balance these differences, trying to minimize the purchase of energy from the grid:

$$P_{\Delta}(t) + P_{BSS}(t) = 0 \tag{5.8}$$

with

$$P_{\Delta}(t) = P_{G}(t) - P_{Load}(t) \tag{5.9}$$

The resulting decision on this objective is defined as

$$P_{BSS}(t) = \begin{cases} 0 < P_{BSS} < P_{charge,max}, & if\, P_{\Delta}(t) > 0 \\ P_{discharge,max} < P_{BSS} < 0, & if\, P_{\Delta}(t) < 0 \\ 0, & otherwise \end{cases} \tag{5.10}$$

The third objective of the storage system behavior uses the variability of the price for electrical energy $p(t)$ to maximise the revenues. Consequently, the storage will charge at low price periods and discharge at high price periods, depending on given price limits p_{upper_limit} and p_{lower_limit} that are assumed on economic objectives:

$$P_{BSS}(t) = \begin{cases} P_{charge,max}, & if p(t) < p_{lower\_limit} \\ 0, & otherwise \\ P_{discharge,max}, & if p(t) > p_{upper\_limit} \end{cases} \tag{5.11}$$

5.4.2.3 The Implemented Behavior

Having three different objective functions, the storage agent implements an advanced behavior structure like a finite state machine. The general structure of the agent's performance during the simulation is pictured in Fig. 5.4.

When the agent is set up in the initialization phase, the information about its objective function, either market price driven, self-consumption maximization or network oriented is taken from the database. After having received the new time step t from the node agent, this is taken into account. If the storage agent behaves

Table 5.4 Dynamic parameters of storage agents

	Variable	Unit
(Dis-)charging power	$P_{BSS}(t)$	kW
State of charge (SoC)	$SoC(t)$	%
Charging level (SoC absolute)	$E_{SoC}(t)$	kWh
Available capacity	$E_{cap,av}(t)$	kWh

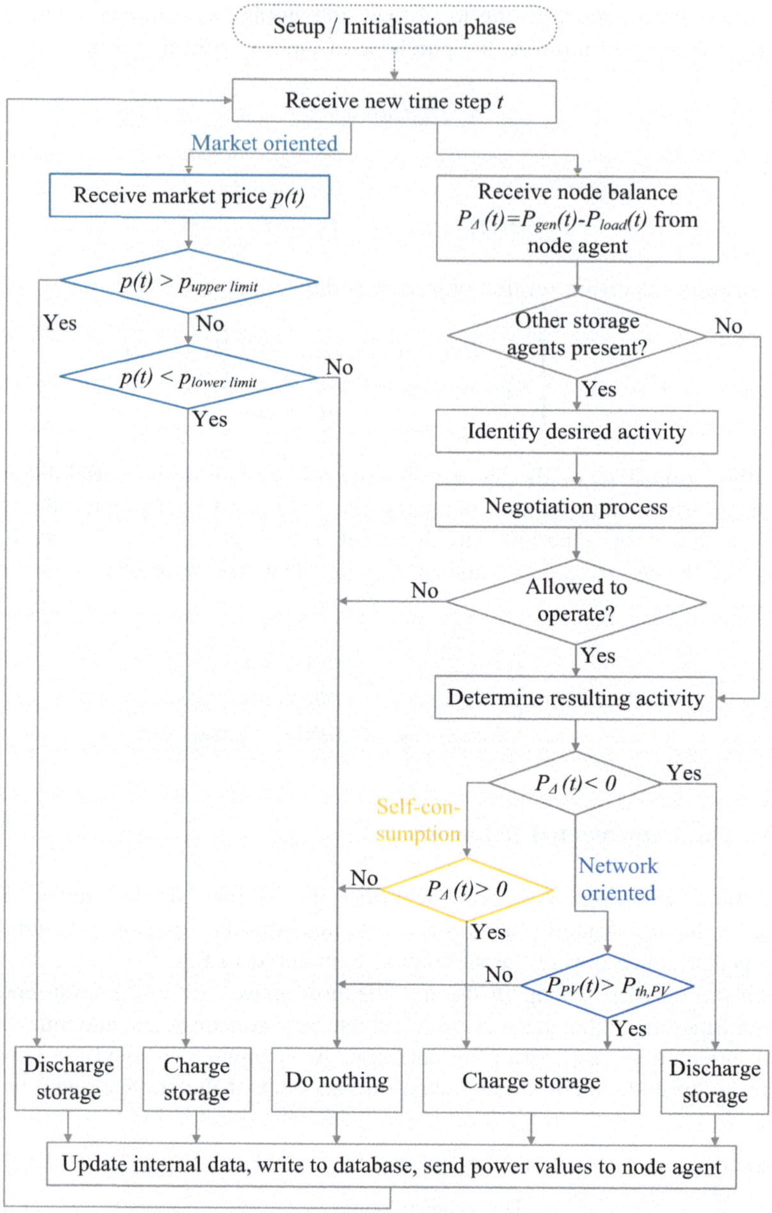

Fig. 5.4 Internal behavior structure of the storage agent

market price driven, it will process the received market price. Depending on the given price limits p_{upper_limit} and p_{lower_limit} the storage system will charge or discharge according to (5.11). If the available capacity for (dis-) charging is insufficient, the power will be reduced (5.12).

$$P_{BSS}(t) = \frac{E_{cap,av}(t)}{\tau} \tag{5.12}$$

Acting either network-oriented or aiming for maximization of the self-consumption, the agent requests the local feed-in $P_{Gen}(t)$ and consumption $P_{Load}(t)$ to derive the nodal balance $P_{\Delta}(t)$. Additionally, the presence of further storage or EV agents is checked. If further storing agents are present, the desired behavior is identified and a negotiation amongst them is started to coordinate their behavior before charging. Being the winner of the negotiation process or the only storage system at the node, the agent determines its charging behavior depending on the considered objective functions which are described with (5.6)–(5.10). Finally, after having updated internal data, the time step results are stored into a database and the node agent is informed about its consumed or injected power.

5.4.2.4 Negotiation Between Two Storage Agents Acting at the Same Node

If more than one storage agent with network oriented behavior is connected to a node, a negotiation process between these agents is essential. Otherwise, with all storage agents behaving uncontrolled in the same way, the grid situation would not be improved but probably worsened. For example, if all storage systems are charging to reduce the feed-in peak of a PV unit, the voltage could exceed the lower voltage interval limit instead of the upper limit in the case of no charging storage systems. The implemented negotiation process between all storage agents that are connected at the same node solves this problem. Because of the storage agents having the same network oriented behavior, they do not react egoistically but cooperatively during the negotiation and demonstrate implicitly the agent basic characteristic of social ability. The implemented message exchange is depicted in Fig. 5.5. Focusing on the interactions between storage agents, the messaging with the other agents assigned to the node is only indicated in the figure. However, the amount of agents, including electric vehicles or other network users is not limited.

After informing the storage agents about the new time step and the reception of request messages, the node agent sends the nodal power balance to the storage agents (1). Without knowing if any further storage agents are present in the current time step, they check the receiver list of the sent propose message (a). If the list contains more than one addressee, a negotiation is started. Otherwise, the proceeding with only one storage agent occurs. One of the storage agents becomes the chief negotiator to coordinate the negotiation (storage agent *SA 1* in Fig. 5.5). It queries the other storage agents (here *SA 2*), whether they want to charge or

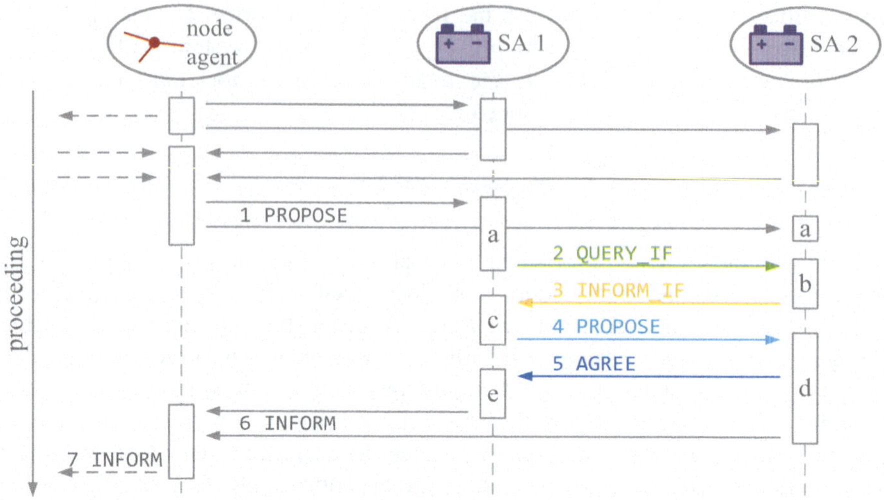

Fig. 5.5 Negotiation process with two storage agents

discharge in the current time step (2). Based on the internal parameters and the current nodal power balance, *SA 2* determines its desired behavior (b) and informs *SA 1* about its intention (3). Additionally, *SA 2* adds a charging priority to the message, which is derived in dependence on the current $SoC(t)$. Also having determined a charging behaviour on its own, *SA 1* evaluates all received intended behaviors with the corresponding charging priorities in a non-discriminating manner (c). Then, the chief negotiator sends messages to all storage agents with the proposed behavior (4). Following the priority ranking, the storage agents are allowed to perform their intended behavior until the common target is reached. On this basis, the storage agents check, if their desires have been considered (d). Although able to behave completely selfish, the agents are implemented to be constructive. This constructive behavior is implicitly assumed in the pursued objective that aims to support the grid. Additionally, this ensures the convergence of the negotiation. However, the chief negotiator waits (e) until all others agents have agreed to the determined solution (5). Then, all storage agents inform the node agent about their deduced power consumption or injection (6). Finally the node agent takes all collected power data and forwards the aggregated values to the grid agent (7).

5.5 Simulation Example

For the demonstration of the multi-agent simulation environment, a low voltage test grid is utilized. It aims for representing the current situation in a small rural settlement or village in Germany and is based on real grid data from internal grid

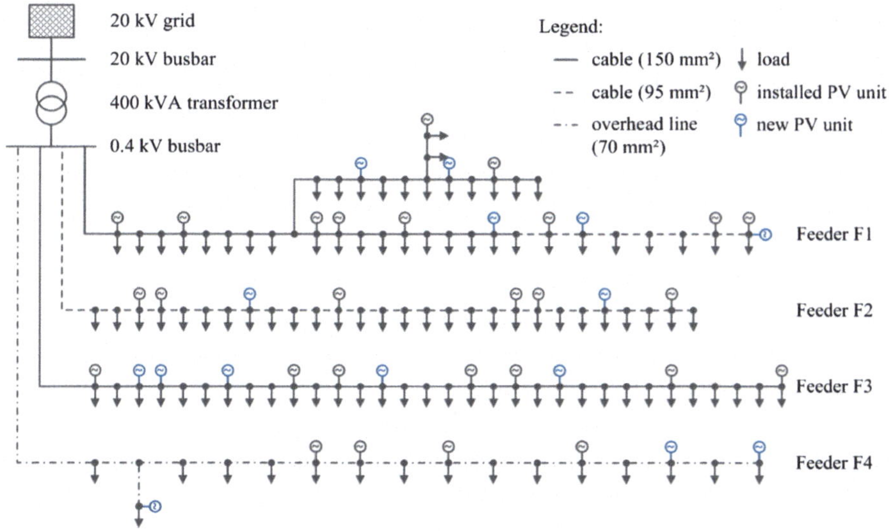

Fig. 5.6 Illustration of the low voltage test grid for the initial and future scenario

Table 5.5 Line parameters in the test grid

	R [Ω/km]	X [Ω/km]	B [μS/km]	I_{max} [A]
Cable 95 mm^2	0.32	0.069	185.4	235
Cable 150 mm^2	0.207	0.069	216.8	300
Overhead line 70 mm^2	0.443	0.072	163.4	195

studies. Historically grown and extended, the grid serves an area with spacious distributed detached houses. This leads to an inhomogeneous application of grid equipment with long feeders, consisting of cables with different cross-sections or overhead lines. The schematic illustration of this grid is given in Fig. 5.6.

Supplied from the 20 kV medium voltage grid via a 400 kVA local transformer, the grid consists of four feeders with different characteristics (line parameters see Table 5.5). The feeder F1 has a total length of about 850 m and an additional branch that has a length of 380 m. The first part is realized as a cable with a cross-section of 150 mm^2, the last part with an older one with a cross-section of only 95 mm^2. Except some longer distances, the average distance between the connected 38 private customers is around 30 m. The overall length of F2 results in 700 m with 28 residential loads being connected to, having an average distance of 25 m between them. The cable type of F2 has a cross-section of 95 mm^2. The 32 households connected to feeder F3, which is designed with a 150 mm^2 cross-section cable, have an average distance of 25 m, too. Feeder F4 supplies a smaller group of 17 customers that are more spatially distributed. Therefore, the average distance between them is about 60 m and the branch has the longest distance of 100 m. In contrast to the other feeders, F4 is realized as a 70 mm^2 overhead line. The installed PV units have a installed rated capacity of either 10 or 30 kW and are operated with $\cos(\varphi) = 1$. Their complete installed capacity is 330 kW. This set of input parameters defines the *initial scenario*.

Originating from the *initial scenario*, the *future scenario* assumes a development of the grid's supply task. Against the background of a furthermore increasing installation of PV units, an additional installation of new PV units with 10 and 30 kW and a power factor of $\cos(\varphi) = 0.95$ resp. $\cos(\varphi) = 0.9$ is assumed for some households in this area.

5.5.1 Conventional Grid Analysis

In the first instance, the test grid's performance at both scenarios is checked and evaluated with the conventional grid planning method. The grid's performance is analyzed for two extreme situations in every supply scenario with commercial network calculation software (DIgSILENT PowerFactory). The maximum load scenario assumes 100% of the rated load power and 0% of PV feed-in. The feed-in scenario assumes 10% of the rated load power and 85% of PV power. The residential loads have a rated consumption of 2 kW at peak load with a power factor of $\cos(\varphi) = 0.97$. The reactive power behaviour of the PV units is set according to the scenario description above.

The grid is able to serve the supply task of the *initial scenario* without any problems. However, the additional PV installation in the *future scenario* causes severe problems. Besides an overloading of the local transformer with 115% of its rated power, over-voltages, up to 1.09 pu, appear in all feeders. Therefore, extensive reinforcement measurements, i.e. transformer exchange and new parallel cables in the feeders, are necessary (see Table 5.6).

5.5.2 Time Series Based Analysis

Now, the test grid's performance is analyzed in the scenarios with the help of time series, which are generated with the developed multi-agent system. While some information, like the grid topology and PV unit nominal power, are already available on the basis of the test grid definition, some agents need additional

Table 5.6 Necessary investment in the grid

	Investment	Subtotal (€)
1 Transformer upgrade	10,000 €/unit	10,000
1450 m of 150 mm^2 cable	60,000 €/km	87,000
Total sum		97,000

Table 5.7 Assumptions for the required parameters of the PV agent

Parameter	Symbol	Assumption
Module azimuth	α_E	Random value based on Gaussian distribution, expectation 0° S, interval [−90°, 90°]
Module elevation	γ_E	Random value based on Gaussian distribution, expectation 38.5°, interval [0°, 60°]
Inverter efficiency	η_{INV}	Random value based on uniform distribution on interval [0.95, 0.99]
Albedo	A	Set to 0.2

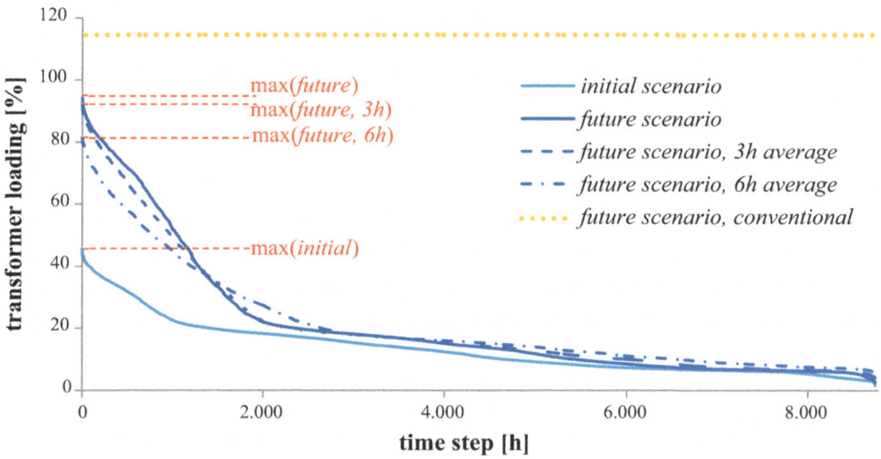

Fig. 5.7 Transformer loading duration curves for the scenarios

information to perform the simulation. The household agents require an annual energy consumption, which is, in this case, homogeneously assumed to be $E_a = 4000$ kWh and the basis for probabilistic hourly power consumption. Similarly, all household agents are not sensitive to the spot market price. Additional parameters for the PV agents are listed in Table 5.7. Because of them being inaccessible to the DSO, they have to be estimated for the simulation.

The simulated time period covers the year 2011. A processing of the resulting time series to duration curves facilitates the derivation of the maximum and minimum loading as well as the analysis of the frequency of occurrence of loading situations. The resulting duration curves for both scenarios are illustrated in the diagram of Fig. 5.7.

In contrast to the depicted loading situation in the conventional analysis with extreme scenarios, the maximum loading of the transformer is reduced significantly in the test grid simulation with the multi-agent system. With the transformer's maximum loading being below 95% of the rated power in the future scenario,

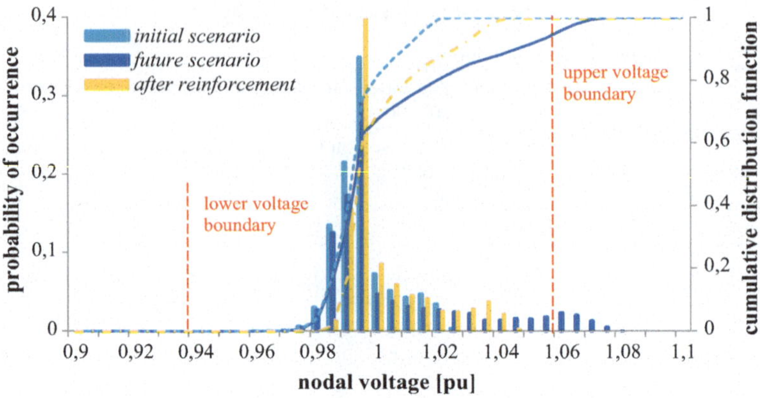

Fig. 5.8 Histogram and distribution function of the nodal voltage at last node of feeder F4

a transformer exchange to a higher rated type is not necessary. If the 6 h average is considered, the maximum loading is even reduced to 81%.

Besides the transformer overloading, the nodal voltages at the feeders' ends exceed the permitted voltage interval in the feed-in situation of the *future scenario* in the conventional grid analysis. The time series analysis reveals that only feeder F4 has voltage problems. The resulting histogram and distribution function of the appearing nodal voltage at the last node in F4 is depicted in Fig. 5.8.

The analysis of the test grid with the multi-agent simulation identifies an overall grid extension necessity of 360 m of new 150 mm^2 cable. Assuming investment costs of 60,000 €/km, the induced investment in the test grid amounts to 21,600 €. That is 75,400 € or 77% less than the resulting investment that can be identified via the conventional extreme scenario based method.

5.5.3 Analysis of the Influence of New Network Users on the Grid

For the demonstration of the objective functions' effects of storage systems, every PV unit in the test grid is equipped with a storage system. A 4 kW battery is installed at each 10 kW PV units and a 12 kW battery at the 30 kW PV units. The analyzed objective functions try either to maximize the self-consumption, to minimize the impact on the grid or to maximize the revenues due to fluctuating market prices. The resulting voltage histogram is depicted in Fig. 5.9, demonstrating the impact of the analyzed objective functions on the grid. Exemplarily it reveals that the market-driven behavior is only a problem in a few hours during a year, where countermeasures can be defined.

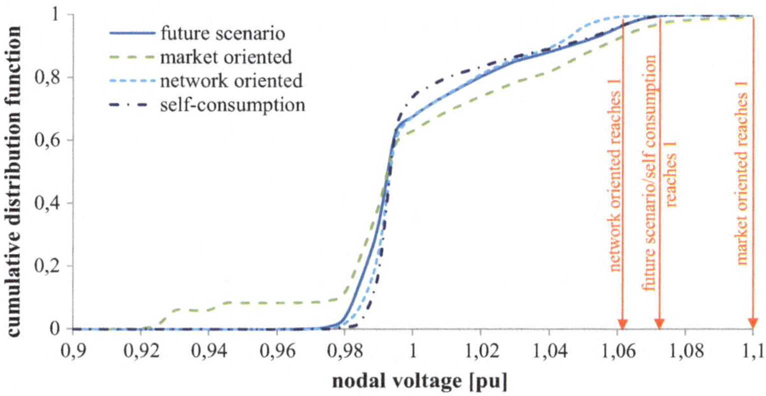

Fig. 5.9 Cumulative distribution function of the voltage at node for different storage objective functions

This analysis can be performed for all innovative network users with different interdependencies and objective functions. Therefore, it allows for the first time a holistic analysis of the impact of innovative and functional dependent grid elements and users on the distribution grid.

References

1. C. Baudot, G. Roupioz, A. Billet, Modernizing distribution network management with linky smart meters—lessons learned in greenlys project, in *Proceedings Of CIRED 2015* (Lyon, France, 15–18 June, 2015)
2. G. Celli et al., A Comparison of distribution network planning solutions: traditional reinforcement versus integration of distributed energy storage, in *IEEE PowerTech 2013* (Grenoble, France, 16–20 June 2013)
3. S. You et al., An overview of trends in distribution network planning: A movement towards smart planning, in *T&D Conference and Exposition, 2014 IEEE PES* (Chicago, USA, 14–17 April, 2014)
4. Cigré Task Force C6.19, *Planning and Optimization Methods for Active Distribution Systems* (2014)
5. A. Keane et al., State-of-the-art techniques and challenges ahead for distributed generation planning and optimization. IEEE Trans. Power Syst. **28**(2), 1493–1502 (2013)
6. P.S. Georgilakis, N.D. Hatziargyriou, Optimal distributed generation placement in power distribution networks: models, methods, and future research. IEEE Trans. Power Syst. **28**(3), 3420–3428 (2013)
7. S. Prabhakar Karthikeyan et al., A review on soft computing techniques for location and sizing of distributed generation systems, in *2012 International Conference on, Computing, Electronics and Electrical Technologies (ICCEET)* (21–22 March, 2012)
8. P. Wiest et al., New Hybrid planning approach for distributions grids with a high penetration of RES, in *Proceedings of CIRED 2015* (Lyon, France, 15–18 June, 2015
9. I. Ziari et al., Optimal distribution network reinforcement considering load growth, line loss, and reliability. IEEE Tran. Power Syst. **28**(2), 587–597 (2013)

10. V. Klonari etal., Probabilistic analysis tool of the voltage profile in low voltage grids, in *Proceedings of CIRED 2015* (Lyon, France, 15–18 June, 2015)
11. D.F. Frame, G.W. Ault, S. Huang, The uncertainties of probabilistic LV network analysis, in *IEEE Power and Energy Society General Meeting 2012* (San Diego, USA, 22–26 July, 2012)
12. C. Engels, L. Jendernalik, M. Osthues, H. Spitzer, 'Smart planning'—an integrated approach for distribution system planning to cope with its future requirements, in *Proceedinds of CIRED* (Stockholm, Sweden, 10–13 June, 2013)
13. R.F. Arritt, R.C. Dugan, Value of sequential-time simulations in distribution planning. IEEE Trans. Ind. Appl. **50**(6), 4216–4220 (2014)
14. E.Tønne, J.A. Foosnæs, T.Pynten, Power system planning in distribution networks today and in the future with Smart Grids, in *Proceedings of CIRED* (Stockholm, Sweden, 10–13 June, 2013)
15. G. Roupioz, X. Robe, F. Gorgette, First use of smart grid data in distribution network planning, in *Proceedings of CIRED* (Stockholm, Sweden, 10–13 June, 2013)
16. M. Nick et al., On the optimal placement of distributed storage systems for voltage control in active distribution networks, in *IEEE PES Innovative Smart Grid Technologies (ISGT Europe)* (Berlin, Germany, 14–17 Oct, 2012)
17. N. Siebert et al., Scheduling demand response and Smart Battery flexibility in a market environment: results from the Reflexe demonstrator project, in *IEEE PowerTech 2015* (Eindhoven, Netherlands, June 29–July 2, 2015)
18. R. Roche et al., A multi-agent model and strategy for residential demand response coordination, in *IEEE PowerTech 2015* (Eindhoven, Netherlands, June 29–July 2, 2015)
19. N.G. Paterakis et al., Distribution system operation enhancement through household consumption coordination in a dynamic pricing environment, in *IEEE PowerTech 2015* (Eindhoven, Netherlands, June 29–July 2, 2015)
20. S. Bashash, H.K. Fathy, Cost-optimal charging of plug-in hybrid electric vehicles under time-varying electricity price signals. IEEE Trans. Intell. Transp. Syst. **15**(5), 1958–1968 (2014)
21. F. Boulaire et al., A hybrid simulation framework to assess the impact of renewable generators on a distribution network, in *Proceedings of the 2012 Winter Simulation Conference (WSC)* (Berlin, Germany, 9–12 Dec, 2012)
22. P. Jahangiri et al., Development of an agent-based distribution test feeder with smart-grid functionality, in *IEEE Power and Energy Society General Meeting 2012* (San Diego, USA, 22–26 July, 2012)
23. C. Chengrui et al., Agent-based simulation of distribution systems with high penetration of photovoltaic generation, in *Power and Energy Society General Meeting, 2011 IEEE* (24–29 July, 2011)
24. J. Hu et al., Multi-agent based modeling for electric vehicle integration in a distribution network operation. Electr. Power Syst. Res. **136**, 341–351 (2016)
25. T. Pinto et al., Smart grid and electricity market joint simulation using complementary multi-agent platforms, in *IEEE PowerTech 2015* (Eindhoven, Netherlands, June 29–July 2, 2015)
26. L. Hattam, D.V. Greetham, Green neighbourhoods in low voltage networks: measuring impact of electric vehicles and photovoltaics on load profiles. J. Mod. Power Syst. Clean Energy **5**(1), 105–116 (2017)
27. E.F. Bompard, B. Han, Market-based control in emerging distribution system operation. IEEE Trans. Power Delivery **28**(4), 2373–2382 (2013)
28. S.D.J McArthur et al., Multi-agent systems for power engineering applications—part I: concepts, approaches, and technical challenges. IEEE Trans. Power Syst. **22**(4), 1743–1752 (2007)
29. C. Rehtanz, *Autonomous Systems and Intelligent Agents in Power System Control and Operation* (Springer, Berlin, New York, 2003). ISBN 3540402020

30. A. Prostejovsky et al., Demonstration of a multi-agent-based control system for active electric power distribution grids, in *IEEE International Workshop on Intelligent Energy Systems (IWIES) 2013* (Vienna, Austria, 14 Nov, 2013)
31. F.I. Hernandez et al., Active power management in multiple microgrids using a multi-agent system with JADE, in *International Conference on Industry Applications (INDUSCON), 2014 11th IEEE/IAS* (Juiz de Fora, Brazil, 7–10 Dec, 2014)
32. E. Polymeneas, M. Benosman, Multi-agent coordination of DG inverters for improving the voltage profile of the distribution grid, in *IEEE PES General Meeting 2014* (National Harbor, USA, 27–31 July, 2014)
33. E.A.M. Klaassen et al., Integration of in-home electricity storage systems in a multi-agent active distribution network, in *IEEE PES General Meeting 2014* (National Harbor, USA, 27–31 July, 2014)
34. I. Pisica et al., A multi-agent model for assessing electricity tariffs, in *Innovative Smart Grid Technologies Conference Europe (ISGT-Europe), 2014 IEEE PES* (Istanbul, Turkey, 12–15 Oct, 2014)
35. G.H. Merabet et al., Applications of multi-agent systems in smart grids: a survey, in *International Conference on Multimedia Computing and Systems (ICMCS) 2014* (Marrakech, Marocco, 14–16 April, 2014)
36. J. Kays, *Agent-based Simulation Environment for Improving the Planning of Distribution Grids*, Ph.D. thesis, TU Dortmund University, Dortmund, Germany, 2014, ISBN 9783868446623
37. A. Seack, *Time-series Based Distribution Grid Planning Considering Interaction of Network Participants with A Multi-agent System*, Ph.D. thesis, TU Dortmund University, Dortmund, Germany, 2016
38. J.Kays, A. Seack, C. Rehtanz, Consideration of smart-meter measurements in a multi-agent simulation environment for improving distribution grid planning, in *Innovative Smart Grid Technologies Conference (ISGT), 2016 IEEE PES* (Minneapolis, USA, 6–9 Sept, 2016)

Chapter 6
Optimal Siting and Sizing of Distributed Generations

Karar Mahmoud and Yorino Naoto

Abstract Recently, the penetration of distributed generations (DG) has been obviously increased in electric distribution networks throughout the world. DGs are small scale generators connected near load centers in networks, thereby avoiding losses in transmission systems and releasing system capacity. At present, there are many types of DG, such as wind power, solar power, fuel cell, biomass, micro-turbines, and diesel engines. DG can play an important role in improving the performance of the networks; therefore, allocating DG optimally is one of the most crucial subjects in DG planning. In this chapter, the DG allocation problem is studied, and an efficient method is presented for accurately solving this optimization problem. The proposed method combines between analytical expressions and an optimal power flow (OPF) algorithm to determine the optimal locations, sizes and the best mix of various DG types for minimizing the total real power loss in electric distribution networks. The proposed analytical expressions are general for directly calculating the optimal sizes of any combination of multi-type DG technologies. The optimal power factors of the various units can be analytically computed, thereby contributing positively to loss reduction. The 69-bus test system is used to test the proposed method. The effectiveness of the proposed method is demonstrated for determining the optimal mix of various combinations of different DG types.

Keywords Electric distribution networks · DG location · DG size
Power loss · Optimal allocation

K. Mahmoud (✉)
Electrical Engineering Department, Aswan University, Aswan, Egypt
e-mail: karar.alnagar@aswu.edu.eg

Y. Naoto
Department of Electrical and Computer Engineering, Hiroshima University,
Hiroshima, Japan
e-mail: yorino@hiroshima-u.ac.jp

F. Shahnia et al. (eds.), *Electric Distribution Network Planning*,
Power Systems, https://doi.org/10.1007/978-981-10-7056-3_6

6.1 Introduction

The needs for reliable and efficient electric distribution networks have motivated the research on renewable energy sources (RES). RESs, such as wind turbines, photovoltaic (PV), and biomass systems, are clean sources with low investment costs [1]. In electric distribution networks, the penetration of distributed generations (DGs) has steadily increased because of their benefits. DG units are allocated near load centers, reducing the stress on the transmission systems and saving costs [2–5].

The introduction of DG units into electric distribution networks has great impacts on the operation, stability, and protection of the systems. These impacts vary depending on the selected locations, sizes and types of the DG units [6–8]. As most of the losses in the entire power systems are normally dissipated in electric distribution networks, considering DG impacts on losses is important when allocating the DG units. The total active losses are significantly affected by DG, where they can be decreased/increased according to the DG allocation [9, 10]. Voltage rise and reverse power flow are common technical problems associated with the integration of DG in electric distribution networks. These technical problems constraints the allowable penetration of DG. An efficient method is required to allocate DG in an optimal way with considering overall system and DG constraints.

The allocation of DG units in MV electric distribution networks is considered an important issue for system planners. The optimal DG allocation aims at determining the best locations and sizes of DGs to optimize the network operation. Recently, many methods have been presented in the literature for solving the optimal DG allocation problem in networks. These methods are categorized as follows: (1) numerical-based methods, (2) heuristic-based methods, and (3) analytical-based methods. Examples of numerical-based methods are gradient search [11], linear programming [12], optimal power flow [13], and exhaustive search [14, 15]. These methods can determine the optimal DG sizes at candidate locations. To determine the optimal locations of DGs, these numerical-based are applied to solve the optimal DG sizes at all possible combinations of DG locations. The heuristic-based methods utilize artificial intelligence algorithms, e.g., genetic algorithms [16, 17], particle swarm optimization [18], harmony search [19], and tabu search [20]. These heuristic-based methods have the ability to deliver near-optimal solutions of DG sizes and locations but require exhaustive computational efforts. Regarding analytical-based methods, they are simple, easy to be applied, and computationally fast. The analytical-based methods simplify the DG allocation problem by considering only uniformly distributed load types or single DG placements [21, 22]. In [23], an analytical method is presented to deliver the optimal locations of DG units, and their sizes are optimally calculated using the Kalman filter method. In [24, 25], a method based on the concept of load centroid to optimally allocate multiple DGs. An analytical method for allocating a single DG unit is proposed in [26], and then it is improved in [27, 28] by considering the reactive power capability of multiple DG units.

This book chapter is on the optimal siting and sizing of multi-type DG units in electric distribution networks. In this chapter, two efficient methods for the optimal allocation of multi-type DG for loss minimization are presented. The first method is based on analytical expressions that directly can calculate the optimal sizes of multi-type DG units and evaluate the corresponding DG benefits. A second hybrid method is presented which combines the first analytical method and an optimal power flow (OPF) algorithm for solving the DG allocation problem. The presented methods are accurate, general for multi-type DGs, and valid for radial and meshed systems. The performance of the proposed methods is tested and validated using the 69-bus test system.

6.2 DG Models

According to the output scheme of DG units, they can be classified to three models: (1) DG Model A, (2) DG Model B, and (3) DG Model C. First, for DG Model A, its active power is not specified and needed to be optimally computed. Second, unlike DG Model A, the variable in the DG Model B is reactive power, not the active power generation. Third, the DG Model C model has two variables includes both active and reactive power generation, which means that this model is more complex than the other two models to be optimally solved. The mathematical representations of these units are described in Fig. 6.1. As seen in the figure, the different RES technologies have their interfaced devices to the main grid. For each DG type, once the state variable (active and/or reactive power generation), the interfaced device and the configuration of DG technology can be properly designed [29, 30]. It is important to mention that considering the reactive power capability of DG is important to simulate the real situation of employing the DG reactive power

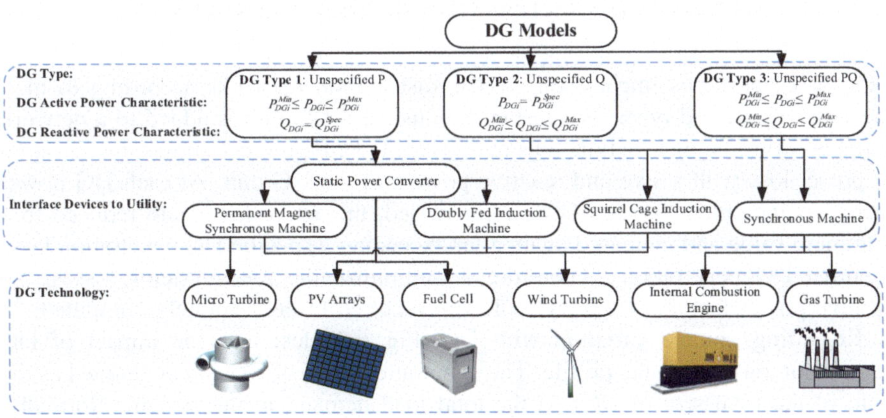

Fig. 6.1 Steady state models of different DG technologies

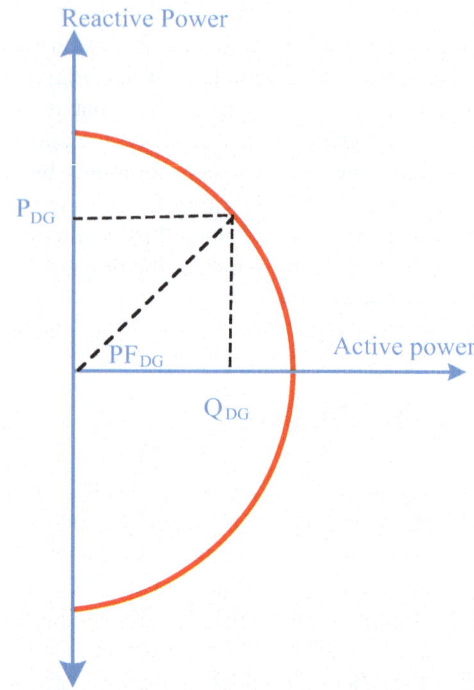

Fig. 6.2 Generation capability curve of PV inverters

injection/rejection for voltage regulations. For example, Fig. 6.2 shows the generation capability curve of PV inverters, where the red circle indicates the rated power of the interfaced inverter.

6.3 DG Impacts on Electric Distribution Networks

DGs have enormous impacts on electric distribution networks according to their locations, sizes, and types. For instance, consider a DG unit is added to a network supplied from the distribution substation (Fig. 6.3). Figure 6.4 shows the variation of power loss with active and reactive powers of the DG unit. At each DG power factor, as the active power of DG is increased, the active losses are reduced to a minimum value and increased again after exceeding a specific DG penetration level (optimal DG penetration). Therefore, to minimize the total systems losses, the optimal power factors of the DG units are required to be accurately computed.

Regarding voltage variation with DG, Fig. 6.5. describes the impact of DG penetration on the voltage profile. The penetration level of DG can be defined as the ratio of the total size of DGs to the total load demand in the system. Normally, the voltage of distribution feeders drops with increasing the distance from the distribution substation. However, if a DG unit with high penetration is added,

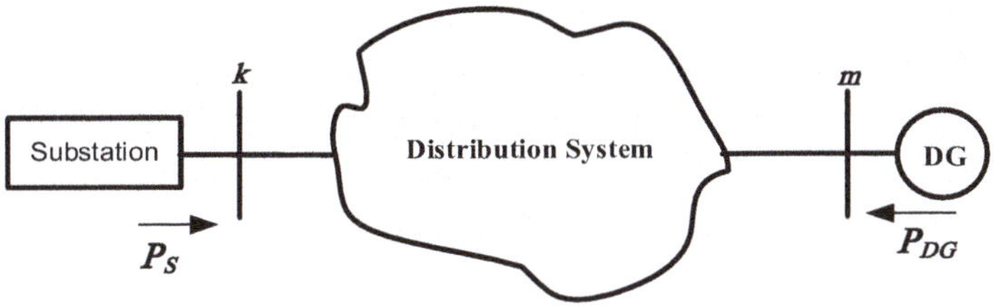

Fig. 6.3 Electric distribution network with DG

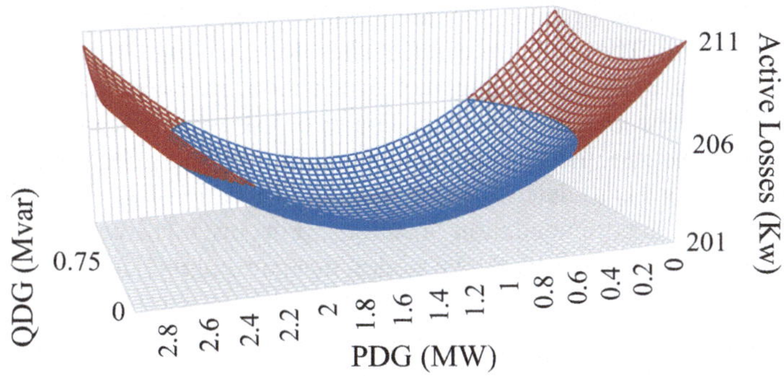

Fig. 6.4 The variation of the active losses with active and reactive powers of DG

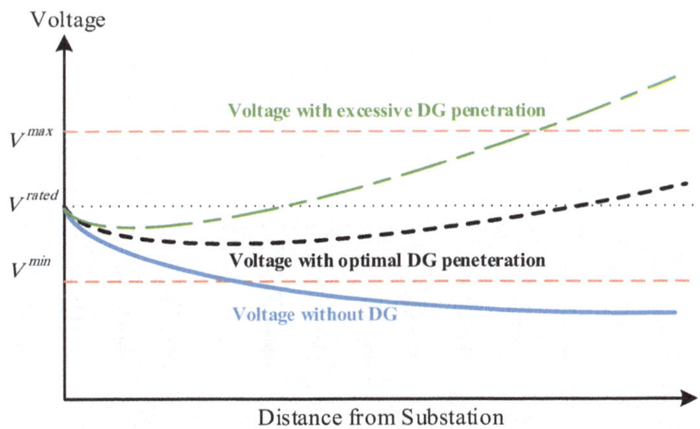

Fig. 6.5 The variation in voltage profile with the penetration level of DG

voltage profile will increase with distance from the substation. For example, the penetration level of the PV units is changed by increasing the number of arrays of the PV units. An optimal penetration of DG is required to ensure that the voltage profiles along the distribution feeder do not exceed lower/upper limits.

Besides losses and voltage profile, overall constraints of the electric distribution network must be considered when allocating DG.

6.4 Description of the DG Allocation Problem

The optimal placement of DG is a complex optimization as the number of alternative solutions (i.e., possible locations and sizes of units) is huge and the electric distribution network is nonlinear. Figure 6.6 shows an example of a network where various components are needed to be placed at some of their recommended sites. These recommended sites for each unit type can be listed according to many factors including fuel distribution, investor strategies, weather conditions (for renewable DG) and etc. Figure 6.7 shows a PV system where its optimal size can be specified by the number of PV models. Therefore, the target of optimal placement is to determine the best set of locations for the various units, according to their recommendation locations. For instance, consider installing N_{DG} DG units in a network with N_B nodes that are eligible for the installation.

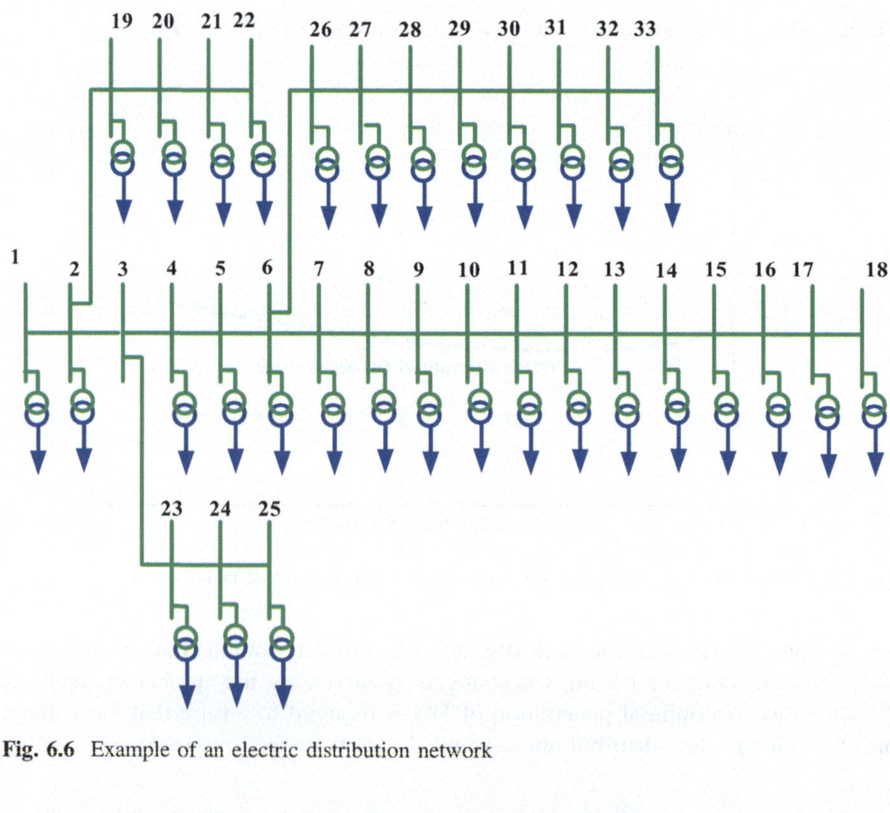

Fig. 6.6 Example of an electric distribution network

Fig. 6.7 PV system model

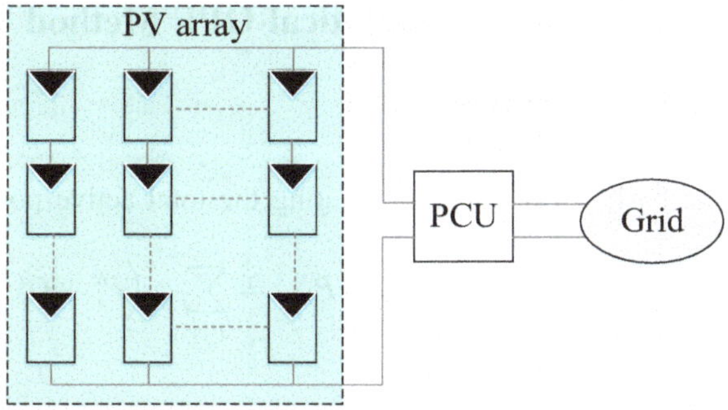

The DG units of type i, whose number is N_{DGi}, can be installed only in their corresponding nodes N_{Bi}. Thus

$$N_{DG} = \sum_{i=1}^{N_{DGT}} N_{DGi}, \quad N_B = \sum_{i=1}^{N_{DGT}} N_{Bi} \tag{6.1}$$

where N_{DGT} is the number of DG types to be installed. The number of possible combination of DG sites in this case can be computed from

$$N_{Com} = \left(\prod_{i=1}^{N_{DGT}} C_{N_{DGi}}^{N_{Bi}} \right) (N_{DG}!) \tag{6.2}$$

To determine the optimal combination, it is required to evaluate all of these possible combinations of the DG sites. Note that the number of the possible combinations is high, especially when placing different DG types in large-scale distribution networks. This excessive number of combinations will not only increase the complexity of the optimization problem but also degrade the computational performance. A fast method is required to determine the optimal combination among all of these combinations. The main requirements of the DG allocation method can be listed as

- Accurate (proper DG locations with optimal capacities).
- Generic formulations for multi-type DG allocation.
- High computational speed (especially when allocating multiple DGs in large-scale systems).

6.5 Combined Analytical-OPF Method

6.5.1 *Losses with DG*

The basic formula for calculating the total active power loss P_{loss} is expressed as

$$P_{loss} = \sum_{j \in \phi} \varphi_j \left(P_j^2 + Q_j^2 \right) \tag{6.3}$$

in which

$$\varphi_j = \frac{R_j}{V_j^2}$$

where P_j and Q_j the active and reactive power flows, respectively, through the distribution line j. ϕ is a set of system lines, V_j is voltage magnitude of the receiving bus of the line, and R_j is line resistance. Consider adding a DG or a capacitor, which injects P_g and/or Q_g, at a certain bus in a network, the variation in the reactive power loss can be linearly estimated, whereas it can be computed by

$$P_{loss,DG} = \sum_{j \in \alpha} \varphi_j \left(P_j^2 + Q_j^2 \right) + \sum_{j \in \beta} \varphi_j \left(\left(P_j - P_g \right)^2 + \left(Q_j - Q_g \right)^2 \right), \quad \alpha \cup \beta = \phi \tag{6.4}$$

where α and β represent two different sets of lines whose power flows are not affected and whose power flows are affected by adding the DG, respectively. The above equation can be modified to be in general form for expressing the effect of integrating multiple DG units and capacitors at a set of locations ψ on reactive power loss as follows

$$P_{loss,DG} = \sum_{j \in \alpha} \varphi_j \left(P_j^2 + Q_j^2 \right) + \sum_{j \in \beta} \varphi_j \left(\left(P_j - \sum_{i \in \psi} \Omega_{ij} P_{gi} \right)^2 + \left(Q_j - \sum_{i \in \psi} \Omega_{ij} Q_{gi} \right)^2 \right) \tag{6.5}$$

The Ω matrix can be built based on the radial structure of networks. The binary matrix Ω for a small-scale system shown in Fig. 6.8. when adding PV and wind units, respectively, at buses 11 and 7 can be expressed as follows

$$\begin{array}{c} \textit{System Buses} \\ \begin{array}{cccccccccccc} 1 & 2 & 3 & 4 & 5 & 6 & 7 & 8 & 9 & 10 & 11 & 12 \end{array} \end{array}$$

$$\Omega = \begin{bmatrix} 1 & 1 & 1 & 0 & 1 & 1 & 1 & 0 & 0 & 0 & 0 & 0 \\ 1 & 1 & 1 & 0 & 0 & 0 & 0 & 0 & 1 & 1 & 1 & 0 \end{bmatrix} \begin{array}{ll} 7 & \textit{Wind Bus} \\ 11 & \textit{PV Bus} \end{array} \tag{6.6}$$

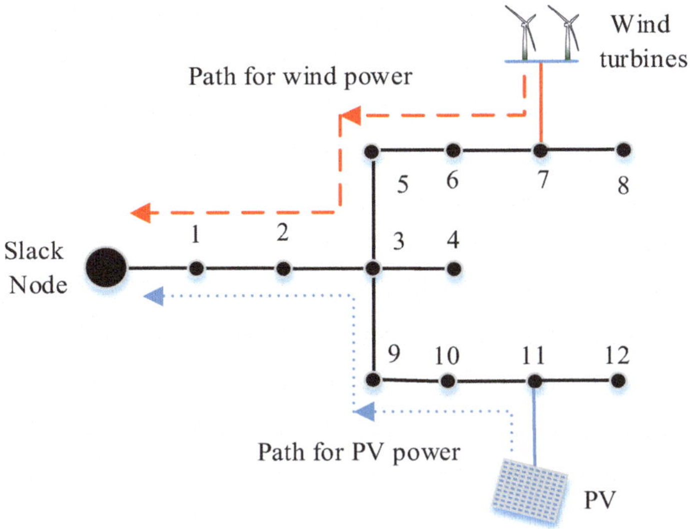

Fig. 6.8 Power flows after adding PV and wind units to an electric distribution network

Equation (6.5) could be expressed in different ways with active or reactive power injection of DG with respect to the power factor PF_g as in (6.7) and (6.8).

$$Q_{gi} = M_{gi} P_{gi} \tag{6.7}$$

where

$$M_{gi} = \frac{\sqrt{1 - PF_{gi}^2}}{PF_{gi}} \tag{6.8}$$

6.5.2 Analytical Expressions for Optimal DG Sizing

The objective of the placement of DG units in electric distribution networks is to minimize the total power loss by selecting best locations and optimal sizes of these units. Since the losses can be represented by (6.5), the objective function can be expressed as minimization of $P_{loss,DG}$. The variable for this optimization problem is the active P_g and reactive Q_g powers of the units. As the variations of $P_{loss,DG}$ with P_g and Q_g are equal to zero at the optimal point,

$$\frac{\partial P_{loss,DG}}{\partial P_{gm}} = 0, \quad \forall m \in \Psi \tag{6.9}$$

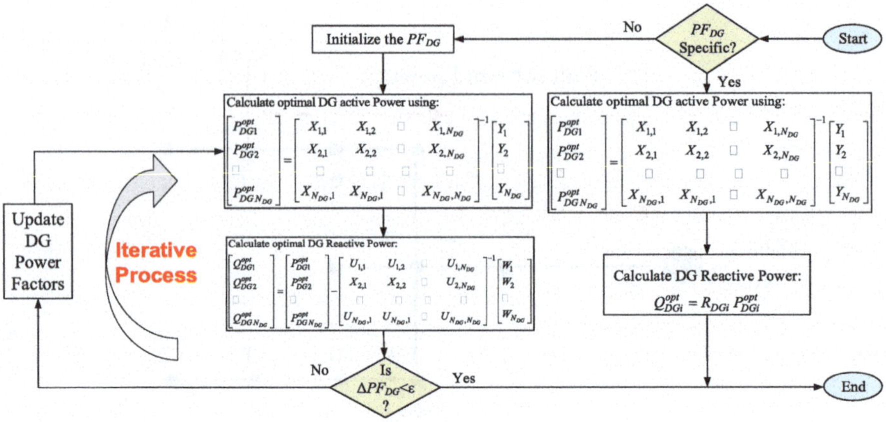

Fig. 6.9 Optimal DG sizing process

$$\frac{\partial P_{loss,DG}}{\partial Q_{gm}} = \frac{\partial P_{loss,DG}}{\partial P_{gm}}, \quad \forall m \in \Psi \tag{6.10}$$

The above two equations can be written for each unit; therefore, their number is equal to double the number of the units to be placed. For the set of (6.9) and (6.10), they can be arranged in matrix form to be as given in (6.11) and (6.12), respectively. With employing the new (6.11) and (6.12), a direct optimal solution of P_g and Q_g for all units (i.e., optimal power factors) can be delivered. Once the combination of the DG and capacitor locations is defined, the optimal sizing for all units can be computed, as shown in Fig. 6.9. The parameters of XYWU matrices in (6.11) and (6.12) can be completely computed directly from the power flow result for the base case without requiring iterative processes. The proposed formulae have been established based on the radial structure of power distribution systems. In order to apply the proposed method to radial systems, it is required to break all system loops and then the DG allocation problem is solved for the resulted radial system [31, 32].

$$\begin{bmatrix} P_{g\Psi_1} \\ P_{g\Psi_2} \\ \vdots \\ P_{g\Psi_N} \end{bmatrix} = \begin{bmatrix} X_{\Psi_1,\Psi_1} & X_{\Psi_1,\Psi_2} & \cdots & X_{\Psi_1,\Psi_3} \\ X_{\Psi_2,\Psi_1} & X_{\Psi_2,\Psi_2} & \cdots & X_{\Psi_2,\Psi_N} \\ \vdots & \vdots & \vdots & \vdots \\ X_{\Psi_N,\Psi_1} & X_{\Psi_N,\Psi_2} & \cdots & X_{\Psi_N,\Psi_N} \end{bmatrix}^{-1} \begin{bmatrix} Y_{\Psi_1} \\ Y_{\Psi_2} \\ \vdots \\ Y_{\Psi_N} \end{bmatrix} \tag{6.11}$$

$$\begin{bmatrix} Q_{g\Psi_1} \\ Q_{g\Psi_2} \\ \vdots \\ Q_{g\Psi_N} \end{bmatrix} = \begin{bmatrix} P_{g\Psi_1} \\ P_{g\Psi_2} \\ \vdots \\ P_{g\Psi_N} \end{bmatrix} - \begin{bmatrix} U_{\Psi_1,\Psi_1} & U_{\Psi_1,\Psi_2} & \cdots & U_{\Psi_1,\Psi_3} \\ U_{\Psi_2,\Psi_1} & U_{\Psi_2,\Psi_2} & \cdots & U_{\Psi_2,\Psi_N} \\ \vdots & \vdots & \vdots & \vdots \\ U_{\Psi_N,\Psi_1} & U_{\Psi_N,\Psi_2} & \cdots & U_{\Psi_N,\Psi_N} \end{bmatrix}^{-1} \begin{bmatrix} W_{\Psi_1} \\ W_{\Psi_2} \\ \vdots \\ W_{\Psi_N} \end{bmatrix} \tag{6.12}$$

where

$$X_{n,m} = \sum_{j\in\beta} \Omega_{nj}\varphi_j\Omega_{mj}(1 + M_{DGm}M_{DGn}), \qquad Y_m = \sum_{j\in\beta} \Omega_{mj}\varphi_j(P_j + M_{DGm}Q_j)$$

$$U_{n,m} = \sum_{j\in\beta} \Omega_{nj}\varphi_j\Omega_{mj}, \qquad W_m = \sum_{j\in\beta} \Omega_{mj}\varphi_j(P_j - Q_j)$$

6.6 Solution Process

The proposed method for determining the optimal mix of DG involves a combination of the proposed analytical expressions and OPF [31–34]. The objective function of OPF is set to be the minimization of the losses with considering equality and inequality constraints (6.13)–(6.17). Since the analytical expressions are general for optimally solving any combination of sites where various units are placed, they can be employed to evaluate all possible combinations of the sites. This evaluation process is essential to select the optimal combination of unit sites (i.e., the optimal mix). The computation burden of the evaluation process is greatly improved using the proposed analytical expressions, as the optimal solution can be directly computed using (6.11) and (6.12). The benefits of employing the OPF formulation are to apply system constraints for the optimal combination obtained by the analytical expressions and slightly correct the unit sizes to the exact optimal solution. The flowchart which illustrates the solution process of the proposed method is given in Fig. 6.10. The backward/forward sweep power flow method presented in [35] is employed as a power flow solver. As clear in the figure, the proposed analytical expressions are needed to calculate the optimal mix, while OPF is employed once for considering system constraints. This combination between these two formulations is efficient; since the proper optimal combination can be obtained with the analytical expressions, and the optimal solution can be accurately computed with including various constraints via OPF.

Minimize:

$$F = \sum_{j=1}^{N_{Line}} P_L^i \tag{6.13}$$

Subject To: (1) Equality constraints

$$P_S - P_D - \sum_{i=1}^{N} |V_j||V_i|(G_{ij}\cos\theta_{ij} + B_{ij}\sin\theta_{ij}) = 0 \tag{6.14}$$

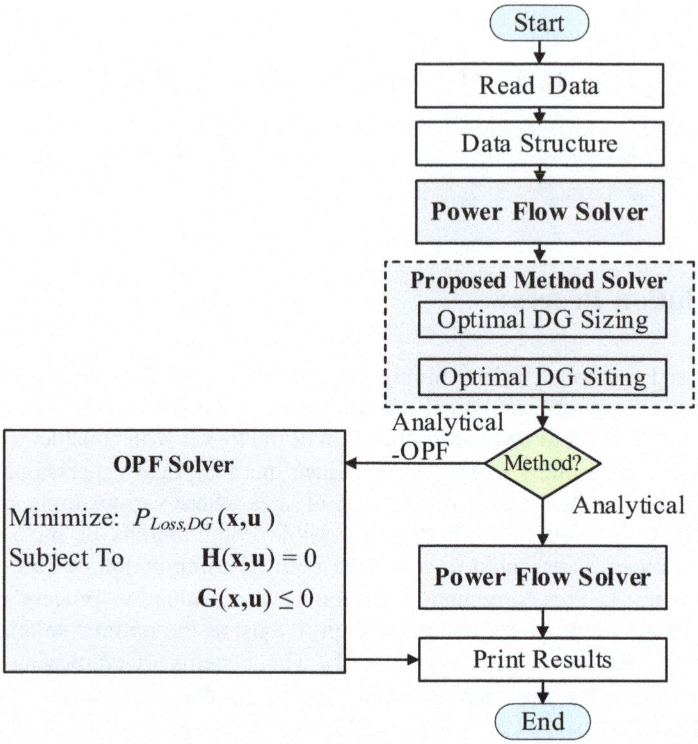

Fig. 6.10 Flowchart of the proposed method for DG allocation

$$Q_S - Q_D - \sum_{i=1}^{N} |V_j||V_i|\left(G_{ij}\sin\theta_{ij} - B_{ij}\cos\theta_{ij}\right) = 0 \tag{6.15}$$

(2) Inequality constraints

$$P_{DGi}^{\min} \le P_{DGi} \le P_{DGi}^{\max} \quad \text{for } i = 1, 2, \ldots N_{DG} \tag{6.16}$$

$$V_i^{\min} \le V_i \le V_i^{\max} \quad \text{for } i = 1, 2, \ldots N \tag{6.17}$$

6.7 Results and Discussions

6.7.1 Assumptions

- A single DG unit can be placed on each bus in the test system;
- The maximum allowed number of DG to be connected is three units;

- The specified generation values of DG Types A and DG type B are considered to be zero.
- For DG Type C, its power factor is equal to 0.90 lagging.
- The maximum penetration of DG is 100%.
- The minimum and maximum limits of voltages are 0.9 and 1.05, respectively.

6.7.2 Test System

The 69-bus test system is used to test out the proposed method for DG placement. Figure 6.11 shows the system, which is a preferable benchmark test system for several allocation approaches, where the data are given in [36]. This system consists of 68 load buses and a slack bus, and the active and reactive power loss in the base case are 225 kW and 102 kvar, respectively. The proposed method has been implemented by C++ programming. This analysis aims at demonstrating the effectiveness of the proposed method for solving the allocation problem of different DG types. Assume that an area is recommended for each type of DG. To do so, the 69-bus distribution test system is divided into four different areas as

- Area-A: This area contains candidate locations of DG type A.
- Area-B: This area contains candidate locations of DG type B.
- Area-C: This area contains candidate locations of DG type C.
- Area-D: No DG is allowed to be installed in this area.

Table 6.1 shows different studied cases with different combinations of DG types. The first case (Case 0) is the base case without DG while the other seven cases (Case 1–Case 7) involve installing three DG units of different types, as illustrated in the table.

Table 6.1 Number of DG for different cases

Cases	DG Type A	DG Type B	DG Type C
Case 0	–	–	–
Case 1	2	1	–
Case 2	2	–	1
Case 3	1	2	–
Case 4	–	2	1
Case 5	–	1	2
Case 6	1	–	2
Case 7	1	1	1

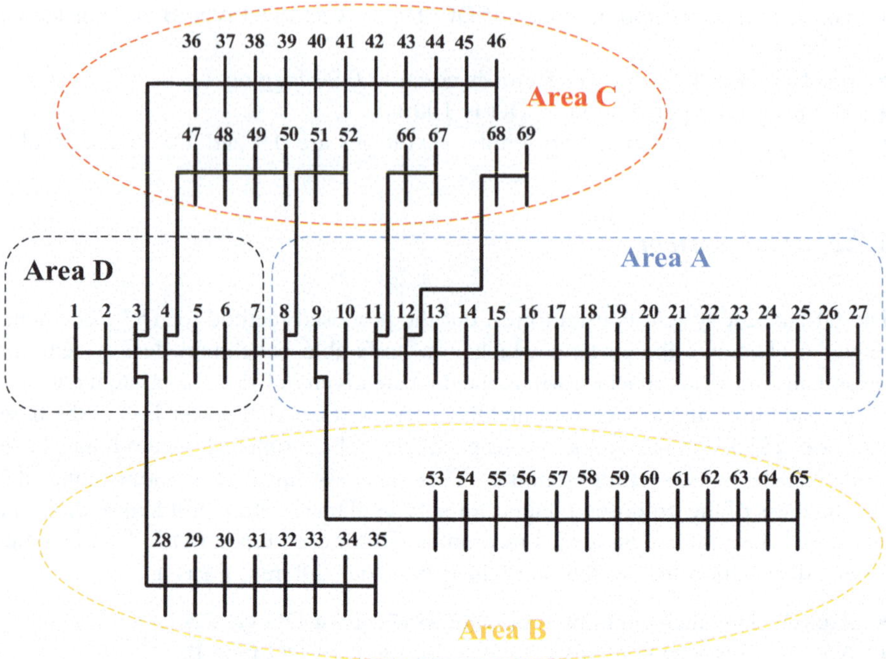

Fig. 6.11 The 69-bus test system

6.7.3 *Analyses*

In this subsection, the benefits of allocating DGs to the test system are discussed. Table 6.2 shows the computed DG sizes and their corresponding buses with using the proposed analytical-OPF method. Note that the calculated DG locations and sizes are almost different for all the cases at which the DG numbers are similar (three units) for the cases. The only difference between the cases is the combinations of the different DG types. These results imply that the type of DG has a significant influence on the DG allocation problem. The major differences between the eight cases can be listed as

- **Total Losses**: Figure 6.12 shows the losses in kW and kVA for the cases after allocating the DG units into the 69-bus test system. It is obvious that there are significant reductions of the total losses for all the cases of DG allocation with respect to the base case, but these reductions are different for the cases. Note that DG type C has a higher effect on loss reduction compared with the other DG types. For example, Case 5 has a high value of loss reduction due to its two DG units of type C. This feature is reasonable since this DG type can inject active and reactive power, thereby contributing heavily in reducing losses. In addition,

Table 6.2 Results for the 69-bus test system

Cases	DG Type A		DG Type B		DG Type C		Total DG size (kVA)
	Bus	Size (kVA)	Bus	Size (kVA)	Bus	Size (kVA)	
Case 0	–	–	–	–	–	–	–
Case 1	9 18	2388.12 451.373	61	1314.6	–	–	4154.0
Case 2	12 21	370.6 312.8	–	–	51	2535.9	3219.2
Case 3	9	2839.5	53 61	652.6 1193.5	–	–	4685.6
Case 4	–	–	61 64	938.9 206.5	51	2640.3	3785.7
Case 5	–	–	61	1138.2	51 68	2007.7 704.3	3850.3
Case 6	21	302.1	–	–	51 66	2129 793.8	3225.2
Case 7	9	2249.6	61	1270.1	68	655.4	4175.2

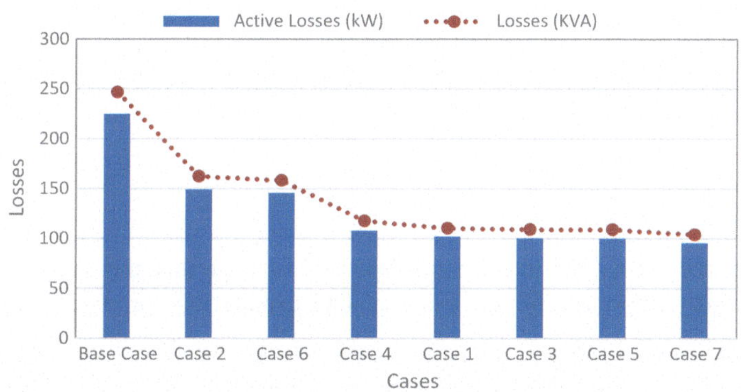

Fig. 6.12 The losses for the different cases

it is noted that Case 7 which involves different three types of DG yields the highest loss reduction.

- **The Total Size of DGs**: The total size of DGs is important in the DG allocation as it can be employed to estimate the installation cost of DG. The higher DG size, the higher cost of DG. The total size of DG units for the case are shown in Table 6.2. It is clear from the table that Case 1 and Case 3 at which DG type C is not included, have the highest capacity. This trend means that the installation costs for these two cases are relatively high. However, this trend can be an advantageous feature if the penetration DG is required to be maximized.
- **Voltage Profile**: Figure 6.13 shows the voltage profile for the different cases. Table 6.3 compares the minimum voltage, maximum voltage, and the value of

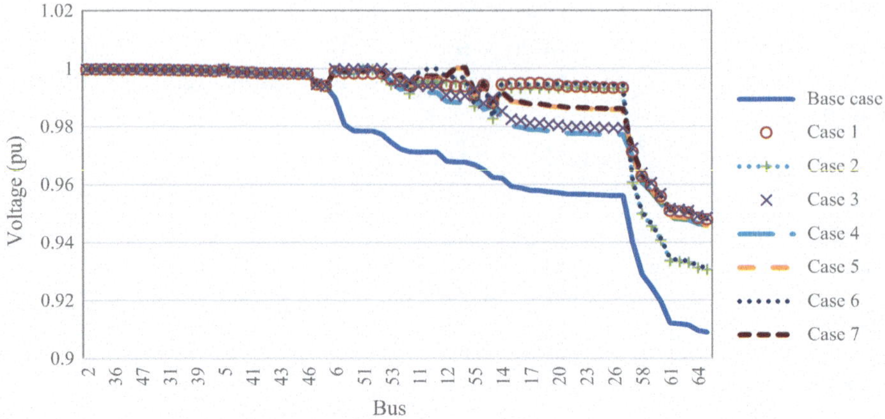

Fig. 6.13 Voltage profile for the different cases

Table 6.3 Voltage for the different cases

Item	Base case	Case 1	Case 2	Case 3	Case 4	Case 5	Case 6	Case 7
Maximum voltage	0.9092	0.9481	0.9307	0.9488	0.9461	0.9461	0.9313	0.9484
Maximum voltage	1.0000	1.0000	1.0000	1.0002	1.0005	1.0006	1.0001	1.0005
Voltage deviation	0.0993	0.0193	0.0345	0.0236	0.0278	0.0224	0.0336	0.0207

voltage deviation for the different cases. It is clear that the voltage profile is significantly improved for the cases of DG installation compared with Case 0.

It is important to mention that the proposed method solves the DG allocation problem optimally, and the total losses are reduced for all cases. The proposed method is general, and so it can be applied for solving other cases and electric distribution networks. The proposed method is a helpful tool for optimizing the networks with DG and selecting the optimal mix of the available DG technologies to maximize benefits. Note that the proposed method is very effective for solving the allocation problem of multi-type DG units compared with existing analytical methods in the literature. This superiority is accomplished as the proposed method has high accuracy rates with fast computational speed, and it can directly compute the optimal power factors of different DG types [31, 32].

References

1. L. Willis, W.G. Scott, *Distributed Power Generation: Planning and Evaluation*, vol. 10 (Marcel Dekker, New York, 2000)
2. O.M. Toledo, D.O. Filho, A.S.A.C. Diniz, J.H. Martins, M.H.M. Vale, Methodology for evaluation of grid-tie connection of distributed energy resources—case study with photovoltaic and energy storage. IEEE Trans. Power Syst. **28**(2), 1132–1139 (2013)
3. A.G. Exposito, A.J. Conejo, C.A. Canizares, *Electric Energy Systems Analysis and Operation* (CRC Press, 2009)
4. X. Zhang, G.G. Karady, S.T. Ariaratnam, Optimal allocation of CHP-based distributed generation on urban energy distribution networks. IEEE Trans. Sustain. Energy **5**(1) (2014)
5. X. Zhang, R. Sharma, Y. He, Optimal energy management of a rural microgrid system using multi-objective optimization, in *Proceedings of IEEE PES Innovative Smart Grid Technology Conference* (Washington DC, USA, 2012), pp. 1–8
6. R.A. Walling, R. Saint, R.C. Dugan, J. Burke, L.A. Kojovic, Summary of distributed resources impact on power delivery systems. IEEE Trans. Power Del. **23**, 1636–1644 (2008)
7. T. Ackermann, V. Knyazkin, Interaction between distributed generation and the distribution network: operation aspects, in *Proceedings on 2002 IEEE T&D Conference* (2002), pp. 1357–1362
8. D. Singh, K.S. Verma, Multiobjective optimization for DG planning with load models. IEEE Trans. Power Syst. **24**(1), 427–436 (2009)
9. Y.M. Atwa, E.F. El-Saadany, M.M.A. Salama, R. Seethapathy, Optimal renewable resources mix for distribution system energy loss minimization. IEEE Trans. Power Syst. **25**(1), 360–370 (2010)
10. P.S. Georgilakis, N.D. Hatziargyriou, Optimal distributed generation placement in power distribution networks: models, methods, and future research. IEEE Trans. Power Systems **28** (3), 3420–3428 (2013)
11. N.S. Rau, Y.-H. Wan, Optimum location of resources in distributed planning. IEEE Trans. Power Syst. **9**(4), 2014–2020 (1994)
12. A. Keane, M. O'Malley, Optimal allocation of embedded generation on distribution networks. IEEE Trans. Power Syst. **20**(3), 1640–1646 (2005)
13. G.P. Harrison, A.R. Wallace, Optimal power flow evaluation of distribution network capacity for the connection of distributed generation. IEE Proc. Gener. Transm. Distrib. **152**(1), 115–122 (2005)
14. D. Zhu, R.P. Broadwater, K.S. Tam, R. Seguin, H. Asgeirsson, Impact of DG placement on reliability and efficiency with time-varying loads. IEEE Trans. Power Syst. **21**(1), 419–427 (2006)
15. D. Singh, R.K. Mirsa, D. Singh, Effect of load models in distributed generation planning. IEEE Trans. Power Syst. **22**(4), 2204–2212 (2007)
16. K. Vinothkumar, M.P. Selvan, Fuzzy embedded genetic algorithm method for distributed generation planning. Electr. Power Compon. Syst. **39**(4), 346–366 (2011)
17. K.H. Kim, Y.J. Lee, S.B. Rhee, S.K. Lee, S.K. You, Dispersed generator placement using fuzzy-GA in distribution systems, in *Proceedings of IEEE Power Engineering Society Summer Meeting* (July 2002), pp. 1148–1153
18. W. Prommee, W. Ongsakul, Optimal multiple distributed generation placement in microgrid system by improved reinitialized social structures particle swarm optimization. Eur. Trans. Electr. Power **21**(1), 489–504 (2011)
19. M.F. Shaaban, Y.M. Atwa, E.F. El-Saadany, DG allocation for benefit maximization in distribution networks. IEEE Trans. Power Syst. **28**(2), 639–649 (2013)
20. M.E.H. Golshan, S.A. Arefifar, Optimal allocation of distributed generation and reactive sources considering tap positions of voltage regulators as control variables. Eur. Trans. Electr. Power **17**(3), 219–239 (2007)

21. H.L. Willis, Analytical methods and rules of thumb for modeling DG-distribution interaction, in *Proceedings of IEEE Power Engineering Society Summer Meeting* (July 2000), pp. 1643–1644
22. C. Wang, M.H. Nehrir, Analytical approaches for optimal placement of distributed generation sources in power systems. IEEE Trans. Power Syst. **19**(4), 2068–2076 (2004)
23. S.-H. Lee, J.-W. Park, Selection of optimal location and size of multiple distributed generations by using Kalman filter algorithm. IEEE Trans. Power Syst. **24**(3), 1393–1400 (2009)
24. T. Xiao-bo, W. Xue-hong, A new method of distributed generation optimal placement based on load centroid, in *2011 Proceedings of IEEE Power and Energy Engineering Conference (APPEEC),* pp. 1–5
25. A. Elmitwally, A new algorithm for allocating multiple distributed generation units based on load centroid concept. Alexandria Eng. J. **52**(4), 655–663 (2013)
26. N. Acharya, P. Mahat, N. Mithulananthan, An analytical approach for DG allocation in primary distribution network. Int. J. Elect. Power Energy Syst. **28**(10), 669–678 (2006)
27. D.Q. Hung, N. Mithulananthan, R.C. Bansal, Analytical expressions for DG allocation in primary distribution networks. IEEE Trans. Energy Convers. **25**(3), 814–820 (2010)
28. D.Q. Hung, N. Mithulananthan, Multiple distributed generators placement in primary distribution networks for loss reduction. IEEE Trans. Ind. Electron. **60**(4), 1700–1708 (2013)
29. M.J.E. Alam, K.M. Muttaqi, D. Sutanto, A three-phase power flow approach for integrated 3-wire MV and 4-wire multigrounded LV networks with rooftop solar PV. IEEE Trans. Power Syst. **28**(2), 1728–1737 (2013)
30. K. Mahmoud, M. Abdel-Akher, Analysis of hybrid photovoltaic and wind energies connected to unbalanced distribution systems, in *Proceedings of 2010 IEEE International Conference on Power and Energy (PEcon),* (Kuala Lumpur, 2010), pp. 79–84
31. K. Mahmoud, N. Yorino, A. Ahmed, Optimal distributed generation allocation in distribution systems for loss minimization. IEEE Trans. Power Syst. **31**(2), 960–969 (2016)
32. K. Mahmoud, N. Yorino, A. Ahmed, Power loss minimization in distribution systems using multiple distributed generations. IEEJ Trans. Elec. Electron. Eng. **10**(5), 521–526 (2015)
33. K. Mahmoud, Y. Naoto, Optimal combination of DG technologies in distribution systems, in *Power and Energy Engineering Conference (APPEEC), 2015 IEEE PES Asia-Pacific* (2015)
34. K. Mahmoud, Optimal integration of DG and capacitors in distribution systems, in *Power Systems Conference (MEPCON), 2016 Eighteenth International Middle Eastern IEEE* (2016), pp. 651–655
35. K. Mahmoud, Y. Naoto, Robust quadratic-based BFS power flow method for multi-phase distribution systems. IET Gener. Transm. Distrib. **10**(9), 2240–2250 (2016)
36. M.E. Baran, F.F. Wu, Optimum sizing of capacitor placed on radial distribution systems. IEEE Trans. Power Del. **4**(1), 735–743 (1989)

Chapter 7
Battery Energy Storage Planning

Mahdi Sedghi, Ali Ahmadian, Ali Elkamel, Masoud Aliakbar Golkar and Michael Fowler

Abstract Rechargeable grid-scale batteries are suitable and mature technology for energy storage in active distribution networks. Battery energy storage (BES) units have many advantages and are used for several purposes in electric systems and distribution grids. They are used not only for peak shaving and voltage regulation, but also for reliability enhancement and dispatching the renewable-based distributed generation (DG) sources. However, BES technologies are still expensive and need to be employed optimally to prevent excess investment cost. Optimal planning of BES is a complex approach that determines the type, location, capacity and power rating of energy storage units. The optimization should handle the uncertain conditions and it requires to develop the appropriate models and methods. There are many effective components that should be addressed. These components influence the results of the optimal planning and make it more complicated. In this chapter the optimal BES planning methodologies are presented. Firstly the optimization problem is formulated considering different economic perspectives. Then the approaches and strategies for solving the combinatorial problem are described. In this way, both the probabilistic and possibilistic methods and models are displayed. In addition, the most important components and factors that affect the

M. Sedghi (✉) · M. A. Golkar
Faculty of Electrical Engineering, K. N. Toosi University of Technology, Tehran, Iran
e-mail: meh.sedghi@gmail.com

M. A. Golkar
e-mail: golkar@kntu.ac.ir

A. Ahmadian
Department of Electrical Engineering, University of Bonab, Bonab, Iran
e-mail: ahmadian@bonabu.ac.ir

A. Elkamel · M. Fowler
Department of Chemical Engineering, University of Waterloo, Waterloo, ON, Canada
e-mail: aelkamel@uwaterloo.ca

M. Fowler
e-mail: mfowler@uwaterloo.ca

F. Shahnia et al. (eds.), *Electric Distribution Network Planning*, Power Systems, https://doi.org/10.1007/978-981-10-7056-3_7

optimal planning are characterized and analyzed, including conventional DGs, renewable-based DGs, capacitor banks, plug-in electric vehicles, etc.

Keywords Battery energy storage · Optimal planning · Active distribution network

7.1 Introduction

The energy storage systems (ESSs) are widely used in active distribution networks for several purposes. The ESSs absorb energy and store it for a period of time and then inject it to supply energy in scheduled times. They can be installed throughout the electric grid and employed in heating and cooling networks, stand-alone and grid-connected electric systems, uninterruptible power supply (UPS) systems, etc. Hence, they provide a valuable opportunity for grid operators to manage the networks optimally. The ESSs can store energy in off-peak hours (when the electricity price is low) and inject it in peak-load times (when the electricity price is high) [1]. So, they are used for the power profile peak shaving and valley filling in efficient manner and such they provide an 'energy arbitrage' service. This service can address the generation, transmission and distribution expansion concerns, and also it defers the reinforcement necessity of the infrastructure. The stored energy enhances the reliability, especially in radial distribution networks and stand-alone systems, as it can restore the loads in outage times [2]. Furthermore, they can provide voltage regulation service and also a valuable tool for grid operators to accommodate the supply and/or demand-side variability [3]. Several technologies have been introduced for energy storage and they can be implemented on large and small scales in distributed and centralized manners within electric systems [4]. While some technologies are mature or in near maturity, some of them are still in the early development stages and will require additional attention before their potential is fully realized [5]. Research and development works are currently underway with the primary goals of realizing technology cost reductions and improving the performance of existing, new and emerging storage technologies. In addition, the non-technical barriers should be addressed for deployment of ESSs by governments and industry stakeholders.

Generally, the ESSs are categorized in thermal and electricity (heat or cold) groups based on their outputs. Both groups can work as generators and consumers, making them useful for energy management. Based on input and output energy types of ESSs, they can provide several services including electricity to electricity, heat to heat, electricity to heat, and heat to electricity applications. Using these services, the ESSs provides a valuable opportunity for energy management improvement both in energy supply and demand sides [6].

Due to their characteristics, each type of the ESSs can be utilized for individual application. For example, the high power ESSs are needed for frequency regulation service to maintain a constant main frequency in utility application, while a more sustained power ESSs are needed for energy management service in which the demand is met and the supply is not wasted. There are six parameters for energy storage systems which make distinguished of different technologies. Specific energy, energy density, specific power, power density, operating cost, efficiency, and cycle life are the main parameters which play important role in technology type choosing for different applications [5]. Specific energy is the energy storage capacity per unit mass while the energy density denotes energy storage capacity per unit volume. The operating cost is the cost of one cycle of charge/discharge in rated power, and it consists of maintenance, heating, labor, etc. Specific power characterizes the capability of the discharge power per unit mass. The efficiency is defined as the ratio of total energy injected to the load over total energy absorbed by the ESS in a single charge/discharge cycle. The cycle life determines the number of charge/discharge cycles which could be fulfilled until the useful capacity of the ESS brings down to 80% of its initial capacity. The importance of every parameter depends on the application.

Among the energy storage technologies, the batteries have been widely considered to be utilized in electric distribution networks. Different battery types consist of lead-acid, NaS, Ni-Cd, lithium-ion, Zn-Br, vanadium redox flow and ZEBRA, some of which are commercialized for utility applications [7]. Each of these battery types have advantages and disadvantages in comparison with each other, making it suitable for individual application.

Although the ESSs have several advantages in distribution networks, however the optimal planning and scheduling of them are the main prerequisites for optimal exploitation of them. Generally, the battery energy storage (BES) can be implemented in the most buses of the distribution networks as the batteries have less environmental and non-technical constraints. However, the electrical considerations such as power follow, power loss, voltage regulation and etc. affect on optimal location of batteries [8]. Similarly, the optimal scheduling of batteries depends on electrical and non-electrical considerations of networks such as power price, power loss, power follow, renewable energies availability, flexible load and generation availability, etc. As a result, the batteries should be optimally scheduled and planned in distribution networks to prevent excess operation and investment costs [9]. For these purposes, the effective components in generation and consumption sides as well as electric grid should be modeled properly. In generation sides the components such as conventional distributed generations (DGs), wind and solar energy based DGs, etc., and in the consumption side the flexible and non-flexible load demands should be modeled appropriately.

The BES planning approach should determine the optimal type, location, size and power rating of the batteries in electric systems and grids. A lot of research has focused on the optimal sizing of BES in electric systems including renewable-based resources [10–14]. The optimization of the BES location, capacity and power rating in distribution network is shown in [15–17] while the optimal typing has been investigated in [7, 9]. With respect to the literature, there are several methods and models for optimal BES planning that is displayed here.

In this chapter, firstly the BES planning problem is defined. It is formulated in two ways, considering the owner of storage that invests in BES. In order to solve this problem, the various methods and techniques employed for optimal BES planning are presented in continue. These methods and techniques include

- Multi-layer methods (including master/slave based strategy and optimal power flow based strategy)
- Probabilistic optimal power flow
- Possibilistic optimal power flow
- Monte Carlo simulation
- Point estimate method
- Fuzzy load flow

After introducing the problem and solution techniques, the results of optimal BES planning presented in literature are discussed and analyzed. The discussion represents how the results of case studies are influenced by the most effective components in active distribution networks. These components which affect the optimal BES planning can be categorized in three main groups of

- The other facilities of distribution network
- Storage capabilities
- Techno-economic factors

The components which belong to these groups are shown in Fig. 7.1. The impact of effective components on BES are comprehensively compared at the end.

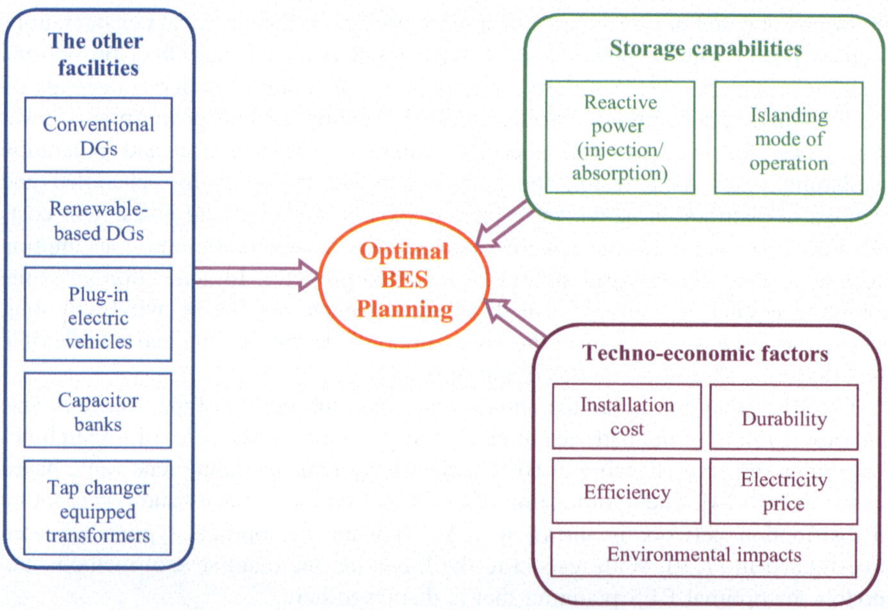

Fig. 7.1 The most effective components which influence the optimal BES planning

7.2 Optimal BES Planning

The long-term BES planning determines the optimal type, location, capacity and power rating of grid-scale batteries. In other words, they are the decision variables in optimal BES planning. The optimal solution should minimize the cost function (or maximize the benefit) considering the technical constraints. The costs usually include the investment cost, operation cost and reliability cost. The optimal solution should satisfy the voltage magnitude constraint, maximum allowed capacity constraint and power balance constraint. It is possible to model the technical constraints as a penalty factor in the objective function. So the unpractical solutions are penalized in this way.

With respect to the economic perspectives, there are two kinds of planning as follow:

- scenario-I: The distribution utility invests in BES; and,
- scenario-II: An independent organization rather than distribution company (DisCo) is the owner of storage.

In the first case, a cost objective function is defined to be minimized subject to the technical constraints. While in the second state, one is faced to two objectives that should maximize the benefits of two different companies simultaneously. Therefore two objective functions are defined in the latter case. Both definitions are presented and formulated below.

7.2.1 *Definitions*

7.2.1.1 Objective Function(s)

In the first scenario a total cost function is determined to be minimized subject to the technical constraints. Generally the objective function consists of investment cost, operation cost, reliability cost and maybe a penalty factor, as

$$\text{Min}.f = IC + OC + RC + \text{M} \times PF \tag{7.1}$$

where M is a large fixed value for penalization of the infeasible solutions.

The investment cost contains three terms: installation cost related to the battery capacity, installation cost related to the battery power rating, and replacement cost of batteries, as

$$IC = \sum_{b=1}^{n_N} \left[x_b \times C_{Cap}^{INS} + x_b \times C_{Pow}^{INS} + x_b \times C_{Batt}^{REP} \right] \tag{7.2}$$

The operation cost includes the cost of purchased power by DisCo and the operation and maintenance (O&M) cost of batteries, over the project period, as

$$OC = C_E^{AR}(T) + \sum_{b=1}^{n_N} \left[x_b \times C_{Batt}^{OM}(T)\right] \tag{7.3}$$

It should be noted that $C_E^{AR}(T)$ contains the cost of power loss as well.

The reliability cost is the cost of outage which arises from all the probable failure events in distribution network. *RC* is a function of energy not supplied (ENS) as well as the type of interrupted loads. It is noticeable that *RC* denotes the cost of ENS in (7.1). So it is possible to combine the costs with the same unit in a single objective function. If the other reliability indices with different units are employed, different weights should be used in the objective function to consider the importance of the different reliability indices.

The penalty factor is corresponding with the number of violated constraints.

In the second scenario, two profit functions i.e. f_{DisCo} and f_{StOwn} are defined. f_{DisCo} and f_{StOwn} are the profit functions of DisCo and storage owner, respectively. The profit of the storage owner is equal to the difference between the income and cost, as

$$\text{Max.} f_{StOwn} = C_{IN}^{ST} - C_{COS}^{ST} \tag{7.4}$$

The storage owner purchases the electrical energy from DisCo in light-load periods, and then, it sells the electricity to the DisCo in peak-load times. So, the income of storage owner can be formulated as

$$C_{IN}^{ST} = C_{Sel}^{ST} - C_{Pur}^{ST} \tag{7.5}$$

The cost of storage owner is obtained from

$$C_{COS}^{ST} = \sum_{b=1}^{n_N} \left[x_b \times C_{Cap}^{INS} + x_b \times C_{Pow}^{INS} + x_b \times C_{Batt}^{REP}\right] + \sum_{b=1}^{n_N} \left[x_b \times C_{Batt}^{OM}(T)\right] \tag{7.6}$$

The profit of DisCo is corresponding with the reduction of the cost due to the installation of storage. Moreover, the DisCo is responsible for satisfying the technical constraints. Hence the penalty factor should be taken into account in DisCo's profit, as shown below.

$$\text{Max.} f_{DisCo} = \frac{C_{withoutST}^{DC} - C_{withST}^{DC}}{1 + PF} \tag{7.7}$$

These cost functions are presented in Eqs. (7.8) and (7.9).

$$C^{DC}_{withoutST} = C^{AR}_{E,withoutST}(T) + RC_{withoutST} \tag{7.8}$$

$$C^{DC}_{withST} = C^{DC}_{E,ST}(T) + C^{AR}_{E,withST}(T) + RC_{withST} \tag{7.9}$$

In fact the DisCo purchases the electricity from the storage owner with a specific tariff.

If all the technical constraints are satisfied, the penalty factor is equal to zero, and hence, $PF = 0$ in (7.7). Otherwise $PF > 0$, so that the profit of the DisCo is reduced considerably.

In scenario-II the Pareto set is used to assign the global best solution similar to the work shown in [18].

It is notable that (7.1) represents a static BES planning. Generally the optimal BES planning may be defined as a multistage expansion planning where the planning horizon can be divided into several stages, so that the elements to be installed in each stage should be determined. In other words, the installation date is a decision variable as well. In this situation, the objective function is defined as

$$F = \sum_{s=1}^{N} f_s \tag{7.10}$$

where

$$f_s = IC(s) + OC(s) + RC(s) + PF \tag{7.11}$$

In this definition, the investment and operation cost formulations are modified as follow:

$$IC = \sum_{b=1}^{n_N} \left[x_{s,b} \times C^{INS}_{Cap}(T_s) + x_{s,b} \times C^{INS}_{Pow}(T_s) + x_{s,b} \times C^{REP}_{Batt}(T_s) \right] \tag{7.12}$$

$$OC(s) = C^{AR}_{E}(T_s) + \sum_{b=1}^{n_N} \left[x_{s,b} \times C^{OM}_{Batt}(T_s) \right] \tag{7.13}$$

Similar to the investment and operation costs, the reliability cost is a function of the number and duration of stages as well. In all the situations, the technical constraints should be satisfied at all the operation times, as formulated below.

7.2.1.2 Technical Constraints

There are several technical constraints that should be considered in optimal storage planning. In an acceptable solution, the voltage level of all the nodes must be limited to the allowed boundary; the constraint of maximum allowed capacity should not be violated for all the equipment; and the active/reactive power balance should be satisfied. These technical constraints are formulated below.

$$V_{\min} \leq V_{n,t} \leq V_{\max}, \quad \forall n \in \mathrm{A_N}, \quad \forall t \in \mathrm{A_T} \tag{7.14}$$

$$S^u_{\min} \leq S^u_t \leq S^u_{\max}, \quad \forall u \in \mathrm{A_{EQ}}, \quad \forall t \in \mathrm{A_T} \tag{7.15}$$

$$\sum_{s \in \mathrm{A_{SS}}} S^{SS}_{s,t} + \sum_{d \in \mathrm{A_{DG}}} S^{DG}_{d,t} = \sum_{k \in \mathrm{A_{ST}}} S^{ST}_{k,t} + \sum_{l \in \mathrm{A_{LD}}} S^{LD}_{l,t} + S^{LOSS}_t, \quad \forall t \in \mathrm{A_T} \tag{7.16}$$

In (7.16), the real part of $S^{ST}_{k,t}$ is a negative value during discharge period. In addition, the state of charge (SOC) of the batteries are updated as

$$SOC_{k,t+1} = \begin{cases} SOC_{k,t} + P^{ST}_{k,t}/\eta_{dis}, & if \quad P^{ST}_{k,t} \leq 0 \\ SOC_{k,t} + \eta_{ch} \times P^{ST}_{k,t}, & otherwise \end{cases} \tag{7.17}$$

7.2.1.3 Penalization

The optimal solution should satisfy the technical constraints. In this way the candidate solutions in which the constraints are violated, should be penalized. The penalty factor is proportional to the total number of violations in every solution. However, the penalization should be corresponding with the probability of the violation, as well. For example, the violations with 1 and 100% probability should be considered in different manner. Hence, the penalty factor is defined as

$$PF = \sum_{v=1}^{n_{VIO}} p(v) \tag{7.18}$$

$p(v)$ is obtained from the probabilistic/possibilistic load flow (PLF) analysis that is described in the next subsections.

7.2.2 *Optimal Power Flow (OPF)*

The type, location, capacity and power rating of energy storage units are the main decision variables in optimal battery planning. However, the long-term optimization should be accomplished considering the optimal charge/discharge cycles. In real conditions an optimal scheduling i.e. OPF is required to be taken into account. Unlike the common OPF which is performed for a snapshot, here the OPF is an uncertain probabilistic/possibilistic approach. So the intermittent renewable power and the stochastic load profiles are taken into consideration in probabilistic/possibilistic optimal power flow (POPF).

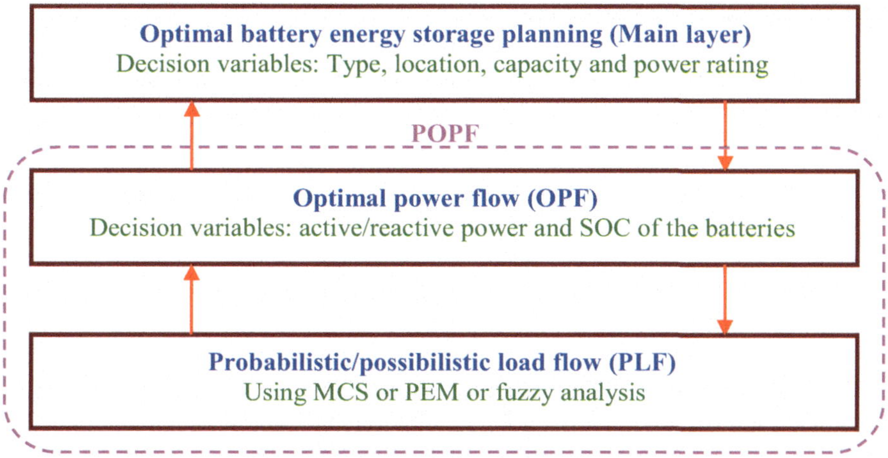

Fig. 7.2 Multi-layer approach for optimal BES planning (master/slave strategy)

In order to employ the POPF in optimal battery planning, there are two strategies as follow: the Master/slave strategy, and, the OPF-based strategy. The master/slave strategy is based on a multi-layer method. The optimal planning is as the main optimization layer that considers the results of POPF as a sub-layer. The main layer determines the optimal type, location, capacity and power rating as decision variables, while the active/reactive power and SOC are the decision variables of the sub-layer. For every given type, location, capacity and power rating, the POPF is executed completely to specify the optimal active/reactive power and SOC.

On the other hand, the implementation of POPF requires executing PLF analysis. Therefore the PLF appears as a sub-layer in POPF as shown in Fig. 7.2.

Similar to the master/slave strategy, the OPF-based strategy contains the PLF analysis as a sub-layer, however, the location, capacity and power rating are not predetermined in a master layer. In OPF-based strategy the POPF is executed considering stochastic active/reactive power and SOC. Then the candidate capacity, power rating and the locations are determined according to Eqs. (7.19)–(7.21).

$$S_{Cap}(n) = \max_{t \in \mathrm{A_T}} [SOC(t, n)], \quad \forall n \in \mathrm{A_N} \tag{7.19}$$

$$P_{Rat}(n) = \max_{t \in \mathrm{A_T}} \left\{ \sqrt{[P(t, n)]^2 + [Q(t, n)]^2} \right\}, \quad \forall n \in \mathrm{A_N} \tag{7.20}$$

$$\mathrm{A_{ST}} = \{ n \in \mathrm{A_N} | S_{Cap}(n) > 0 \} \tag{7.21}$$

So the candidate location, capacity and power rating are functions of the POPF program. In both strategies it is required to execute the probabilistic or possibilistic load flow analyses that are described in the next subsections.

7.2.2.1 Probabilistic Load Flow

The probabilistic load flow (PLF) is based on the basic load flow analysis whose main equations are

$$P_p = V_p \sum_{q=1}^{n} V_q (G_{pq} \cos(\delta_{pq}) + B_{pq} \sin(\delta_{pq})), \quad p = 1, 2, \ldots, n \tag{7.22}$$

$$Q_p = V_p \sum_{q=1}^{n} V_q (G_{pq} \sin(\delta_{pq}) - B_{pq} \cos(\delta_{pq})), \quad p = 1, 2, \ldots, n \tag{7.23}$$

There are several load flow methods to solve the above equations in active distribution networks as shown in [19]. However, in PLF the active and reactive power as well as the voltage and current magnitudes and angles are stochastic variables [20].

The methodologies for PLF can be categorized to three types of analytical methods, numerical methods and approximate methods. The analytical methods are very complex, while the numerical ones are much simpler but time consuming. The approximate methods are not very complex and does not take a long time, however, the results are less precise compared with the other methods. Despite of this fact, the accuracy of the approximate methods is usually acceptable in optimal storage planning. Here, two well-known methods i.e. Monte Carlo simulation (MCS) and point estimate method (PEM) are briefly presented.

7.2.2.2 Monte Carlo Simulation (MCS)

MCS, as a numerical method, is the simplest approach for PLF analysis. It is based on deterministic load flow and random sampling. Random samples are selected considering the probability distribution function (PDF) of input stochastic variables. The load flow analysis is frequently executed considering the selected samples and the results are saved. Finally the PDF of output stochastic variables are estimated. Figure 7.3 represents the flowchart of the MCS implementation. Although the MCS is very simple and accurate, but it needs a large computation effort. So the approximate methods are more suitable in real conditions.

7.2.2.3 Point Estimate Method (PEM)

The PEM has been used widely for PLF analysis in electric system studies [21]. It is based on Taylor series expansion. Suppose that the load flow nonlinear functions are presented as $Y = h(X)$. The PDF of input and output variables are shown with $f_X(x)$ and $f_Y(y)$, respectively. $f_X(x)$ and h are known functions, while the goal is to determine the PDF $f_Y(y)$.

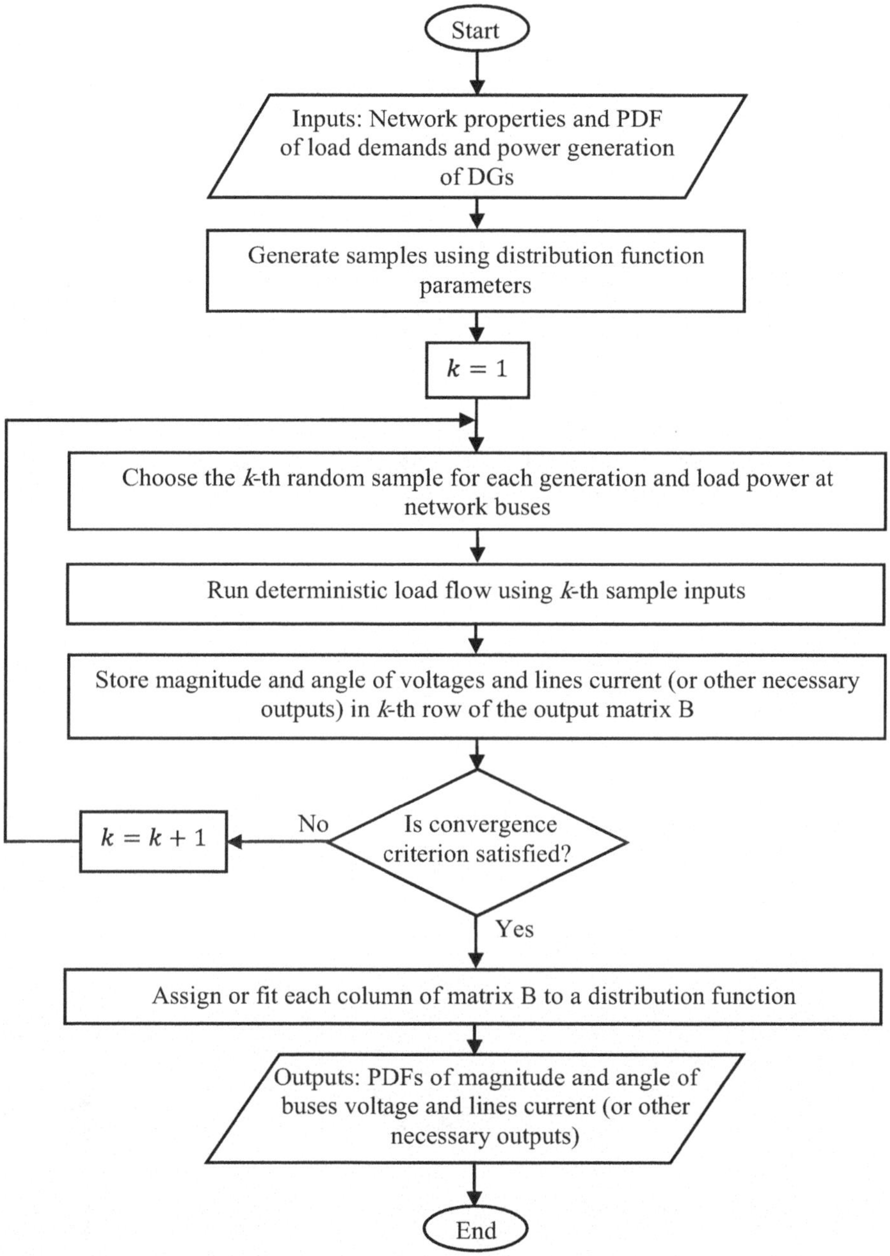

Fig. 7.3 The flowchart of the MCS implementation for PLF analysis

The PDF $f_X(x)$ is constricted in n points i.e. x_i (i = 1, 2, …, n), so that it can be written as

$$f_X(x) = \sum_{i=1}^{n} P_i\,\delta(x - x_i) \tag{7.24}$$

In n-PEM, n points are considered for contraction. The more the number of points, the more accurate the results would be.

As an example, in 2-PEM the PDF $f_X(x)$ is approximated according to

$$f_X(x) \cong P_1\,\delta(x - x_1) + P_2\,\delta(x - x_2) \tag{7.25}$$

This approach is illustrated in Fig. 7.4.

In order to establish the approximation, the first three moments of the actual function and the approximated function should be equal with each other [22].

The jth central moment of $f_X(x)$ is defined as

$$M_j(X) = \int_{-\infty}^{+\infty} (x - \mu_X)^j f_X(x)dx, \quad j = 1, 2, \ldots \tag{7.26}$$

To normalize the approximated function, variable ξ_i is defined as

$$\xi_i = \frac{x_i - \mu_X}{\sigma_X}, \quad i = 1, 2 \tag{7.27}$$

The moments of $f_X(x)$ are then

$$M_0 = P_1 + P_2 = 1 \tag{7.28}$$

$$M_1 = \xi_1 P_1 - \xi_2 P_2 = 0 \tag{7.29}$$

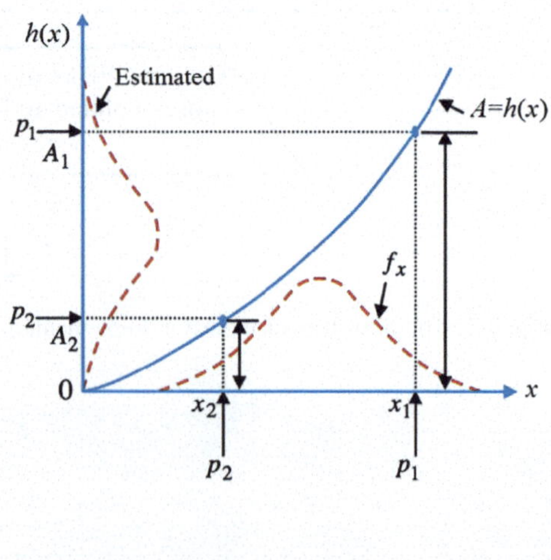

Fig. 7.4 The 2-PEM approach for PLF analysis

$$M_2 = \sigma_X^2(\xi_1^2 P_1 + \xi_2^2 P_2) = \sigma_X^2 \tag{7.30}$$

$$M_3 = \sigma_X^3(\xi_1^3 P_1 - \xi_2^3 P_2) = \lambda_X \sigma_X^3 \tag{7.31}$$

Using the Taylor series expansion of $h(X)$ about μ_X, and considering the approximation shown in (7.25), one gets [22]

$$P_1 + P_2 = M_0(X) = 1 \tag{7.32}$$

$$P_1\xi_1 + P_2\xi_2 = \frac{M_1(X)}{\sigma_X} = \lambda_{X,1} \tag{7.33}$$

$$P_1\xi_1^2 + P_2\xi_2^2 = \frac{M_2(X)}{\sigma_X^2} = \lambda_{X,2} \tag{7.34}$$

$$P_1\xi_1^3 + P_2\xi_2^3 = \frac{M_3(X)}{\sigma_X^3} = \lambda_{X,3} \tag{7.35}$$

Solving the system of four equations with four unknowns results in

$$\xi_1 = \frac{\lambda_{X,3}}{2} + \sqrt{1 + (\frac{\lambda_{X,3}}{2})^2} \tag{7.36}$$

$$\xi_2 = \frac{\lambda_{X,3}}{2} - \sqrt{1 + (\frac{\lambda_{X,3}}{2})^2} \tag{7.37}$$

$$P_1 = -\frac{\frac{\lambda_{X,3}}{2} - \sqrt{1 + (\frac{\lambda_{X,3}}{2})^2}}{2\sqrt{1 + (\frac{\lambda_{X,3}}{2})^2}} \tag{7.38}$$

$$P_2 = \frac{\frac{\lambda_{X,3}}{2} + \sqrt{1 + (\frac{\lambda_{X,3}}{2})^2}}{2\sqrt{1 + (\frac{\lambda_{X,3}}{2})^2}} \tag{7.39}$$

The locations of x_i (i = 1,2) are calculated using (7.27). Finally the kth moment of Y is determined as

$$\mathrm{E}(Y^k) \cong P_1[h(x_1)]^k + P_2[h(x_2)]^k \tag{7.40}$$

The mean value and variance of Y are obtained from

$$\mathrm{E}(Y) \cong P_1 h(x_1) + P_2 h(x_2) \tag{7.41}$$

$$\sigma_Y^2 = \mathrm{E}(Y^2) - [\mathrm{E}(Y)]^2 \tag{7.42}$$

Similarly, in n-PEM ($n > 2$), the locations x_i ($i > 2$) and the corresponding probabilities P_i ($i > 2$) are estimable. Then the kth moment of Y is calculated as

$$\mathrm{E}(Y^k) \cong \sum_{i=1}^{n} \left\{ P_i [h(x_i)]^k \right\} \tag{7.43}$$

The PDF of the output stochastic variables are extracted using the moments of Y. The maximum entropy and gram-charlier are the most well-known methods employed for this estimation [23]. The flowchart of 2-PEM application is represented in Fig. 7.5.

7.2.2.4 Possibilistic OPF

Another POPF i.e. possibilistic OPF is based on fuzzy load flow (FLF) analysis. It is useful when there is no exact model or precise information about the uncertain parameters. Implementation of the FLF is simpler than PLF analysis. Similar to the probabilistic OPF, the possibilistic OPF can be handled using master/slave or OPF-based strategies, however, the FLF layer is used instead of the PLF layer.

The FLF analysis is based on the conventional load flow methods e.g. backward/forward sweep load flow, but the input and output variables are fuzzy values. Hence it is required to use fuzzy operators in computations [8]. The fuzzy values can be defined using several types of membership functions. The triangular and trapezoidal membership functions are the most common functions which has been widely used in FLF analyses. These membership functions are shown in Fig. 7.6.

A triangular fuzzy value is introduced using three parameters i.e. f_1, f_2 and f_3.

If $\widetilde{A} = (a_1, a_2, a_3)$ and $\widetilde{B} = (b_1, b_2, b_3)$ are two fuzzy numbers, then the main fuzzy operators are defined as follow [8]:

$$\widetilde{A} \pm \widetilde{B} = (a_1 \pm b_1, a_2 \pm b_2, a_3 \pm b_3) \tag{7.44}$$

$$\widetilde{A} \,.\, \widetilde{B} = (a_1 b_1, a_2 b_2, a_3 b_3) \tag{7.45}$$

$$\widetilde{A} / \widetilde{B} = (a_1/b_3, a_2/b_2, a_3/b_1) \tag{7.46}$$

The trapezoidal fuzzy values are shown using four parameters i.e. l_1, l_2, l_3 and l_4.

Suppose that $\widetilde{C} = [c_1, c_2, c_3, c_4]$ and $\widetilde{D} = [d_1, d_2, d_3, d_4]$ are two trapezoidal fuzzy numbers, then the main fuzzy operators are defined as follow [24]:

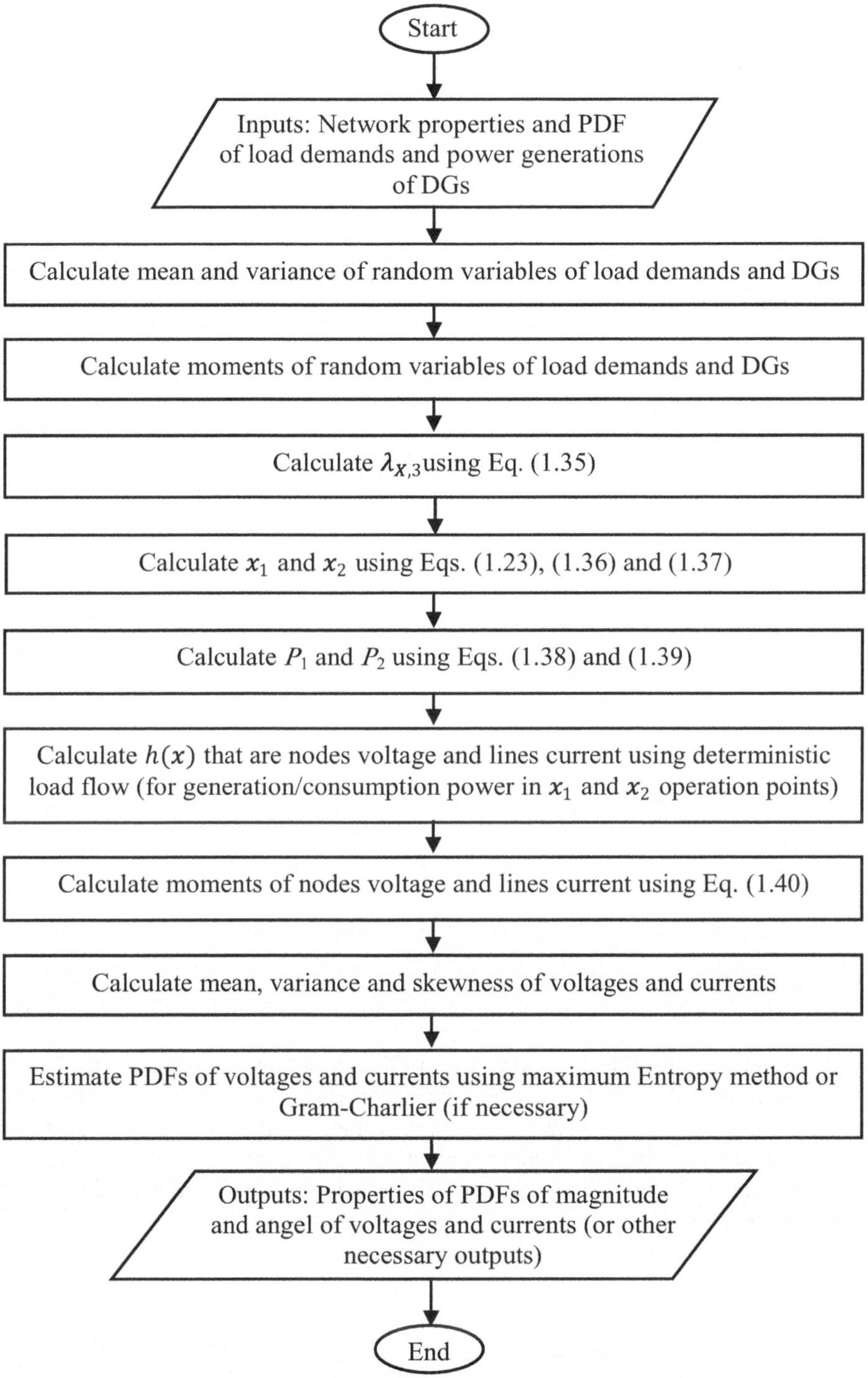

Fig. 7.5 The flowchart of 2-PEM implementation for PLF analysis

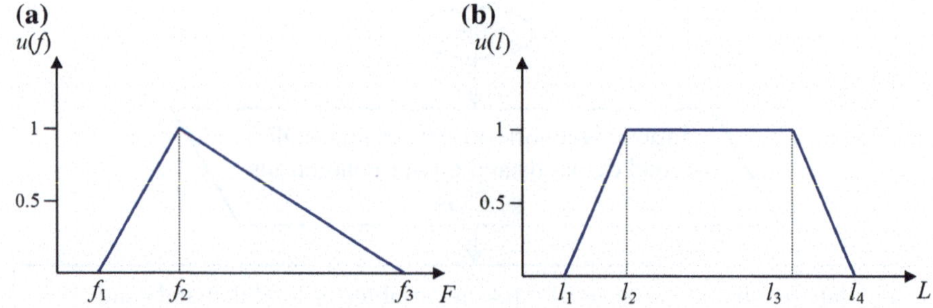

Fig. 7.6 **a** The triangular and **b** the trapezoidal membership functions used in FLF analysis

Fig. 7.7 Multi-linearization scheme of membership function for FLF analysis

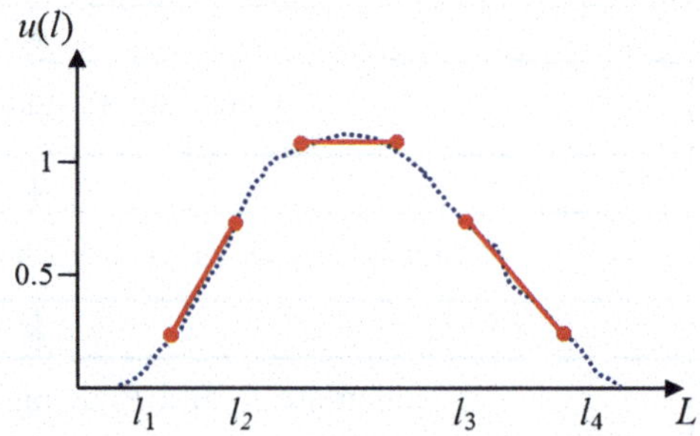

$$\widetilde{C} \pm \widetilde{D} = [c_1 \pm d_1,\ c_2 \pm d_2,\ c_3 \pm d_3,\ c_4 \pm d_4] \tag{7.47}$$

$$\begin{aligned} \widetilde{C}.\widetilde{D} = [&\min(c_1d_1, c_1d_4, c_4d_1, c_4d_4), \min(c_2d_2, c_2d_3, c_3d_2, c_3d_3), \\ &\max(c_2d_2, c_2d_3, c_3d_2, c_3d_3), \max(c_1d_1, c_1d_4, c_4d_1, c_4d_4)] \end{aligned} \tag{7.48}$$

$$\begin{aligned} \widetilde{C}/\widetilde{D} = [&\min(c_1/d_1, c_1/d_4, c_4/d_1, c_4/d_4),\ \min(c_2/d_2, c_2/d_3, c_3/d_2, c_3/d_3), \\ &\max(c_2/d_2, c_2/d_3, c_3/d_2, c_3/d_3),\ \max(c_1/d_1, c_1/d_4, c_4/d_1, c_4/d_4)] \end{aligned} \tag{7.49}$$

When the membership functions are nonlinear and complex, the computation effort increases. Therefore the membership functions are simplified using linearization techniques. It is usually preferred to linearize the membership functions about several points, but not about a single point [25]. In this way the simplification becomes more accurate. Figure 7.7 represents a multi-linearization scheme for a fuzzy membership function.

7.3 Effective Components

As shown in Fig. 7.1, there are several components that affect the optimal BES scheduling and planning. These components provide opportunities and threats for battery storage and they influence the penetration of battery units in active distribution networks. They are described and analyzed below.

7.3.1 Conventional DGs

Conventional dispatchable DGs such as micro-turbines and diesel generators can be used for several purposes such as peak shaving, reliability enhancement and voltage regulation. In fact, they are rivals for storage units in grid-connected as well as stand-alone electric systems. However, they are basically different technologies with various costs and benefits. In active distribution networks, the storage units usually absorb active power in light-load times and they inject the power to the grid in peak-load periods. While the conventional DGs are used only to inject the power in normal-load and/or peak-load times [26–28]. So storage units are more flexible to smooth the load profile. Moreover, in presence of renewable-based DGs, the excess energy can be stored in storage units, while without storage it should be injected to the upstream network. In such times, the electrical energy price may be low and it is not optimal to sell the energy to the grid.

In stand-alone systems, diesel generators are widely used to meet load and generation especially in peak-load periods. If the renewable-based DGs are employed in such systems, the excess energy should be curtailed in off-peak times when the load is less than the generated power. So in this case, the batteries are much more beneficial [29].

Using the BES usually increases the power loss in distribution network due to the energy storage efficiency which is less than 100% in actual conditions [8]. However, the cost of power loss is reduced thanks to the cheap electrical energy in light-load periods [17].

In the competition between the BES and conventional DGs, the most important factors are the investment cost, fuel price and the electrical energy price in light-load times. The energy storage is an expensive technology with higher cost for installation/replacement. However, it is more economical than DGs if the fuel price increases or the electrical energy price reduces in light-load times.

Finally, from the environmental point of view, employing BES instead of conventional DGs reduces the pollution. The reason is that it reduces the required capacity of the fossil fuel power plants and provides the techno-economic conditions for increasing the penetration of renewable energy.

7.3.2 Renewable-Based DGs

Renewable energies such as wind and solar are intermittent resources which cannot be dispatched without energy storage units. The combination of renewable-based DG and BES is very interesting for stand-alone electric systems in remote areas. Additionally it can be beneficial in grid-connected systems for peak shaving, reliability enhancement, voltage regulation and dispatching the intermittent power. The required capacity of storage increases as the penetration of renewable energy increases in active distribution networks [9, 17]. Traditionally using the BES in grid-connected systems is taken into attention if the penetration of renewable-based DGs is high [15, 30–32]. The reason is that the energy storage technology is very expensive. In distribution networks without intermittent energy resources, it is not economical to install BES units, unless they are used for several objectives simultaneously [27].

7.3.3 Plug-in Electric Vehicles (PEVs)

The PEVs can be connected to distribution networks in grid-to-vehicle (G2V) or vehicle-to-grid (V2G) modes of operation. In uncoordinated G2V manner, they are appeared as new uncertain loads, so that more storage units are required to support the grid. However, less capacity of storage is needed, if the PEVs are coordinated in G2V mode of operation. The reason is that the coordinated PEVs are usually charged in light-load periods when the storage units should be charged as well. So the grid capacity should be optimally apportioned between the PEVs and storage. In the case of smart V2G, the PEVs can play the role of storage in distribution network and the storage application is challenged critically.

In near future, both G2V and V2G modes of operation are possible, as a part of vehicles are plugged into the grid in uncoordinated manner [28]. Hence, in order to reinforce the network, the BES units should be employed beside the DG resources, as a global optimal solution [8]. In these situations, the share of BES and DG is a function of PEVs penetration as shown typically in Fig. 7.8.

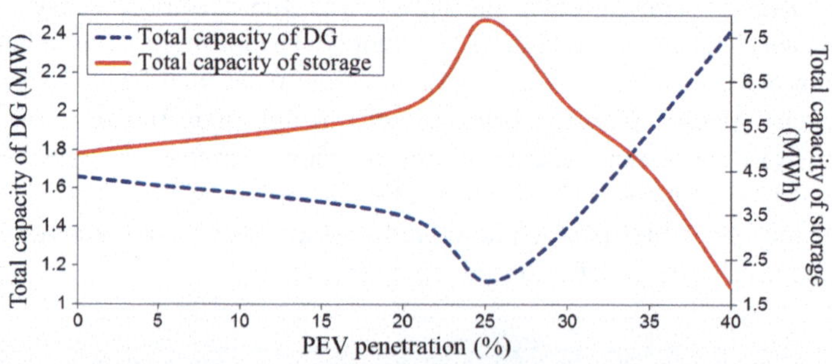

Fig. 7.8 The optimal portion of BES and DG as a function of PEVs penetration [33]

As can be seen, the optimal capacity of BES and DG is a nonlinear function of the PEVs penetration. With increasing the penetration of PEVs up to 25%, the capacity of storage should be increased, and after that, the storage capacity is reduced. The growing of BES and DG capacity are in reverse with each other. The reason is that the V2G (with a high penetration) can treat as stationary storage unit, and hence, more DG capacity (instead of storage) is required to support the grid against those PEVs that are operated in uncoordinated manner.

In addition, the uncertainty behaviour of PEVs is another effective factor which influences the optimal capacity of storage units. Let's consider the following two scenarios:

- scenario-A: Optimal planning of BES considering the deterministic model of PEVs; and,
- scenario-B: Optimal planning of BES considering the uncertain model of PEVs.

The optimal capacity of storage in scenario-B is more than that in scenario-A [8, 33]. So the uncertain manner of PEVs results in more required BES capacity in active distribution network.

7.3.4 Tap-Changer Equipped Transformers

The BES units can be used to inject the power to the upstream grid and sell the electrical energy if it is beneficial considering the electricity price profile. However, reverse power to high voltage (HV) network results in raising the voltage level in medium voltage (MV) or low voltage (LV) feeders. Therefore the number or capacity of BES units is limited due to the violation of technical constraints [34]. In this state, absorbing more reactive power in BES units increases the power rating cost (but not necessarily the capacity cost) and it is not efficient enough in reverse power flow conditions. Reducing the tap position of HV/MV transformers, which are equipped to on-load tap-changer, is more effective to decrease the voltage level of the grid. As a result, the penetration of BES units is allowed to be increased thanks to regulating the tap position. The on-load tap changer and the BES units embedded in distribution grid should be operated in coordinated manner.

It is notable that using tap changer for voltage regulation reduces the lifetime of the HV/MV transformer, and hence, the reliability of the system decreases [35]. The failure event, due to the unsuccessful change of tap position, causes a wide outage in distribution network.

7.3.5 Capacitor Banks

Capacitor banks have been used widely in passive distribution networks to increase the voltage level of long MV feeders. However, in modern distribution networks containing DG sources and flexible active/reactive power flow capability of storage units, it may not be needed to install capacitor banks seriously. Without capacitor banks, more storage capacity is required and the cost function increases [35]. Consequently the combination of storage and capacitor is more beneficial. In this case, the optimal siting and sizing of both technologies i.e. the BES units and capacitors, should be performed simultaneously.

7.3.6 Battery Capabilities

BES units have two important capabilities which are very helpful for distribution grids. These capabilities influence not only the conditions of the network, but also the optimal operation, siting and sizing of storage units. These effects are presented below.

7.3.6.1 Reactive Power

The grid-scale BES units can inject/absorb not only the active power, but also the reactive power. So they can be operated in four zones of active-reactive power surface [36]. However, the electrical current of the battery and the converter which connect the dc source to the ac grid, is limited. The apparent current has two components on the real and imaginary axes, as shown in Fig. 7.9.

The apparent current can be stated as

$$I_{st}^2 = I_{st,re}^2 + I_{st,im}^2 \tag{7.50}$$

For a given voltage, the allowed active-reactive power zone where the storage can be operated, is inside a circle with the radius of apparent power, as shown in Fig. 7.10.

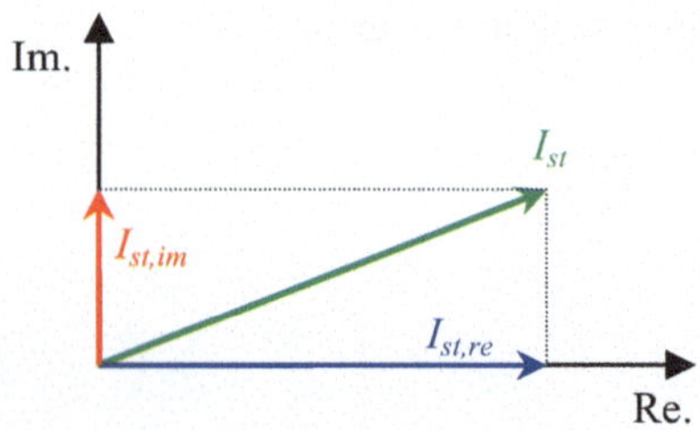

Fig. 7.9 The vector of storage current (I_{st}) and its components on real (Re.) and imaginary (Im.) axes

Fig. 7.10 The active/reactive power limit in a BES unit

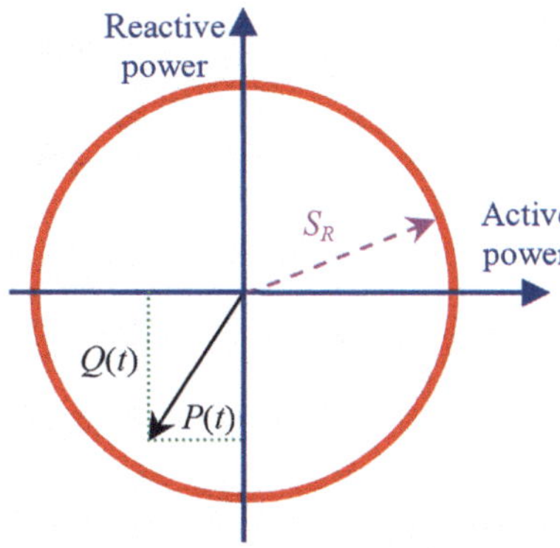

According to this figure, the maximum allowed value of $Q(t)$ is a function of $P(t)$.

If the BES is not allowed to inject/absorb reactive power, it is forced to regulate the voltage only using the active power injection/absorption. As a result, more number or capacity of BES is necessary [17]. So utilizing the reactive power capability reduces the required capacity of storage, and hence, the investment cost decreases.

The BES units that are allowed to exchange reactive power with the network, are preferred to be located near the weak nodes from voltage level point of view. In such nodes, the variance of voltage is usually high, and the flexible active/reactive capability of BES improves the voltage regulation.

7.3.6.2 Islanding Mode Operation

The distributed energy storage units are capable to reduce the ENS within a distribution grid. When a device e.g. a feeder section or transformer is failed, it is disconnected from the grid using normally closed switches (NCSs). In this case, the BES unit can be employed to restore a part of the load in islanding mode of operation, as shown in Fig. 7.11.

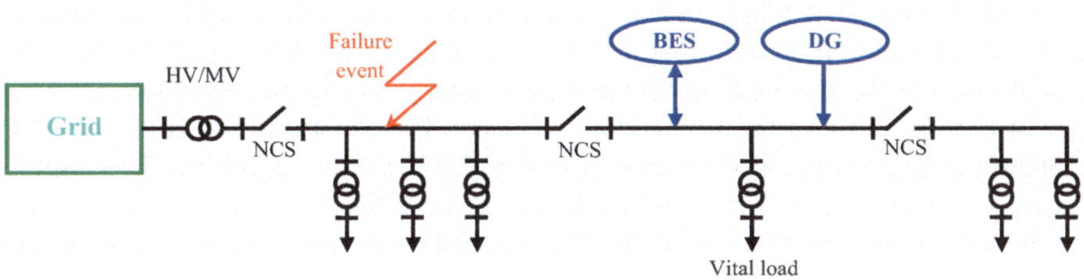

Fig. 7.11 Islanding mode operation of storage in order to reduce ENS cost

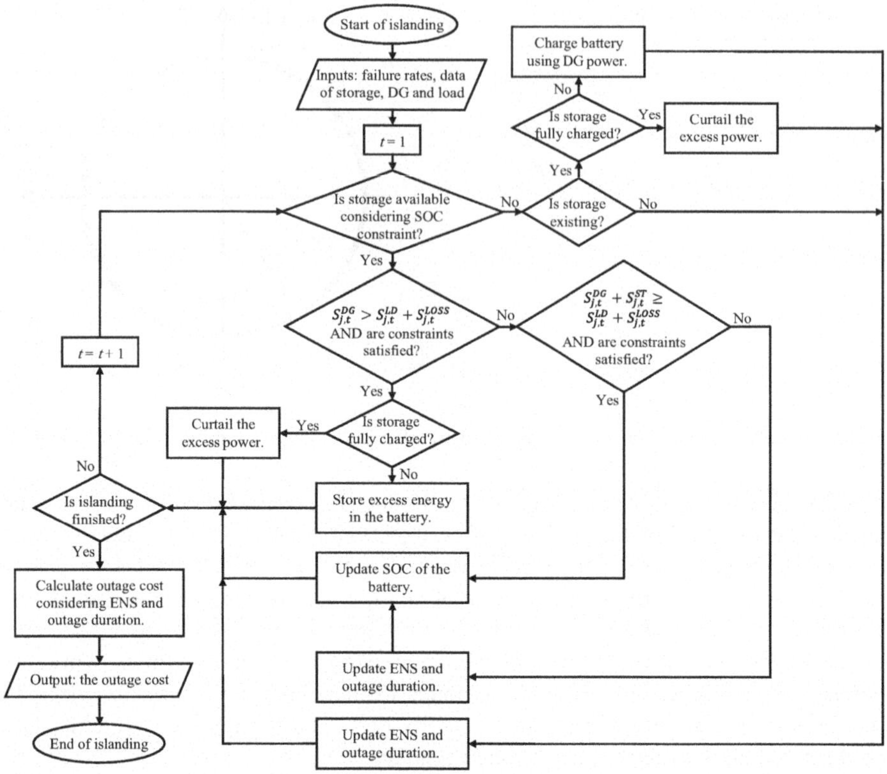

Fig. 7.12 The flowchart of BES operation algorithm in islanding mode with renewable-based DG (in *j*th node of the grid)

The islanding mode operation of BES may take place for several hours during which the islanding mode algorithm is used for operation. The goal of algorithm is to minimize the outage cost. If the storage SOC is not enough to restore the vital load, DG sources inject the power to the islanded area. Due to their intermittency, the renewable-based DGs cannot restore the loads alone. Therefore they should be coordinated with the BES units if possible. The excess energy of the DG sources can be used to charge the BES, so that the battery is sustained to feed the loads in the next hours of islanding operation. If the battery is fully charged, the excess energy of DG should be curtailed. All the technical constraints as well as the SOC limitations must be satisfied within the islanding mode operation, otherwise the load cannot be restored and the ENS is not reduced. The algorithm which is used for islanding mode of operation is shown in Fig. 7.12. In this algorithm the solution feasibility in contingent cases and adequacy of installed devices in system's contingencies are evaluated to consider the cost of ENS properly in the planning algorithm.

The capability of storage to reduce the ENS, influences both the optimal siting and sizing of battery. The ENS cost is a function of failure rate and load type. The storage is preferred to be placed at the end of long feeders where the outage is more probable compared with the other locations. On the other hand, the outage cost in commercial and industrial nodes is more than that in residential regions. Hence the storage is suggested to be installed near the vital loads. Consequently the final decision should be taken considering all the effective factors.

When the islanding mode operation is taken into account in addition to the other objectives e.g. peak shaving and voltage regulation, the optimal capacity of battery increases and the cost function is reduced [17]. As a result, the application of BES is justified and it becomes more beneficial. It should be noted that using BES in grid-connected systems is not appropriate from economic perspectives, if it is employed only for reliability enhancement in islanding mode of operation [27].

When the reliability enhancement is considered as an objective beside the other objectives e.g. peak shaving, the optimal power flow of the battery is changed as well. Consider the following scenarios:

- scenario-A: Charge/discharge battery optimally only for peak shaving; and,
- scenario-B: Charge/discharge battery optimally not only for peak shaving, but also for reliability improvement.

In scenario-A the battery is charged/discharged to minimize the cost of purchased energy and power loss. In scenario-B, it is tried not only to minimize the cost of purchased energy and power loss, but also to increase the average SOC as much as possible. In the latter strategy, more back-up energy is available to support the grid against the probable outages. For example, Fig. 7.13 represents the average SOC of a typical battery in two scenarios.

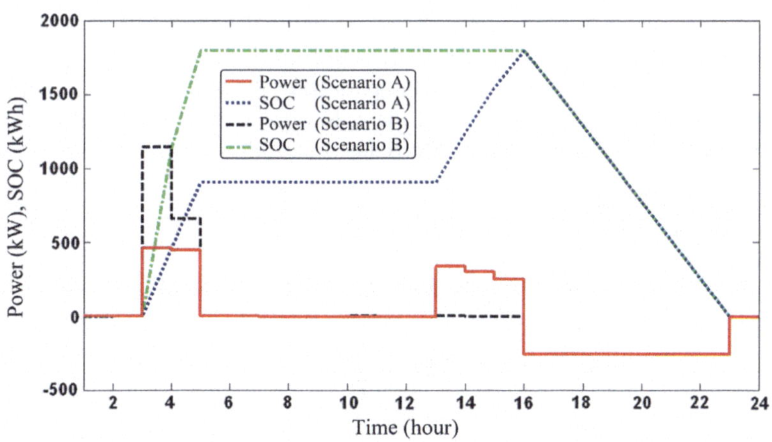

Fig. 7.13 A typical charge/discharge power and SOC of a battery in two scenarios

As can be seen, the battery is charged faster in scenario-B, and hence, the surface below the SOC profile in scenario-B is more than that in scenario-A. Therefore more back-up energy is available in scenario-B.

7.3.7 Techno-economic Factors

There are some important factors that influence the optimal planning of BES units in grid-connected systems. These factors are summarized as follow:

- Installation cost;
- Durability;
- Efficiency;
- Electricity price; and,
- Environmental impact.

The installation cost is the most important factor that determines the usefulness of the grid-scale batteries. It is an uncertain parameter in real conditions and also in near future. Unlike the uncertain conditions of the other effective components e.g. the power of renewable-based DGs, the uncertainty of the investment cost reduces the optimal capacity of batteries [9]. In other words, the uncertain economic conditions challenge the justification of BES. The durability of battery is usually corresponded to the investment cost, as it affects the replacement cost. Development of battery technology results in more durability of the energy storage. So the penetration of batteries increases in the grid-connected systems as well as the stand-alone electric systems.

The storage efficiency is the overall efficiency of the battery and converter. It is an important parameter which determines the power loss of energy storage. However, the impact of storage efficiency on optimal planning is less significant than the installation cost and the durability effects [9].

The electricity price is a very important parameter that makes the BES applicable or inapplicable from economic point of view. In order to characterize the impact of electricity price, two parameters are introduced as follow:

- Price factor (PRF); and,
- Average price $(\overline{P})$.

The price factor and the average price are defined in Eqs. (7.51) and (7.52):

$$PRF = \frac{\overline{C}_e^{PL}}{\overline{C}_e^{LL}} \tag{7.51}$$

$$\overline{P} = \frac{\sum_{t=1}^{T} C_e(t)}{T_1} \tag{7.52}$$

If the battery is used only for peak shaving, *PRF* and $\overline{P}$ should be large enough to justify the storage application in distribution grid. The optimal capacity of battery rises as the price factor and average price increase.

The environmental impact of batteries is the other factor that influences the penetration of BES in distribution networks. The grid-scale batteries become more popular if they address the environmental concerns.

The impact of effective components on BES are summarized in Table 7.1.

Table 7.1 The impact of effective components on BES

Effective component/factor		Influence on BES	Description
Increasing conventional DGs		Reducing the optimal penetration of BES	Dispatchable DGs are competitors for BES
Increasing renewable-based DGs		Increasing the optimal penetration of BES	BES makes the intermittent power of DGs dispatchable to support the grid
Increasing PEVs penetration	Uncoordinated G2V	Increasing the optimal penetration of BES	BES is required seriously
	Coordinated G2V	Reducing the optimal penetration of BES	Coordinated G2V reduces the light-load times
	Smart V2G	Reducing the optimal penetration of BES	Smart V2G is a serious competitors for BES
	Increasing uncertainty of PEVs behavior	Increasing the optimal penetration of BES	More capacity of BES is required in uncertain technical conditions
On-load tap changer availability		Increasing the optimal penetration of BES	Being more important in reverse power conditions
Using capacitor banks		Reducing the optimal penetration of BES	Preventing excess investment cost
Reactive power capability		Reducing the optimal capacity of BES	Preventing excess investment cost
Islanding mode capability			Also effective on the optimal location of BES considerably

(continued)

Table 7.1 (continued)

Effective component/factor		Influence on BES	Description
		Increasing the optimal capacity of BES	
Investment cost	Installation cost increment	Reducing the optimal penetration of BES	Being much more important than BES efficiency
	Increasing uncertainty	Reducing the optimal penetration of BES	Indicating the uncertain economic conditions
Durability increment		Increasing the optimal penetration of BES	Reducing the replacement cost
Efficiency increment		Reducing the power loss of BES	Not very effective on BES penetration compared with the other factors e.g. installation cost and durability
Electricity price	Average price increment	Increasing the optimal penetration of BES	The BES penetration increases if both of them increase
	Price factor increment		
Environmental concerns increment		Increasing the optimal penetration of BES	Depending on the environmental impacts of BES and renewable energy penetration growth

Appendix

The notations used throughout this chapter are listed below:

f objective function

IC investment cost

OC operation cost

RC reliability cost

PF penalty factor

M a large fixed value

x_b binary decision variable associated to installation of the bth battery unit

n_N number of all the nodes in distribution network

C_{Cap}^{INS} the installation cost related to the battery capacity

C_{Pow}^{INS} the installation cost related to the battery power rating

C_{Batt}^{REP} the replacement cost of batteries

C_E^{AR}	the cost of purchased active/reactive power at the high voltage/medium voltage (HV/MV) substations
T	period of project
C_{Batt}^{OM}	the O&M cost of batteries
C_{IN}^{ST}	the income of the storage owner
C_{COS}^{ST}	the cost of the storage owner
C_{Sel}^{ST}	the cost of power which the storage injects to the grid
C_{Pur}^{ST}	the cost of power that the storage absorbs from the grid
$C_{withoutST}^{DC}$	the cost function of the DisCo when no storage unit is installed in the network
C_{withST}^{DC}	the cost function of DisCo after installing the storage units
$C_{E,WithoutST}^{AR}$	the cost of purchased active/reactive power at the HV/MV substations without any embedded storage unit
$RC_{withoutST}$	the reliability cost without storage
$C_{E,WithST}^{AR}$	the cost of purchased active/reactive power at the HV/MV substations with embedded storage units
RC_{withST}	the reliability cost with storage
$C_{E,ST}^{DC}$	the cost of purchased power from storage units
F	the objective function of the multistage expansion planning
f_s	the objective function of the sth stage
N	number of all the stages
$x_{s,b}$	binary decision variable associated to installation of the bth battery unit at sth stage
T_s	duration of the sth stage
$C_{E,s}^{AR}$	the cost of purchased active/reactive power in HV/MV substations at sth stage
$V_{n,t}$	the voltage magnitude in nth node of the network at tth hour
$V_{\min}$	the allowed minimum voltage magnitude
$V_{\max}$	the allowed maximum voltage magnitude
S_t^u	the power of equipment u at tth hour
$S_{\min}^u$	the allowed minimum power of equipment u
$S_{\max}^u$	the allowed maximum power of equipment u
$S_{s,t}^{SS}$	the power of the sth HV/MV substation at tth hour
$S_{d,t}^{DG}$	the generated power of dth DG unit at tth hour
$S_{k,t}^{ST}$	the charge power of the kth storage unit at tth hour
$S_{l,t}^{LD}$	the load power of the lth node at tth hour
S_t^{LOSS}	the total power loss in distribution network at tth hour
$\mathrm{A_{EQ}}$	the set of all the equipment
$\mathrm{A_N}$	the set of all the nodes
$\mathrm{A_T}$	the set of all the hours
$\mathrm{A_{ST}}$	the set of all the installed storage units
$\mathrm{A_{LD}}$	the set of all the load nodes
$SOC_{k,t}$	the SOC of the kth storage unit at time t

$P_{k,t}^{ST}$	the active power of the kth storage unit at time t
η_{ch}	the charge efficiency of the storage units
η_{dis}	the discharge efficiency of the storage units
$p(v)$	the probability of the vth violation
n_{VIO}	the number of all probable violations in a candidate solution
$S_{Cap}(n)$	the battery capacity in nth node
$P_{Rat}(n)$	the power rating of storage in nth node
$SOC(t, n)$	the SOC of battery in nth node at time t
$P(t, n)$	the active power of the battery in nth node at time t
$Q(t, n)$	the reactive power of the battery in nth node at time t
P_p	the active power of pth node
Q_p	the reactive power of pth node
V_p	the voltage magnitude in pth node
δ_{pq}	the angle between the voltages of the nodes p and q
G_{pq}	the real elements of the grid's admittance matrix
B_{pq}	the imaginary elements of the grid's admittance matrix
P_i	the corresponding probability of x_i
$\delta(.)$	Dirac's delta function
μ_X	the mean value of X
σ_X	the standard deviation of X
λ_X	the skewness of X
$I_{st,re}$	the real component of I_{st}
$I_{st,im}$	the imaginary component of I_{st}
S_R	the apparent power of storage
$P(t)$	the active power of storage at time t
$Q(t)$	reactive power of storage at time t
$\overline{C}_e^{PL}$	the average electricity price in peak-load period
$\overline{C}_e^{LL}$	the average electricity price in light-load times
$C_e(t)$	the electricity price at tth hour
T_1	the number of hours over one year

References

1. P. Poonpun, W.T. Jewell, Analysis of the cost per kilowatt hour to store electricity. IEEE Trans. Energy Conv. **23**(2), 529–534 (2008)
2. E. Naderi, I. Kiaei, M.R. Haghifam, NaS technology allocation for improving reliability of DG-enhanced distribution networks, in *Proceedings of IEEE International Conference on Probabilistic Methods Applied to Power Systems*, Singapore (2010), pp. 148–153
3. M.N. Kabir, Y. Mishra, G. Ledwich, Z. Xu, R.C. Bansal, Improving voltage profile of residential distribution systems using rooftop PVs and battery energy storage systems. Appl. Energy **134**, 290–300 (2014)
4. B. Zakeri, S. Sanna, Electrical energy storage systems: a comparative life cycle cost analysis. Renew. Sustain. Energy Rev. **42**, 569–596 (2015)

5. X. Luo, J. Wang, M. Dooner, J. Clarke, Overview of current development in electrical energy storage technologies and the application potential in power system operation. Appl. Energy **137**, 511–536 (2015)
6. Technology Roadmap Energy storage, [Online] Available: https://www.iea.org/publications
7. M. Daghi, M. Sedghi, M. Aliakbar-Golkar, Optimal battery planning in grid connected distributed generation systems considering different technologies, in *Proceedings of 20th Iranian Electrical Power Distribution Conference*, Zahedan, Iran (2015), pp. 138–142
8. A. Ahmadian, M. Sedghi, M. Aliakbar-Golkar, Fuzzy load modeling of plug-in electric vehicles for optimal storage and DG planning in active distribution network. IEEE Trans. Veh. Technol. **66**(5), 3622–3631 (2017)
9. M. Daghi, M. Sedghi, A. Ahmadian, M. Aliakbar-Golkar, Factor analysis based optimal storage planning in active distribution network considering different battery technologies. Appl. Energy **183**, 456–469 (2016)
10. R. Anindita, S.B. Kedare, S. Bandyopadhyay, Optimum sizing of wind-battery systems incorporating resource uncertainty. Appl. Energy **87**, 2712–2727 (2010)
11. O. Ekren, B.Y. Ekren, Size optimization of a PV/wind hybrid energy conversion system with battery storage using simulated annealing. Appl. Energy **87**, 92–98 (2010)
12. J.M. Lujano-Rojas, R. Dufo-Lopez, J.L. Bernal-Agustin, Optimal sizing of small wind/battery systems considering the DC bus voltage stability effect on energy capture, wind speed variability, and load uncertainty. Appl. Energy **93**, 404–412 (2012)
13. V. Carpentiero, R. Langella, A. Testa, Hybrid wind-diesel stand-alone system sizing accounting for component expected life and fuel price uncertainty. Electr. Power Syst. Res. **88**, 69–77 (2012)
14. J. Wang, F. Yang, Optimal capacity allocation of standalone wind/solar/battery hybrid power system based on improved particle swarm optimization algorithm. IET Renew. Power Gener. **7**(5), 443–448 (2013)
15. Y.M. Atwa, E.F. El-Saadany, Optimal allocation of ESS in distribution systems with a high penetration of wind energy. IEEE Trans. Power Syst. **25**(4), 1815–1822 (2010)
16. J. Tant, F. Geth, D. Six, P. Tant, J. Driesen, Multiobjective battery storage to improve PV integration in residential distribution grids. IEEE Trans. Sustain. Energy **4**(1), 182–191 (2013)
17. M. Sedghi, A. Ahmadian, M. Aliakbar-Golkar, Optimal storage planning in active distribution network considering uncertainty of wind power distributed generation. IEEE Trans. Power Syst. **31**(1), 304–316 (2016)
18. M.A. Abido, Multiobjective particle swarm optimization for environmental/economic dispatch problem. Elect. Power Syst. Res. **79**(7), 1105–1113 (2009)
19. M. Sedghi, M. Aliakbar-Golkar, Analysis and comparison of load flow methods for distribution networks considering distributed generation. Int. J. Smart Elect. Eng. **1**(1), 27–32 (2012)
20. P. Chen, Z. Chen, B. Bak-Jensen, Probabilistic load flow: a review, in *Proceedings of 3rd International Electric Utility Deregulation and Restructuring and Power technology conference*, Nanjing, China (2008), pp. 1586–1591
21. S. Chun-Lien, Probabilistic load-flow computation using point estimate method. IEEE Trans. Power Syst. **20**(4), 1843–1851 (2005)
22. G. Verbic, C.A. Canizares, Probabilistic optimal power flow in electricity markets based on a two-point estimate method. IEEE Trans. Power Syst. **21**(4), 1883–1893 (2006)
23. T. Williams, C. Crawford, Probabilistic load flow modeling comparing maximum entropy and Gram-Charlier probability density function reconstructions. IEEE Trans. Power Syst. **28**(1), 272–280 (2013)
24. Y. Deng, X. Ren, Fuzzy modeling of capacitor switching for radial distribution systems, in *Proceedings of IEEE Power Engineering Society Winter Meeting*, Columbus, OH, USA (2001), pp. 830–834
25. J. Hao, L. Shi, G. Xu, Y. Xie, Study on the fuzzy AC power flow model, in *Proceedings of 5th World Congress Intelligent Control and Automation*, Hangzhou, China (2004), pp. 5092–5096

26. W. Ouyang, H. Cheng, X. Zhang, L. Yao, Distribution network planning method considering distributed generation for peak cutting. Energy Convers. Manage. **51**(12), 2394–2401 (2010)
27. M. Sedghi, M. Aliakbar-Golkar, M.R. Haghifam, Distribution network expansion considering distributed generation and storage units using modified PSO algorithm. Elect. Power Energy Syst. **52**, 221–230 (2013)
28. M. Sedghi, M. Aliakbar-Golkar, Optimal storage scheduling in distribution network considering fuzzy model of PEVs, in *Proceedings of 18th Conference on Electric Power Distribution* 30 Apr–1 May 2013, pp. 1–6
29. D. Suchitra, R. Jegatheesan, M. Umamaheswara Reddy, T.J. Deepika, Optimal sizing for stand-alone hybrid PV-wind power supply system using PSO, in *Proceedings of International Conference on Swarm, Evolutionary and Memetic Computing* (2013), pp. 617–629
30. M. Ghofrani, A. Arabali, M. Etezadi-Amoli, M.S. Fadali, A framework for optimal placement of energy storage units within a power system with high wind penetration. IEEE Trans. Sustain. Energy **4**(2), 434–442 (2013)
31. M. Ghofrani, A. Arabali, M. Etezadi-Amoli, M.S. Fadali, Energy storage application for performance enhancement of wind integration. IEEE Trans. Power Syst. **28**(4), 4803–4811 (2013)
32. H. Kihara, A. Yokoyama, K.M. Liyanage, H. Sakuma, Optimal placement and control of BESS for a distribution system integrated with PV systems. Int. Council Elect. Eng. **1**(3), 298–303 (2011)
33. M. Sedghi, *Optimal Battery Planning in Active Distribution Networks Considering Plug-in Electric Vehicles Uncertainty*, Ph.D. thesis, K. N. Toosi University of Technonlgy, Tehran, Iran, 2015
34. A. Ahmadian, M. Sedghi, M. Aliakbar-Golkar, A. Elkamel, M. Fowler, Optimal probabilistic based storage planning in tap-changer equipped distribution network including PEVs, capacitor banks and WDGs: a case study for Iran. Energy **112**, 984–997 (2016)
35. O. Anuta, N. Wade, J. McWilliams, Coordinated operation of energy storage and on-load tap changer on a UK 11 kV distribution network, in *Proceedings of 22nd International Conference on Electricity Distribution (CIRED)*, Stockholm, Sweden, June 2013, pp. 1–4
36. A. Gabash, P. Li, Flexible optimal operation of battery storage systems for energy supply networks. IEEE Trans. Power Syst. **28**(3), 2788–2797 (2013)

Chapter 8
Optimal Distributed Generation Placement Problem for Power and Energy Loss Minimization

Aggelos S. Bouhouras, Paschalis A. Gkaidatzis and Dimitris P. Labridis

Abstract This chapter introduces the Optimal Distributed Generation Placement problem towards power and energy loss minimization. Several solving methods are applied in order for the most suitable to emerge. Apart from technical and DG constraints, recent raised issues due to high Distributed Generation penetration like the reverse power flow effect is considered as well. The load and generation variability and their impact in integrating Renewable Energy Sources are examined, aided by the use of Capacity Factors implementation. In addition, the impact of Optimal Distributed Generation Placement problem in conjunction with Network Reconfiguration and Optimal Energy Storage Systems Placement is introduced aiming to examine how joined management schemes could be efficiently combined in order to maximize the potential loss and energy reduction.

Keywords ODGP · Loss minimization · Distributed generation Optimization · Heuristics · Reverse power flow · Capacity factors Load/generation variability · ESS · Network reconfiguration

A. S. Bouhouras (✉) · P. A. Gkaidatzis · D. P. Labridis
School of Electrical and Computer Engineering,
Aristotle University of Thessaloniki, AUTH, Thessaloniki, Greece
e-mail: abouhou@ece.auth.gr; abouchou@teiwm.gr

P. A. Gkaidatzis
e-mail: pgkaidat@ece.auth.gr

D. P. Labridis
e-mail: labridis@auth.gr

A. S. Bouhouras
Department of Electrical Engineering, University of Applied Sciences,
TEIWM, Kozani, Greece

F. Shahnia et al. (eds.), *Electric Distribution Network Planning*,
Power Systems, https://doi.org/10.1007/978-981-10-7056-3_8

8.1 Introduction

The penetration of Distributed Generation (DG) units in Electric Distribution Networks has been considered as an efficient way to exploit the benefits of sustainable energy, promoted by distributed energy resources. In most cases, appropriate consideration of DG installation can highly benefit the electric distribution networks in terms of loss reduction, voltage-profile, and reliability improvement [1, 2]. However, high penetration of DG units could potentially cause problems to several operational characteristics, especially due to reverse power flow, leading to excessive losses and feeder overloading [3, 4]. Regarding DG placement, the final decision lies on the owners and investors, depending on site and fuel availability or climatic conditions. Notwithstanding the merits of installing DG and exploiting it in order to solve networks problems, the fact remains that, in most cases, the Distribution Network Operator (DNO) has neither significant control, nor influence over the DG location and size. Still, DG placement affects critically the operation of a network, below a certain limit. Thus, optimization tools which provide both optimal locations and capacity of DG units to be installed should be highly appreciated by DNOs. The Optimal Distributed Generation Placement (ODGP) problem generally deals with the determination of the location and appropriate sizing of DG units to be installed into existing electric distribution networks, subject to networks' and DG operational, as well as investment constraints.

In this chapter, a comparative analysis of several promising methods is initially presented, such as analytical or heuristic ones and their merits and drawbacks are pointed out, when contemplating power loss reduction via ODGP approach. The DG units are considered capable of both active and reactive power production. Secondly, using the most suitable of them, the ODGP towards *power loss reduction* is solved by taking into consideration a possible reverse power flow effect [5, 6]. Thirdly, as a first step for the integration of an optimal combination of Renewable Energy Resources (RESs) into an electric distribution network, a method is demonstrated, considering concurrently the geographical characteristics of the area, where the examined network is placed, the different weather conditions and the availability of RESs, by the introduction of Capacity Factors (CFs), while trying to keep problem complexity at a minimum.

Furthermore, the ODGP towards *energy loss reduction* is coped, initially taking into account the impact of load composition variation while considering the DG units having constant power output, and then with variable power output resembling the function of several RESs such as Wind Turbines, or Photovoltaics. Finally, the cooperation of ODGP with Network Reconfiguration (NR) towards power loss reduction is presented and an initial effort regarding the cooperation of ODGP with the Optimal Energy Storage System Placement (OESSP) problem.

8.2 ODGP Towards Power Loss Minimization—Problem Formulation

8.2.1 *Objective Function—Constraints*

The ODGP problem is a mixed-integer-non-linear-constrained (MINLC) optimization problem; mixed integer because both the power of the DGs installed (sizing) and their position (siting) are requisites; non-linear, due to the power flow equations needed to solve the problem. As an optimization problem, various objectives can be found in literature, such as cost minimization, benefit maximization, greenhouse gas emission reduction, either solved individually, or as a multi-objective approach [7–9]. In this section, power loss minimization is to be contemplated, formulated as

$$F_{loss} = \min \sum_{k=1}^{n_l} g_{i,j} \left[V_i^2 + V_j^2 - 2V_i V_j \cos(\theta_i - \theta_j) \right] \tag{8.1}$$

where:

$g_{i,j}$ is the conductance between buses i and j, respectively,
n_l is the total number of branches of the network,
V_i, V_j are the voltage magnitudes of buses i and j, respectively, and
θ_i, θ_j are the voltage angles of buses i and j, respectively.

The constraints of the problem can be separated to obligatory and occasional. As *obligatory constraints*, the power flow Eqs. (8.2a), (8.2b) and the technical constraints of the electric distribution network (8.3), (8.4) are considered, as they must always be met. They are expressed as:

Power Flow Constraints:

$$P_{G,i} - P_{D,i} - \sum_{j=1}^{n_b} |V_i||V_j||Y_{i,j}| \cos(\varphi_{i,j} - \theta_i + \theta_j) = 0 \tag{8.2a}$$

$$Q_{G,i} - Q_{D,i} + \sum_{j=1}^{n_b} |V_i||V_j||Y_{i,j}| \sin(\varphi_{i,j} - \theta_i + \theta_j) = 0 \tag{8.2b}$$

DN Contraints:

$$V_i^{min} \leq V_i \leq V_i^{max} \tag{8.3}$$

$$S_k \leq S_k^{max} \tag{8.4}$$

where:

$P_{G,i}$, $Q_{G,i}$	is the active and reactive power generation at bus i, respectively,
$P_{D,i}$, $Q_{D,i}$	is the active and reactive power demand at bus i, respectively,
n_b	is the total number of buses of the network,
$Y_{i,j}$	is the magnitude of bus admittance element i,j,
$\varphi_{i,j}$	is the angle of bus admittance element i,j
V_i^{min}, V_i^{max}	are the voltage lower and upper limits of bus i, respectively, and
S_k^{max}	is the thermal limit of line k, by terms of apparent power.

As *occasional constraints*, technical constraints regarding DG units and/or their penetration level are considered. They are classified as occasional because they can be present on occasion, defined by the aspect of the problem examined and not necessarily mandatory, as the previous ones. They can be expressed as:

DG constraints:

$$S_{min}^{DG} \leq S_m^{DG} \leq S_{max}^{DG} \tag{8.5}$$

$$pf_{min}^{DG} \leq pf_m^{DG} \leq pf_{max}^{DG} \tag{8.6}$$

Penetration Constraints:

$$\sum_{m=1}^{n_{DG}} S_m^{DG} \leq \eta \cdot S_{Total}^{Load} \tag{8.7}$$

where:

S_m^{DG}	is the power of a DG unit,
pf_m^{DG}	is the power factor of a DG unit,
S_{min}^{DG}, S_{max}^{DG}	are the limits of power for a DG unit, respectively,
pf_{min}^{DG}, pf_{max}^{DG}	are the limits of power factor of a DG unit, respectively,
n_{DG}	is the total number of DG units to be installed,
η	is a percentage indicating the desired DG penetration level, and
S_{Total}^{Load}	is the total load installed in the DN.

8.2.2 *Penalty Function—Terms*

In general, constrained problems can be solved using deterministic, or stochastic algorithms. However, deterministic approaches such as feasible direction and generalized gradient descent, require strong mathematical properties of the objective function, such as continuity and differentiability. Moreover, solving the ODGP problem by analytical methods could prove to be complex and time-consuming [10], or be restrained to solutions including only one DG unit. In cases, where these properties are absent, evolutionary computation offers reliable alternative methods. Since most evolutionary approaches were primarily designed to address unconstrained problems, constrained handling techniques are usually required to detect only feasible solutions. The most common of those techniques is the use of a penalty function. In spite of its drawbacks, it performs rather efficiently, provided a proper calibration of the penalty parameters is undertaken [11, 12]. According to this approach, the constraints expressed via penalty terms are incorporated into the objective function in order to formulate the penalty function that penalizes any infeasible solutions as:

$$P(x) = f(x) + \Omega(x) \tag{8.8}$$

$$\Omega(x) = \rho\left\{g^2(x) + [max(0, h(x))]^2\right\} \tag{8.9}$$

where:

- $P(x)$ is the Penalty function,
- $f(x)$ is the objective function, in this case the F_{loss}, as expressed in (8.1),
- $\Omega(x)$ is the penalty term,
- ρ is the penalty factor,
- $g(x)$ refers to the equality constraints, in this case as defined in (8.2a), (8.2b), and
- $h(x)$ refers to the inequality constraints, in this case as defined in (8.3)–(8.7).

Thus, in case of the ODGP problem and using, for the sake of argument, only the obligatory constraints, the updated Penalty Function could be expressed as:

$$P(x) = \min(F_{loss} + \Omega_P + \Omega_Q + \Omega_V + \Omega_L) \tag{8.10}$$

where Ω_P and Ω_Q refer to the equality constraints of

$$\Omega_P = \rho_P \sum_{i=1}^{n_b} \left\{ P_{G,i} - P_{D,i} - \sum_{j=1}^{n_b} |V_i||V_j||Y_{i,j}| \cos(\varphi_{i,j} - \theta_i + \theta_j) \right\} \tag{8.11a}$$

$$\Omega_Q = \rho_Q \sum_{i=1}^{n_b} \left\{ Q_{G,i} - Q_{D,i} + \sum_{j=1}^{n_b} |V_i||V_j||Y_{i,j}| \sin(\varphi_{i,j} - \theta_i + \theta_j) \right\} \tag{8.11b}$$

and Ω_V and Ω_L to inequality constraints of

$$\Omega_V = \rho_V \sum_{i=1}^{n_b} \left\{\max\left(0, V_i^{min} - V_i\right)\right\}^2 + \rho_V \sum_{i=1}^{n_b} \left\{\max\left(0, V_i - V_i^{max}\right)\right\}^2 \tag{8.12}$$

$$\Omega_L = \rho_L \sum_{k=1}^{n_l} \left\{\max\left(0, S_k - S_k^{max}\right)\right\}^2 \tag{8.13}$$

As can be easily deduced, any other constraints such as (8.5), (8.6) or (8.7) can be incorporated in (8.10) via the same process.

8.3 ODGP Towards Power Loss Minimization—Solving Methods

According to a literature survey on the subject, a great amount of scientific research has been undertaken with the aim of solving the ODGP problem [13]. Several promising methods have emerged such as analytical ones [14–17], heuristics [5, 6, 18–33], or combination of the above, solving siting and sizing individually, but in a sequential order [34–36].

However, as stated earlier, ODGP is a MINLC optimization problem. The conventional approaches utilizing analytical methods could be intricate and time-consuming in this case, or restricted to solving for just one DG unit being placed. Therefore, over the last few decades, Heuristics such as Particle Swarm Optimization (PSO) [5, 6, 18], Genetic Algorithm (GA) [9, 19, 20], Artificial Bee Colony (ABC) [21–23], Cuckoo Search (CS) [24–27], and Harmony Search (HS) [28–30] have been implemented. Moreover, they have proved quite promising and still evolving in this field. Some additional mentions could be, for example, Bacterial Foraging Optimization Algorithm (BFOA) [31], Ant-Lion Optimization (ALO) [32], Grey Wolf Optimization (GWO) [33], and many more [13], noting more advancement in solving the ODGP problem. In this section, a small comparative analysis will take place in order to determine the most suitable method to solve the ODGP problem. Three versions of PSO, namely, the Local, Global and Unified PSO, GA, ABC, CS and HS methods is compared and evaluated. As analytical methods, the ones presented in [15], namely, Improved Analytical (IA) method, Loss Sensitivity Factor (LSF) method and Exhaustive Load Flow (ELF) method is also demonstrated.

8.3.1 Analytical Methods

For calculating the losses for those methods, instead of (8.1) the exact loss formula is utilized, expressed as:

$$F_{loss} = \sum_{i=1}^{n_b} \sum_{j=1}^{n_b} \left[\alpha_{ij} \left(P_i P_j + Q_i Q_j \right) + \beta_{ij} \left(Q_i P_j - P_i Q_j \right) \right] \tag{8.14}$$

where:

$$\alpha_{ij} = \frac{r_{ij}}{V_i V_j}\cos(\theta_i - \theta_j), \quad \beta_{ij} = \frac{r_{ij}}{V_i V_j}\sin(\theta_i - \theta_j) \tag{8.15a, b}$$

where:

$r_{ij} + jx_{ij} = Z_{ij}$ is the *ij*th element of the impedance matrix,
P_i and P_j are the active power injections at *i*th and *j*th buses, respectively, and
Q_i and Q_j are the reactive power injections at *i*th and *j*th buses, respectively.

8.3.1.1 IA Method

In IA different formulas are formed according to the DG type to be used, i.e. injecting only active, and/or reactive, or both. The advantage of the method is that load flow is required only twice: once at the initial state of the electric distribution network and once the DG is in place. The drawback is that only a single DG unit is placed at a time.

8.3.1.2 LSF Method

LSF is based on the linearization of the power flow Eqs. (8.2a), (8.2b). It is most appropriate for locating the most suitable buses to host DG units by ranking them according to their LSF values. Then, a DG unit is placed at the bus with the highest priority and its size is calculated by increasing it in small steps and running load flow. The merits of the method are its simplicity and directness. However, as in IA method, only a single DG unit is placed at a time and naturally after the first time, the solution is biased since some DG units have been already installed.

8.3.1.3 ELF Method

ELF method, also known as repeated load flow solution, requires excessive computational time since all buses are considered in calculation; however, it can lead to a completely optimal solution. Also, as the number of DG units to be installed increases to more than one, so does the computational load and indeed does so in an exponential rate.

8.3.2 Heuristic Methods

8.3.2.1 GPSO, LPSO, UPSO Methods

PSO was introduced by Eberhart and Kennedy [37]. It was inspired by the social behavior of bird flocking. A swarm of particles is assigned to explore the solution space in order to retrieve the optimal solution. Their movement in the solution space is defined by three key elements:

1. their personal knowledge of the solution space, represented by the Personal Best parameter,
2. the social knowledge gained by exchanging information among a group of particles, represented by the Social Best parameter, and finally,
3. their current movement on the solution space, represented by the previous gained velocity.

Regarding the Social Best, when the information exchange takes place amongst all particles within the swarm, it is called Global Best, and the respective algorithm Global PSO (GPSO), whereas if it takes place among smaller formations, called neighborhoods, it is called Local Best and the respective algorithm Local PSO (LPSO).

Concerning GPSO, because the particles are instantly aware of the swarm's best position, rapid convergence is achieved, therefore better exploitation of the knowledge gathered regarding the solution space. However, this happens at the expense of exploration of the solution space, thus resulting in probable local minima entrapment and therefore not achieving a near-optimal solution.

In contrast, in LPSO the formation of overlapping particle neighborhoods and the information exchange within them enables for better exploration of the solution space [38]. However, this happens at the expense of exploitation, thus longer convergence, since the information exchange is distilled among the various neighborhoods, instead of the whole swarm.

Therefore, under GPSO, or LPSO, the algorithm is biased towards exploitation, or exploration, respectively. UPSO, introduced by Parsopoulos and Vrahatis [39], has been developed as an attempt to harness their merits and, at the same time, aiming to neutralize their flaws. In this chapter, the UPSO's Swarm Partitioning scheme is applied for merging the two versions of PSO, as the most promising [40]. A generic flowchart for PSO is presented in Fig. 8.1.

8.3.2.2 GA Method

GA was introduced by Holland [41]. A simulation of the three fundamental genetic processes comprises the technique, namely selection, crossover and mutation. A group of chromosomes is designated in the solution space, here considered a genetic pool. The most fitted are selected as parents to form the next generation.

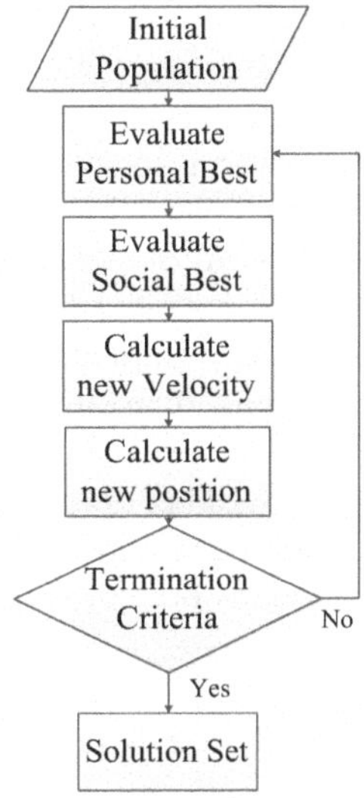

Fig. 8.1 PSO flowchart

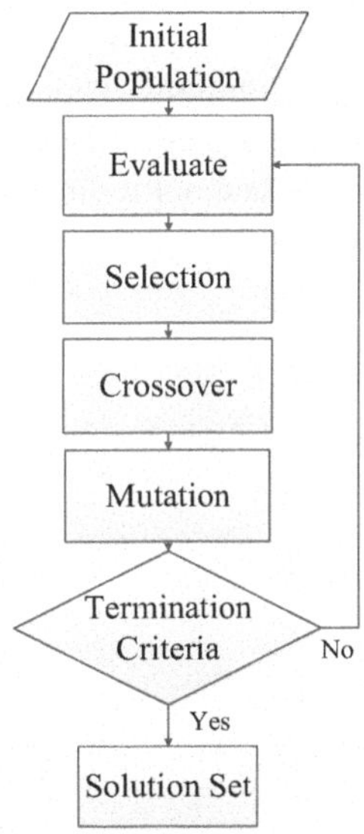

Fig. 8.2 GA flowchart

In this chapter, the roulette-wheel selection scheme is applied. The parents are stochastically combined to breed offspring that bear combinations of their chromosomes. In addition, a mutation process takes place, where stochastically several parts of the offspring's chromosomes are altered. Finally, the best among both parents and offspring are chosen to constitute the next generation of chromosomes, as presented in Fig. 8.2.

8.3.2.3 ABC Method

ABC method was proposed by Karaboga [42]. It was inspired by the intelligent way bee swarms locate and harness their food. The candidate solution space in that case is represented by places of potential food sources. The bee colony, divided into employed, onlooker and scout bees, spreads across it. Employed bees target and exploit potential food position and inform the onlookers for more potential food sites. The employed bees then are trying to determine the food potential of those positions. If an employed bee's position does not represent a good solution, then the bee turns into a scout and starts exploring the solution space. The number of the employed bees is equal to the number of food sources, each of which also represents a site, being exploited at the moment or to the number of solutions in the population, as presented in Fig. 8.3.

8.3.2.4 CS Method

CS method was first introduced by Yang and Deb [43]. It was inspired by the way some cuckoo species lay their eggs in the nests of other host birds (of other species), to be nurtured. Each egg in the nest represents a solution, and cuckoo eggs represent new solutions. The aim is to use the new and potentially better solutions (cuckoos) to replace the least suitable solutions in the nests. In each iteration, one cuckoo egg is laid randomly in a selected nest; The nests with high quality eggs will carry over to the nest generation; Then, in the remaining least suitable nests, a discovery

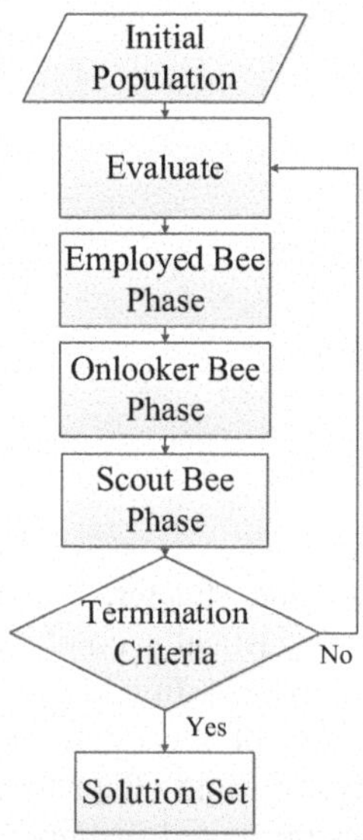

Fig. 8.3 ABC flowchart

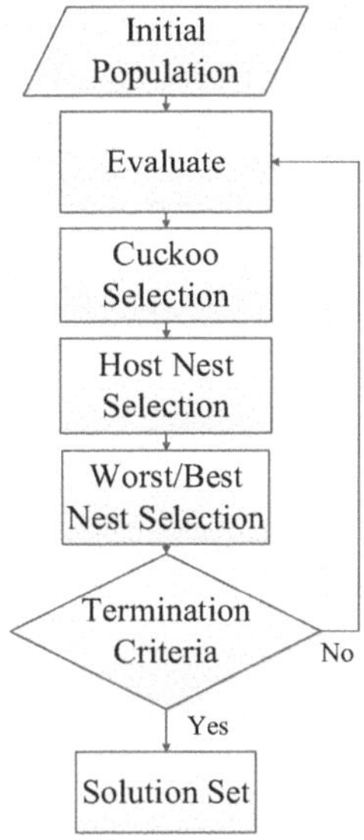

Fig. 8.4 CS flowchart

operation takes place by the host birds, stochastically retrieving cuckoo laid eggs and discarding them, therefore ignoring them from further calculations. A generic flowchart is presented in Fig. 8.4.

8.3.2.5 HS

HS method, introduced in [44], is inspired by the improvisation process of jazz musicians. Improvisation is a process of searching for the most appropriate harmony by trying various combinations of rhythms, under the following three rules:

1. playing any existing rhythm from the memory;
2. playing an altered rhythm from the memory;
3. playing a random rhythm from the possible range.

HS simulates this procedure as:

1. choosing any value from the HS memory;
2. choosing an altered value from the HS memory;
3. choosing a random value from the possible value range.

A generic flowchart is presented in Fig. 8.5.

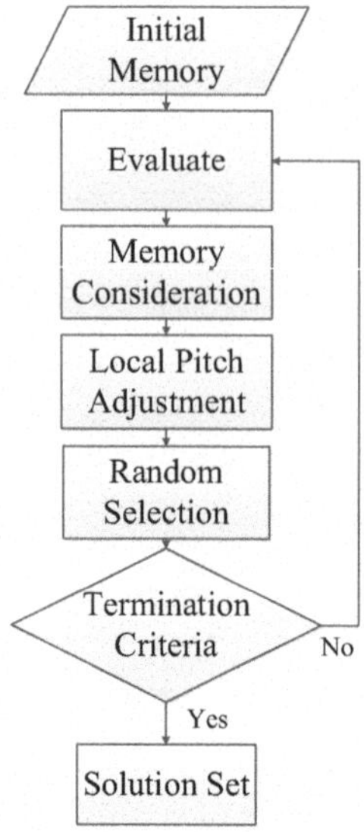

Fig. 8.5 HS flowchart

8.3.3 Heuristic Methods Evaluation

For the evaluation of all aforementioned solution techniques, the typical 33-bus system [45] has been employed, as depicted in Fig. 8.6. It is a radial electric distribution network and has a total load of 3.72 MW and 2.38 MVAr, presenting initial power loss of 211 kW. Due to their stochastic nature, the methods have been

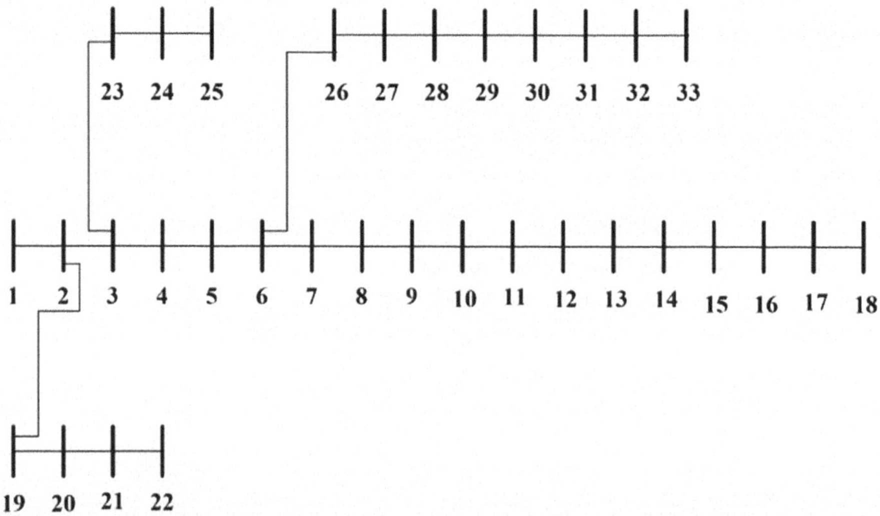

Fig. 8.6 The 33-bus system

applied 1000 times each, and within an ample time of 1000 iterations. Also, they have been let unrestrained in terms of number of DG units, so as to deduce how close the optimal solution they could arrive; with the actual optimal solution being the one with DG units installed in all nodes with nominal capacity equal to the nodes' respective load. The installed DG units are considered capable of injecting active power and injecting/consuming reactive power. Similar results have been extracted from implementation on other networks, such as the typical 16, 30 and 69-bus systems.

In Table 8.1 solution-related properties for all examined heuristic techniques are presented: the minimum loss achieved by each technique, the loss reduction percentage, the number of DG units installed along with the total DG installation size in MVA, as provided by the best solution among the 1000 trials that each technique has reached. Due to the ample time given, every technique has achieved a significant loss reduction in both systems and the differences are virtually slim, though, GA seems to be slightly in a bit of a disadvantage, as also confirmed by Fig. 8.7, where the mean Bus Voltage profile for all the examined heuristic techniques is shown.

In Table 8.2 convergence related properties are presented, i.e. the average execution time of one trial, the iteration number required for each technique to

Table 8.1 Heuristics' solution performance comparison

Method	Minimum power loss (kW)	Power loss reduction (%)	Total DG no.	Total size of installed DG units (MVA) P + *j*Q
GPSO	0.34	99.84	20	3.64 + *j*2.32
LPSO	0.22	99.89	20	3.55 + *j*2.21
UPSO	0.13	99.94	22	3.66 + *j*3.26
GA	10.77	94.89	21	3.00 + *j*0.71
ABC	0.52	99.75	17	3.73 + *j*2.28
CS	0.48	99.77	20	3.82 + *j*2.32
HS	2.67	98.74	19	3.30 + *j*2.08

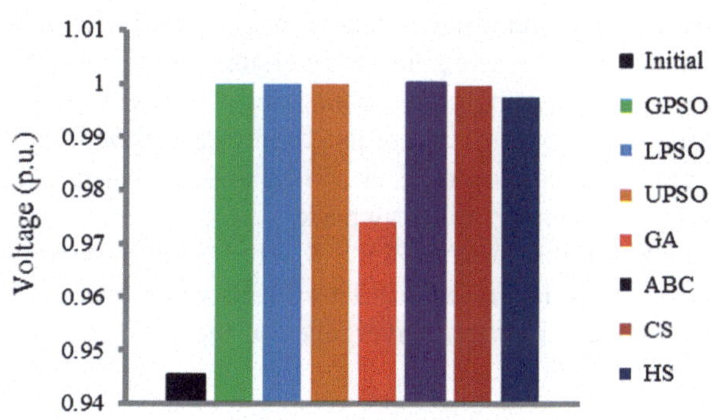

Fig. 8.7 Average voltage

Table 8.2 Heuristics' convergence performance comparison

Method	10% tolerance iteration	1% tolerance iteration	0.1% tolerance iteration	93.22% loss reduction iteration	Average execution time (min)
GPSO	832	983	999	6	6.7
LPSO	845	983	999	7	6.9
UPSO	339	839	900	5	6.9
GA	747	975	999	900	7.3
ABC	846	987	999	19	13.7
CS	909	992	999	43	12.5
HS	684	945	996	9	3.8

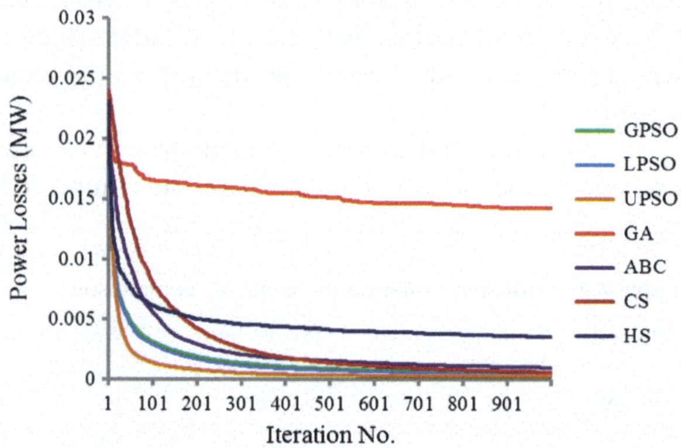

Fig. 8.8 Average convergence

reach in average within 10, 1 and 0.1% tolerance of its final optimal solution, respectively, e.g. for UPSO given that its optimal solution is 0.133 kW, the 10% tolerance is 0.146 kW power loss.

In terms of execution time, evidently HS proves to be the fastest with less than four minutes execution time. Moreover, the iteration number needed for each technique to reach a certain amount of loss reduction is also presented in Table 8.2. It is set to 93.22%, regarding the average loss reduction achieved by the least performing technique, namely being GA. Although all the techniques perform rather well, it seems, the PSO versions, and especially UPSO, performs better than the rest, regarding convergence and iteration steps, reaching their final solution in the least amount of iterations. Therefore, although UPSO is not as efficient as HS in terms of execution time, it can be argued that it can be applied for less iterations, thus overcoming this drawback.

This is illustrated in Fig. 8.8, where each technique's average convergence of the 1000 trials is presented. This is also confirmed by Fig. 8.9, where again each technique's average convergence of the 1000 trials is presented, but in a margin of

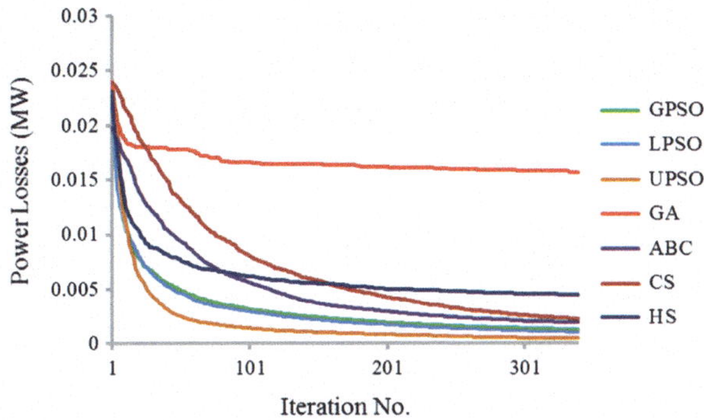

Fig. 8.9 Average convergence zoom-in

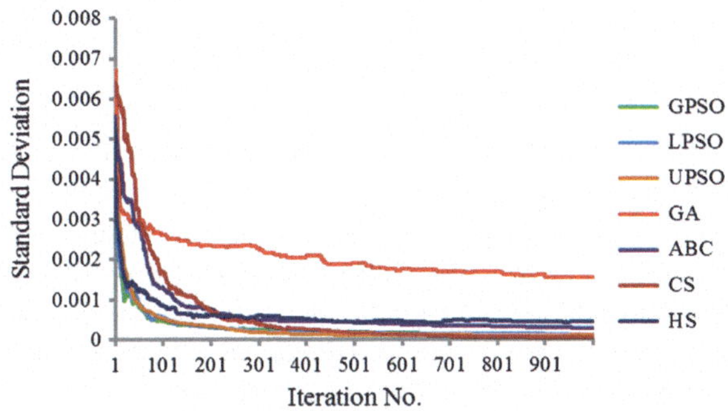

Fig. 8.10 Convergence's deviation

less than 1000 iterations, and specifically, within the 10% iteration tolerance of the best performing technique, being UPSO.

In addition, as shown in Fig. 8.10, the PSO versions, and especially UPSO, have the lowest convergence of standard deviation along 1000 iterations, meaning that their 1000 trials do not deviate far from each other, ensuring the robustness of their solution process and even that less trials are possible.

8.3.4 Heuristic Versus Analytical Methods Evaluation

For a more direct evaluation comparison of the most prominent Heuristic technique, i.e. UPSO, with the analytical methods presented in this section (IA, LSF, ELF) again the typical 33-bus system is employed. Three DG units are considered for installation and capable of injecting only active power. In Table 8.3 the solutions reached by the four methods are presented.

Based on the results of the previous section, UPSO has been applied 50 times and with 400 iterations. As can be deduced, UPSO performs rather better than the analytical ones, in terms of optimal solution, but rather poorly in terms of execution time. However, as evidenced in the precious section, a Heuristic method is able to

Table 8.3 Heuristics versus analytical methods solution comparison

Method	Minimum power loss (kW)	Power loss reduction (%)	DG position	DG size (kW)	Total DG installed (MW)	Time (s)
UPSO	77.9	65.50	13 24 30	802 1092 1054	2.95	70
IA	81.05	61.62	6 12 31	900 900 720	2.52	0.4
LSF	85.07	59.72	18 25 33	720 900 810	2.43	0.23
ELF	74.27	64.83	13 24 30	900 900 900	2.7	3.06

perform with the same efficiency, regardless of the considered number of DGs for installation, whereas the analytical ones would be either restricted to a small number of DGs, as in this case, or be biased since installing one DG unit at a time alters the electric distribution network every time. Moreover, since, the ODGP problem addresses primarily network planning and operational issues, it can be argued that time is not considered as important as finding the optimal solution, resulting in giving priority to the latter.

In conclusion, it is indicated that when contemplating ODGP towards power loss with a small amount of DG units to be installed, an analytical method might prove a better option than a heuristic one, in terms of time with only a minor setback in terms of optimal solution. When, an optimal or near optimal solution is required and more DG units should be considered for installation, a heuristic method would prove more suitable.

8.4 ODGP Towards Power Loss Minimization—Reverse Power Flow

Integration of DG in existing electric distribution networks has been discussed and studied thoroughly during the last years as a measure of reducing grid's power loss. However, the possible impacts of Reverse Power Flow (RPF), caused by extended DG penetration, on solving the ODGP have not been fully considered.

While reaching optimal solutions for the ODGP problem, recent and forthcoming massive DG integration brings to light RPF considerations, i.e. power flow pushed upstream of the network and on neighbouring networks. So far, literature solves the ODGP problem, towards different optimization functions, either without considering possible RPF to adjacent grids, or by simply not allowing it.

However, this solution approach could be proved inadequate; on the one hand, if RPF is ignored, unfair power displacement to neighbouring grids may occur, or not recognised; on the other hand, if it is strictly prohibited, it might lead to biased and sub-optimal solutions, since there are indications that when RPF is included during the planning process, it could lead to different ODGP solutions that can further reduce power loss [46–49].

In this section, it is shown that the RPF effect can be integrated into the ODGP problem as an occasional constraint imposed on the Slack Bus itself. As an alternative, an intermediate bus between the Slack Bus and the rest of the electric distribution network might be inserted, and imposing the constraint on the total power that flows through it, via its adjacent branches, modifying the network slightly [50]. The constraint and the corresponding penalty term can be expressed as

$$P_{perm} \leq \eta_{RPF}^{\%} \cdot P_{init} \tag{8.16}$$

$$\Omega_{RPF} = \rho_{RPF}\left[\max\left(0, |P_{perm}| - \eta_{RPF}^{\%} \cdot |P_{init}|\right)\right]^2 \tag{8.17}$$

where:

P_{init} is the initial power flowing through the Slack to the network
P_{perm} is the permitted power flowing through the Slack Bus to/from the network
$\eta_{RPF}^{\%}$ is the percentage of the allowed RPF, with respect to the initial Slack Bus flowing power.

Results from implementation on the typical 30 and 33-bus systems [51], a radial and a meshed electric distribution network, respectively, are shown in Figs. 8.11, 8.12, 8.13 and 8.14. Power loss reduction is the objective function, while gradually increasing RPF percentages are considered, and therefore the total permitted DG capacity to be installed is accomplished. A total number of seven DG units capable only of active power was considered for both examined networks. Furthermore, it should be stressed that the 30-bus system has already a 100% DG penetration with respect to installed load, whereas the 33-bus system has none. Overall, increasing the RPF results in reduced power loss savings, as can be deduced by Figs. 8.11

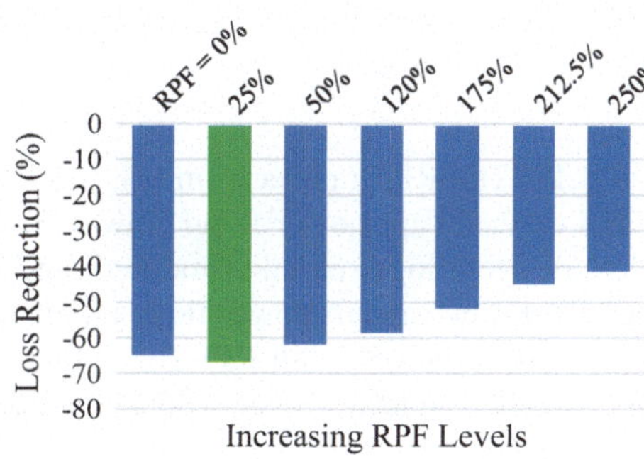

Fig. 8.11 RPF impact on loss reduction (%) in the 33-bus system

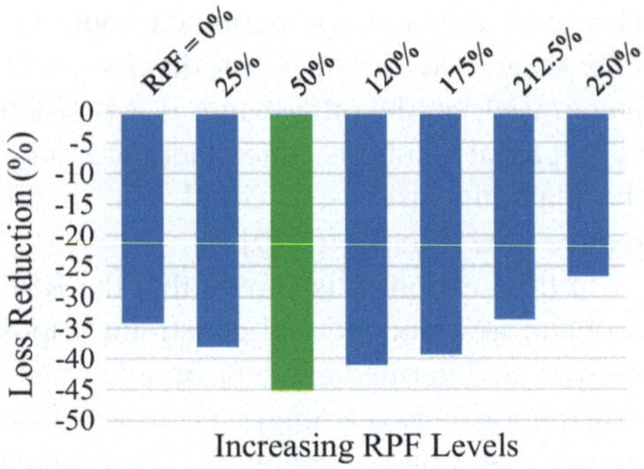

Fig. 8.12 RPF impact on loss reduction (%) in the 30-bus system

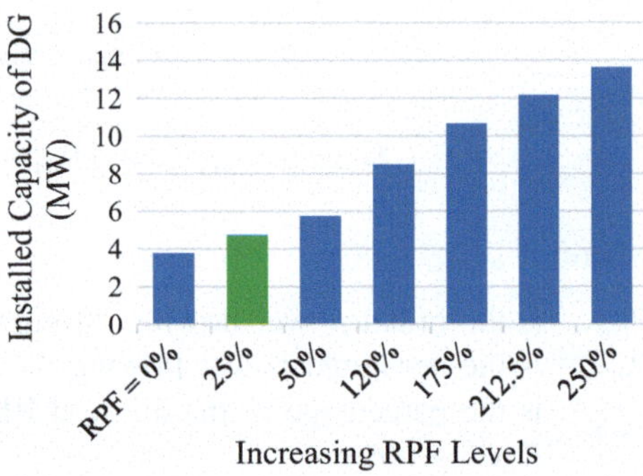

Fig. 8.13 Installed DG capacity in the 33-bus system

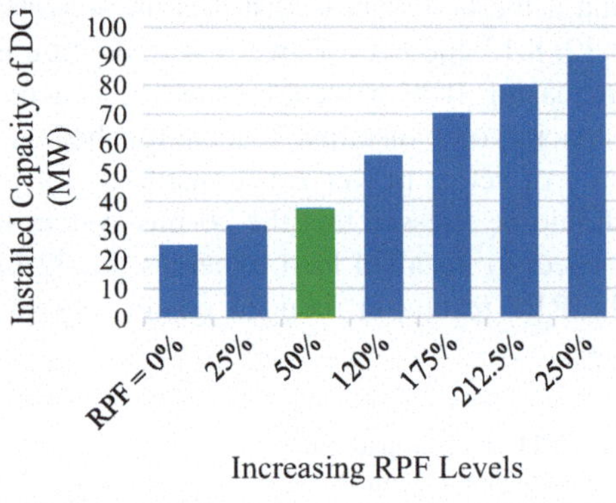

Fig. 8.14 Installed DG capacity in the 30-bus system

and 8.12. The RPF ranges from 0% (no RPF allowed) up to equal to 250% of the initial downstream power flow. However, it is also deducted that regardless of the network's topology (radial or meshed) and with RPF percentage from 25 to 50% of the initial downstream power flow, an even better loss reduction is achieved, when compared to the one with RPF 0%. Moreover, for these RPF percentages the total DG installed reaches over 100% penetration in both networks, as shown in

Figs. 8.13 and 8.14; in the 33-bus system, for the best loss reduction for 25% RPF a total over 4 MW DG is installed, whereas its total installed load is 3.72 MW and in the 30-bus system, a nearly additional 40 MW DG is installed, in spite of already having achieved 100% DG penetration.

In conclusion, RPF existence up to a certain level will not necessarily affect negatively the ODGP, when considering power loss reduction. Additionally, it might lead to solutions with greater loss reduction and DG penetration over than 100%. This could benefit other operational aspects of the network, e.g. reliability improvement and environmental benefits under the installation of Renewable Energy Sources.

8.5 ODGP Towards Power Loss Minimization—Renewable Energy Sources

When examining ODGP towards power loss minimization, it is rather difficult to examine Renewable Energy Sources (RESs) directly and their installation in an electric distribution network, while keeping problem complexity at minimum, since their most distinctive feature, stochasticity, is dependent on time. In ODGP towards power loss minimization only a single state or snapshot of the network is taken into account. In examining different types of DGs, the most direct distinction from the network's point of view are:

- type 1: DG injecting only active power,
- type 2: DG injecting only reactive power,
- type 3: DG injecting active power and injecting/consuming reactive power.

With that in mind, apart from optimal site and size, an aspect of optimal mix of DG types can be added in the problem formulation. However, the question remains if it would be possible to examine the integration of RESs in an electric distribution network even in the current stage, without integrating on time, in other words if an Optimal Renewable Energy Sources Placement (ORESP) problem can be contemplated. To that end, several alternatives are offered.

More specifically, one alternative refers to solving the problem separately, i.e., to find the optimal siting and sizing of DG units in a network, as an ordinary ODGP problem, and then to determine the RES type, e.g. Photovoltaic or Wind Turbine [52]. However, no mix of RESs is examined for penetration and an impartial solution might not be achievable. If an optimal mix is required, an investigation regarding the different impacts of DGs on power quality and reliability must be performed; the DG penetration level could be limited by harmonic distortion because of the nonlinear current injected by inverter-based DG units, as well as by protection coordination constraints because of the variation in fault current caused by synchronous-based DG units [53]. Another approach is to implement the ELFs concept for each bus and each technology of DG [54], though either a possible needless computational effort would take place, or the solution could just rely on approximations. A considerable contribution at this field has been accomplished in [55], in which the optimal mix of DGs of different technologies has been achieved, via stochastic models of wind speed and solar irradiance. The candidate nodes for DG installation are predefined however and thus, only their size is estimated. Additionally, the different DG technologies could be divided according to their power output, i.e. whether they can control active/reactive power independently (PQ mode, or constant power factor mode), or active power and voltage (PV mode, or variable reactive power mode) [56]. In this latter approach, a simultaneous solution regarding number, siting and sizing is achieved, under a multi-objective function that includes active/reactive power loss minimization and voltage profile improvement. However, the aforementioned distinction between different DG technologies might not be quite so accurate.

Finally, in this section, the concept of Capacity Factors is implemented [57]. The basic issue in ORESP is the variations regarding RESs' power output; it is related to their technology and the natural resources' potential and they have to be taken into account. For example, solar irradiance, wind speed and water availability are expected to vary among candidate nodes within an electric distribution network, and these variations could have a significant impact on the optimal siting and sizing of such RESs, especially when a mix of different RESs is examined. The network's nodes are divided into groups, representing different areas with different natural characteristics and resources, and hence with different CFs. The values of these CFs express the potential of the respective natural resources available in that node. Thus, according to their positions, all RESs are assigned their CFs. These CFs are then included as an additional occasional constraint in the problem formulation, as

$$\Omega_{CF} = \rho_{CF} \sum_{l=1}^{n_{res}} CF_l \tag{8.18}$$

where n_{res} the number of RESs, and ρ_{CF} the corresponding penalty factor.

For example, let three RESs technologies to be considered, e.g. Photovoltaic (PV), Wind Turbine (WT) and Hydro-Plant (HP), for installation in the typical 69-bus system [58]. In order to assign potential for the local natural resources at each node, the electric distribution network in question has been divided into three areas, as depicted in Fig. 8.15. At each area, each of the nodes is assigned a value for the CFs of the three technologies examined, i.e. PV, WT, and HP, respectively. Assuming that each area is relatively small, it is evident that the nodes within that area share the same value of the CFs of their respective technologies.

In Table 8.4 a set of typical CFs values for each technology is assigned to each area, whereas in Tables 8.5 and 8.6 the results of the proposed method are presented. The results of the ODGP implementation on the same electric distribution network are also presented for comparison and LPSO was used in both approaches. Five DG units were considered for installation for the ODGP problem, whereas 5 RESs units for each technology in the ORESP problem. It can be concluded that via

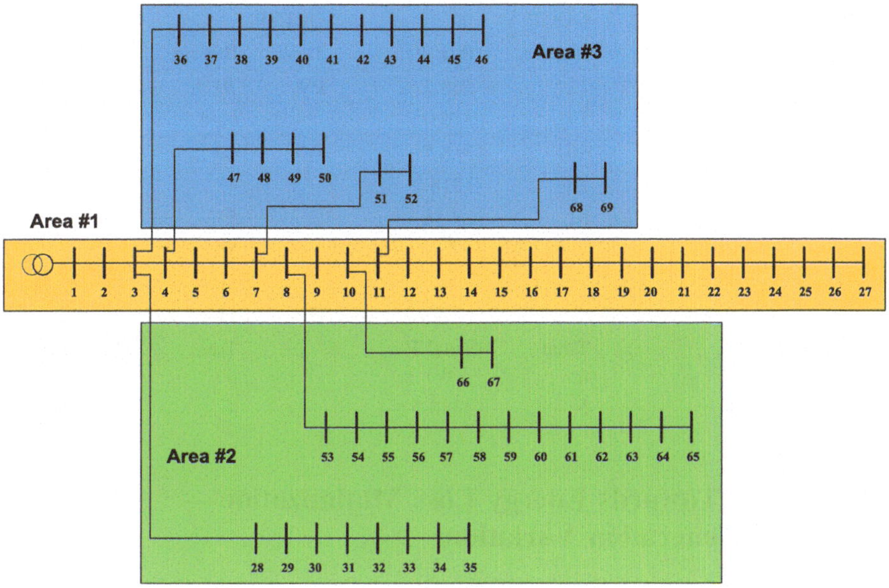

Fig. 8.15 The 69-bus system divided into CF areas

Table 8.4 CFs values

Area	RES type		
	PV	WT	HP
#1	0.10	0.00	0.42
#2	0.10	0.25	0.00
#3	0.15	0.12	0.00

Table 8.5 Overall results

	Initial losses (kW)	Minimum losses (kW)	Loss reduction (%)
ODGP	602.2	148.4	75.357
ORESP	602.2	169.4	71.8698

this method the different geographical characteristics of an area and different weather conditions leading to availability of RESs, can be taken into account all at once, by the aid of CFs. Moreover, the corresponding ORESP problem can be solved while keeping its complexity at minimum and an optimal solution, in terms not only of siting and sizing, but also of RES type, is achievable.

Table 8.6 Detail results

ODGP		ORESP		
Bus no.	P (kW)	Type	Bus no.	P (kW)
12	503.2	PV	20	420.3
19	376.0		61	23.0
40	718.5	WT	40	723.2
53	1718.8		45	580.8
61	29.48		53	1458.1
			56	226.0
			59	57.7
		HP	12	283.5
Total no.	Total P (kW)		Total no.	Total P (kW)
5	3346.5		8	3778.2

8.6 ODGP Towards Energy Loss Minimization—Load/Generation Variation

Although many issues can be examined in ODGP, such as power loss minimization or reduction, reverse power flow, voltage stability and reliability improvement, the approach remains incomplete without the time variable. If time is taken into account though, the problem becomes more complex and time-consuming, than it already is. Thus, a solving method able to prove the Golden Section between quick convergence time and optimal solution will become more than useful, as that examined in Sect. 8.3. Thus, the analysis of ODGP towards power loss minimization is useful and important and also a significant step before examining ODGP towards energy loss minimization.

Still, an electric distribution network's load does not remain constant, but varies over time. Furthermore, the stochasticity of RESs' generation, and their impact on a network cannot be examined, when only a single snapshot of the latter is considered.

For the ODGP towards energy loss, an energy loss minimization objective function could be

$$F_{eloss} = \sum_{\Delta t=1}^{t} \sum_{k=1}^{n_l} g_{i,j} \left[V_i^2 + V_j^2 - 2V_i V_j \cos\left(\theta_i - \theta_j\right) \right] \tag{8.19}$$

where Δt is the time interval and t is the time period examined.

Regarding the constraints, they remain the same, with the following exception: the single constraint value retrieved from the single snapshot's load flow analysis in power loss minimization approach, now in the energy loss minimization approach, is replaced by the maximum absolute value retrieved from the time period examined t, in order to maintain the same order of magnitude in the penalty function.

The impact of DG units on energy losses depends on the specific characteristics of the network, such as demand profile, topology, as well as the relative location of the generators and whether their output is considered constant or variable. Incorporating these complexities into an optimization framework for energy loss minimization is a challenge that has only been partially addressed by a few studies [59]. In [47] the analysis regarding load and DG power output variations relies on uniformly distributed loads while these variations refer to a typical daily pattern for both. Moreover, only the optimal siting of DG units is examined, and one DG unit is considered for installation. In [60] the case of one wind power unit under both power output and load demand variations is examined. The analysis yields the optimal node for the wind power unit installation by considering a sequential analysis with only one candidate node for DG installation at a time and concludes that subject to load variations, the optimal location is different when compared to the operational snapshot. In [55] a probabilistic technique is proposed for optimally allocating different types of DG technologies. The technique is based on generating a probabilistic generation-load model. Beta and Rayleigh Probability Density Functions (PDFs) are used for simulating solar irradiance and wind speed uncertainty, respectively, while IEEE-RTS for the load profile. However, the positions of the DGs are predetermined, as the number of DGs as well. Other approaches incorporating load or DG power variations, as in [61], may provide biased solutions since the installation nodes are predetermined. Furthermore, the analysis in [62] concludes that the power analysis of one load snapshot is not necessarily adequate for the overall operation of the electric distribution network in [63] a two-stage method of optimal siting and sizing of DG units is proposed. Finally, in [64], a method to address and evaluate the economic benefits of RESs is proposed, when applied to networks, but the candidate buses are predetermined and the number of DGs for each type is limited and predefined.

8.6.1 Load Variation

In electric distribution networks, the loads are highly distributed and quite variable. Thus, detailed modelling is not possible, as yet, and even more difficult due to the absence of available real data. Thus, mathematical methods are resorted to formulate the load variations. In a first approach, the load of a test network, like the ones examined so far in this chapter, could be stochastically altered, in order to create different snapshots of a network, or even more, the load in each node could be stochastically altered, regarding the current/original load value of the network, either as its average, or its maximum value.

If the load of an electric distribution network, or even its load composition, is considered as an average snapshot of the network, then load variations or even load composition variations could be constructed via a uniform distribution, within a 20 and 50% range of the original snapshot. The loading condition of the IEEE-24 bus Reliability Test System [65] can be studied as a base case, in order to justify the

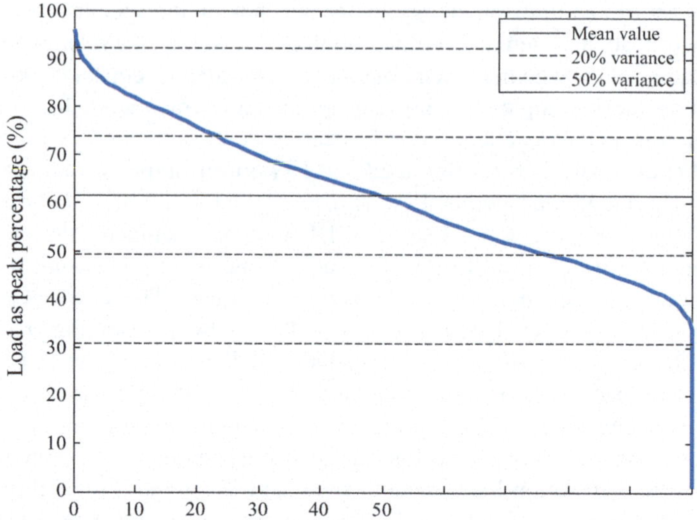

Fig. 8.16 Load duration curve for the typical IEEE-24 bus reliability test system

load variations modelling for the present analysis. Its hourly, daily, and weekly peak load factors were used to construct the annual load curve. These peak load factors have been selected in order to capture the loading conditions that yield the highest annual energy losses and moreover in order to justify the upper limit for the load variations (i.e. 50%) adopted in this analysis. This selected variance could cover a loading composition for the network that refers to the highest load demands that are expected within a one year time period. Subsequently, the annual load curve is transformed into a cumulative power curve, as shown by the blue line in Fig. 8.16 to investigate the loading variability. In the same figure, the mean annual power, 61.45% of the annual peak power, along with the 20 and 50% variance limits are also marked in continuous, dotted, and dashed grey lines, respectively. It is calculated that the 20 and 50% variance cover the 55.08 and 99.40% of the total annual loading levels, respectively. As proved by Fig. 8.16, the majority of loading conditions of the network during a one year time period could be captured by load variations up to 50% of the average load composition of the network.

If the ODGP towards power loss minimization is solved for every snapshot created, then it would provide an optimal solution for each and every snapshot. When examining the overall results, it is deduced that some buses appear more frequently than others, i.e. appear in most of the snapshot's solutions. This suggests that some buses emerge as the most critical for siting DG units. Moreover, this means that the siting stage of the ODGP is insensitive to the load variations, or even to the load composition variations. Thus, the two stages of ODGP, siting and sizing, can be examined separately. In Fig. 8.17 the results from the implementation on the 33-bus system are presented, for 2000 snapshots within 20%, and 6000 snapshots within 50% range of the original load composition solved. The related frequency of

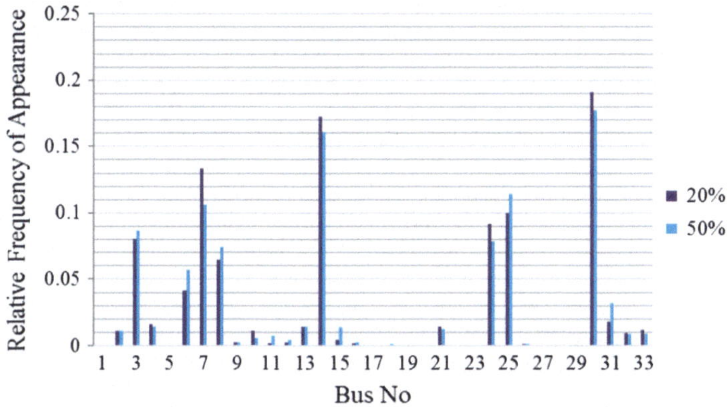

Fig. 8.17 Relative frequency of appearance of buses in the 33-bus system

Fig. 8.18 Energy loss reduction comparison for 1000 snapshots on the 33-bus system

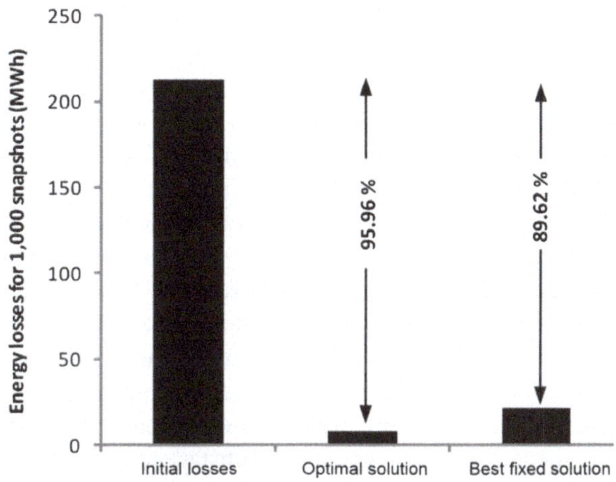

appearance of each bus for each case (20 and 50%, respectively) is shown. It appears that buses No. 3, 7, 14, 24, 25 and 30 for both 20 and 50% range variations emerge as the most prominent for DG installation.

In addition, if their average active and reactive power from the snapshot solutions are to be taken into account for these prominent/critical buses, then they can present a fixed but adequate solution for the ODGP problem towards energy loss minimization [66]. In Fig. 8.18, for instance, results from 1000 snapshots within 20% range applied to the 33-bus system are presented. The total energy losses for the 1000 snapshots without any DG installed is compared to the losses obtained from the optimal solutions of every snapshot and the fixed solution from the most prominent buses along with their respective average active and reactive power. It is demonstrated that the fixed solution's energy loss reduction is very close to the loss reduction sum of the optimal solutions of all the snapshots, diverging only slightly

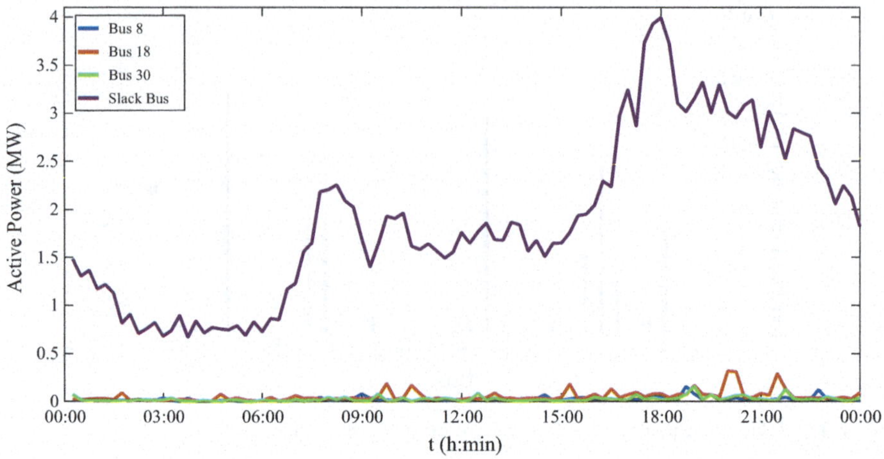

Fig. 8.19 Daily load profile of various buses of the 33-bus system

by 6% from it. Hence, a very first estimation, if not an adequate solution, regarding the ODGP towards energy loss is provided.

An alternative approach to load variations would be to consider that the electric distribution network's snapshot at hand is the peak load for the time period examined. Moreover, instead of stochastically reproducing load snapshots to create a load profile, the network itself can be combined with time-series of standard load profiles, either real, or synthesized via load forecasting techniques. Thus, in a straightforward approach, each bus's load could be multiplied with a normalized standard load profile and so creating the desired snapshots. However, since the loads in a network do not necessarily change simultaneously or present the same pattern, and most importantly the standard load profiles are more or less measurements of the DNO on the substations within its purview and not on the load buses themselves, a more elaborate scheme can be contemplated. It could be theorized that the total network's load follows the standard load profile's pattern and each bus's load changes in such a way, so that this can be achieved. An example can be seen in Fig. 8.19, where the load profile of three buses is shown along the total load profile of the network, as seen from the Slack Bus for a daily time period of hourly quarter's intervals, from the implementation of this method on the 33-bus system. The buses' load follow their own individual patterns, while the total network load profile is seen from the Slack Bus, and that is observed by the DNO. Thus, a more realistic approach of the problem has been achieved.

Table 8.7 ODGP of different technologies—energy loss reduction results

Technology	Energy loss reduction (%)
DG	82.9874
PV	36.5701
WT	50.2236

Table 8.8 ODGP of different DG technologies—detailed solution results

DG			PV			WT		
Bus no	P (kW)	Q (kVar)	Bus no	P (kW)	Q (kVar)	Bus no	P (kW)	Q (kVar)
3	248	78.4	3	281.6	81.5	3	395.8	78.6
6	281	94	6	550	204.3	6	358.7	71.9
11	222.6	59.1	9	96.8	22.2	11	312.7	64.1
16	217.6	50.9	11	214.8	46.7	16	387.7	63.2
31	206.1	109.2	16	262.5	52.2	30	388.9	152.3
Total no.	Total P (kW)	Total Q (kVar)	Total no.	Total P (kW)	Total Q (kVar)	Total no.	Total P (kW)	Total Q (kVar)
5	1175.5	391.6	5	1405.7	406.9	5	1843.8	430.1

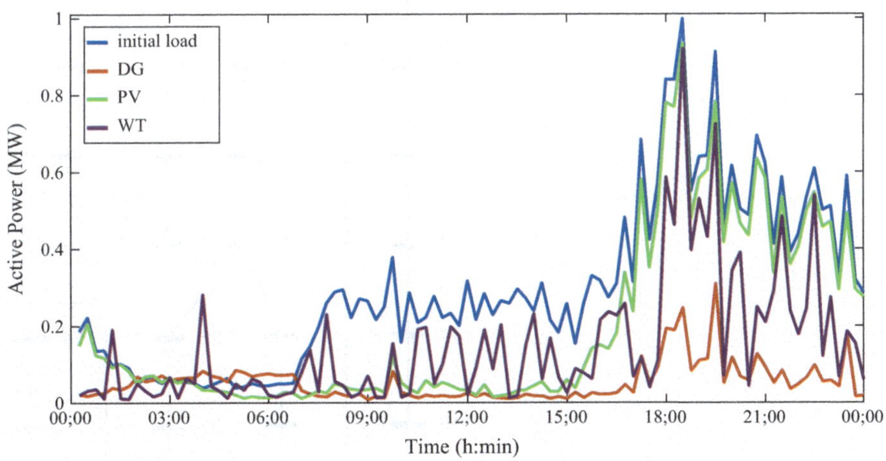

Fig. 8.20 Daily load curve without any DG (initial load), and with generic DG, PV and WT

8.6.2 Load/Generation Variation

With respect to RESs' generation, there are data available both from DG stations and mathematical tools, such as the Weibull distribution for wind speed, or a

Beta PDF for solar irradiance modelling. In spite of the more realistic approach of the problem, it is of interest that still the siting stage of the ODGP is insensitive to the load variation, load composition variation and perhaps DG technology, indicating that it is more network-topology oriented. Furthermore, as expected, the energy loss reduction is DG technology dependent. These can be seen in Tables 8.7 and 8.8, where results from an application on the 33-bus system are presented. A daily period of hourly quarter's intervals is considered and five DG units capable of injecting active power with a maximum power factor of 0.95 leading/lagging. DG units of constant power output, PVs and WTs as renewable technologies have been applied. In case of PVs real data were used, whereas for WTs synthesized data were obtained. The corresponding load curves can be seen in Fig. 8.20.

Additionally, with respect to an optimal mix of DG technologies, e.g. PVs and WTs, it can be argued that the approach proposed in 8.5, is not that far from reality.

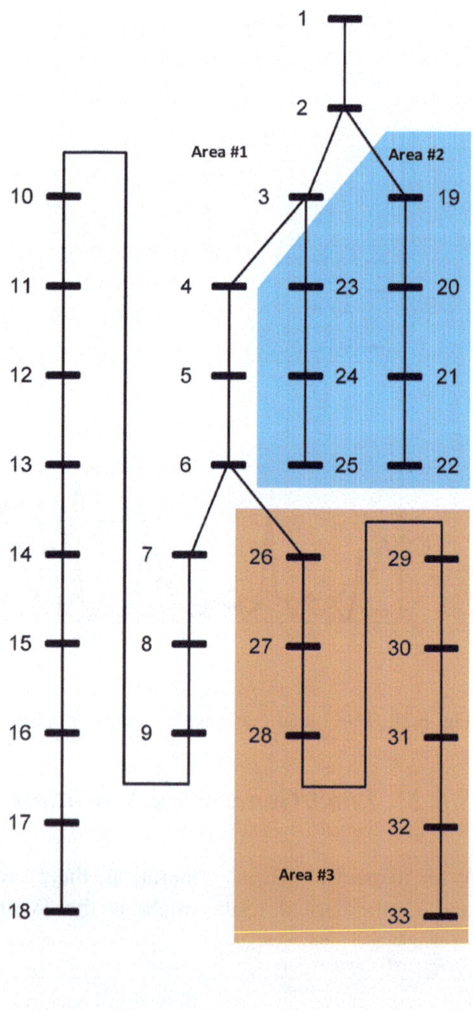

Fig. 8.21 The 33-bus system divided into three CF areas

Table 8.9 ORESP towards energy loss minimization using a realistic approach and CF method

ORESP—realistic approach			ORESP—CF on peak load		
Energy loss reduction (%)	45.1869		Energy loss reduction (%)	28.102	
PV			PV		
Bus no.	P (kW)	Q (kVar)	Bus no.	P (kW)	Q (kVar)
3	40.4	33.9			
6	440.9	73.1	6	725.7	321.4
11	265.1	66.9	11	424.5	124.3
16	270.7	55.2	16	514.7	82.6
Total no.	Total P (kW)	Total Q (kVar)	Total no.	Total P (kW)	Total Q (kVar)
4	1017.1	229.1	3	1664.9	528.3
WT			WT		
Bus no.	P (kW)	Q (kVar)	Bus no.	P (kW)	Q (kVar)
2	0	63.8			
23	667	41.1	23	193.6	25.2
30	1046.5	149	30	239.7	95.7
Total no.	Total P (kW)	Total Q (kVar)	Total no.	Total P (kW)	Total Q (kVar)
3	1713.5	253.9	2	433.3	120.9

More specifically, the 33-bus system is divided in three areas of different weather and geographical potentials, as depicted in Fig. 8.21, where in area #1 and #2 sun and wind potential are dominant, respectively, and in area #3 they are competitive. As earlier, same DG operation regarding active/reactive power and load profile is assumed. As can be deduced from Table 8.9, if the method developed in Sect. 8.5 is performed for the peak load of the network, the solutions reached are a bit different, though comparable. It should be stressed, however, that the analysis is performed in a short time scale, i.e. a daily load curve. However, if the time scale is extended to a whole year, or years, the solutions might bear more resemblance.

8.7 Combination of ODGP with Other Problems

8.7.1 ODGP and NR

In ODGP the siting and sizing of DG units is the objective whereas in Network Reconfiguration (NR) an alternative layout is the objective in order to redistribute the power flow. Both techniques are established as efficient, regarding power loss reduction.

Despite the significant contribution of each technique towards loss reduction, when applied individually, it seems that there are quite few studies that try to examine the potentials of a combined approach under an efficient application order for them [67–69]. Both power loss reduction techniques, when applied individually, affect either the load composition of the electric distribution network (the net power of the nodes that host DG units is altered in ODGP) or its layout (a reconfigured topology after the NR application). Thus, when both techniques are applied, the application order is highly possible to have an impact on the final solution regarding the overall amount of loss reduction. If it is assumed that the highest possible loss reduction refers to the ideal 100%, then the contribution of each technique towards such a solution is affected by the order of their application. For instance, ODGP could theoretically yield a solution with 100% power loss reduction in the ideal case with one DG unit with power injection equal to the local load installed at each node. In this latter case, the further application of NR would be meaningless. On the other hand, if the ODGP problem refers to the more realistic case of limited available DG units to be optimally sized and sited, then the application of the NR technique could yield additional loss reduction and further improve the solution.

If the opposite application order is examined then it is interesting to investigate how both the siting and sizing of the available DG units to be penetrated in the electric distribution network would be affected, given that the ODGP problem will now be applied to an altered network, i.e. with a reconfigured topology while keeping the same load composition.

Let us consider the following three scenarios of the solving order of ODGP and NR:

- scenario-1: NR solved first, then ODGP,
- scenario-2: ODGP first, then NR, and,
- scenario-3: both ODGP and NR are concurrently solved.

The results, when implemented in the 69-bus system are presented in Tables 8.10, 8.11, and 8.12, whereas in Fig. 8.22 the 69-bus system along with its tie-switches is depicted. Seven DG units are considered for installation and capable

Table 8.10 Scenario-1: NR 1st, ODGP 2nd

NR applied	Initial losses (kW)	Sectionalizers open	Tie switches closed	Loss reduction %	Final losses (kW)
	229.8	14,58,62	Tie3-Tie5	54.7	104.1
ODGP applied	Initial losses (kW)	Nodes to host DG units	Active power of each DG unit (kW)	Reactive power of each DG unit (kVar)	Loss reduction % and final losses (kW)
	104.1	5	901.7	189.2	93.65%
		9	241.6	177.2	6.6
		12	427.4	299.6	
		22	338.3	226.6	
		40	0	536.4	
		53	1416.1	938.2	
		56	318.5	226.7	

Table 8.11 Scenario-2: ODGP 1st, NR 2nd

ODGP applied	Initial losses (kW)	Nodes to host DG units	Active power of each DG unit (kW)	Reactive power of each DG unit (kVar)	Loss reduction % and final losses (kW)
	229.8	2 3 9 12 19 40 53	0 539 0 501.2 380.8 717 1674	-53.2 340 184.9 279.8 251.7 512 1178.8	97.35% 6.1
NR applied	Initial losses (kW)	Sectionalizers open	Tie switches closed	Loss reduction %	Final losses (kW)
	6.1	–	–	0	6.1

Table 8.12 Scenario-3: ODGP NR concurrently

Candidate node to host DG units	Active power of each DG unit (kW)	Reactive power of each DG unit (kVar)	Sectionalizers open	Tie switches closed	Loss reduction %—final losses (kW)
57	2021.5	849.8	20,42,46,58,61	Tie1-tie5	68.28% 72.9

of both active and reactive power generation. The UPSO algorithm, as presented in Sect. 8.3, is utilized. The first scenario seems to be advantageous since the switching operations rely on the already existent tie-switches and that results in lower required DG capacity for power loss minimization. In the second scenario, it is highly possible to be unable to apply the NR technique, especially if the ODGP technique performs quite well under high power loss reduction by the installation of the proposed DG units. Finally, in the third scenario since both techniques are considered concurrently, the problem's complexity increases exponentially, thus the algorithm seems unable to provide an adequate solution. It is yet to be investigated, whether the worth of a better solution in this case is overweighted by the increased computational burden [70–72].

8.7.2 ODGP and OESSP

ODGP can be targeted towards energy loss reduction, due to the nature of DG units since they produce electricity even for a certain time period. ESSs, though, present an entirely different complexion. Moreover, because ESSs' integration in a more massive or industrial scale is still in its infancy, cost is still and a more important

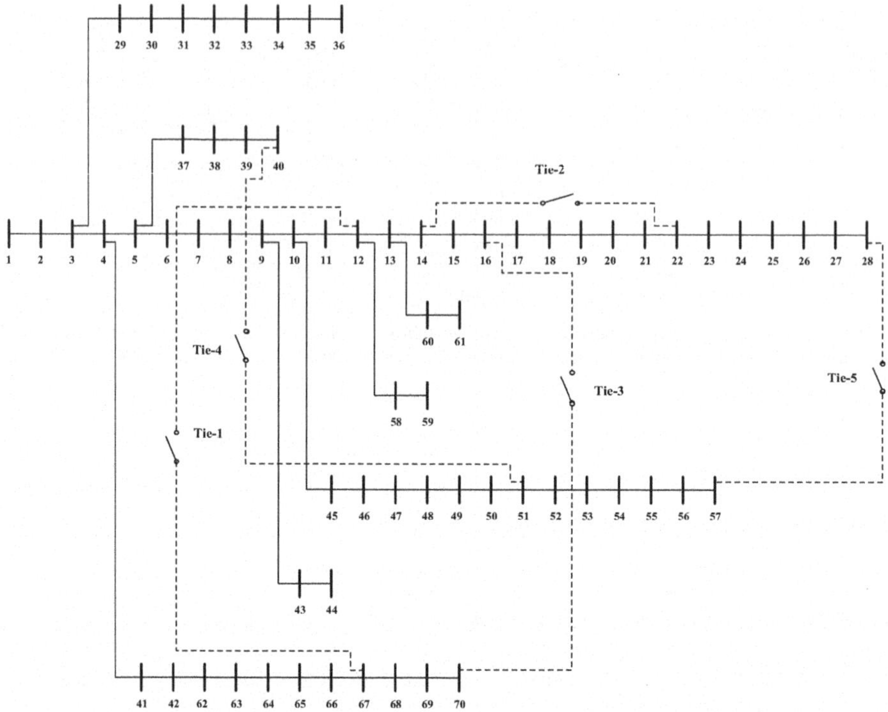

Fig. 8.22 The 69-bus system depicted with its tie-switches

Table 8.13 Energy loss reduction from installing DG units along with ESSs

	Energy loss (MWh)	Energy loss reduction (%)
Initial	2.7647	–
DG	1.5593	43.5997
LS	1.5606	43.5526
EM = 1 MWh	1.5532	43.8203
EM = 2.5 MWh	1.5449	44.1205
EM = 5 MWh	1.5353	44.4678

issue. Thus, when dealing with the OESSP problem, cost or profit objective functions are considered [73]. Furthermore, it might not be of significant aid towards energy loss reduction. For example, using the load/generation tools available from Sect. 8.6.2, two modes of ESSs can be added: load smoothing (LS) and energy management (EM). The former is used in order to smooth out any abrupt spikes in load curves and the latter in storing energy during one time period and providing it at another. PVs and WTs have also been utilised, for a more thorough approach, provided from the example in Sect. 8.6.2. For LS, the ESSs are considered to be installed in the buses where the PVs and WTs have been installed,

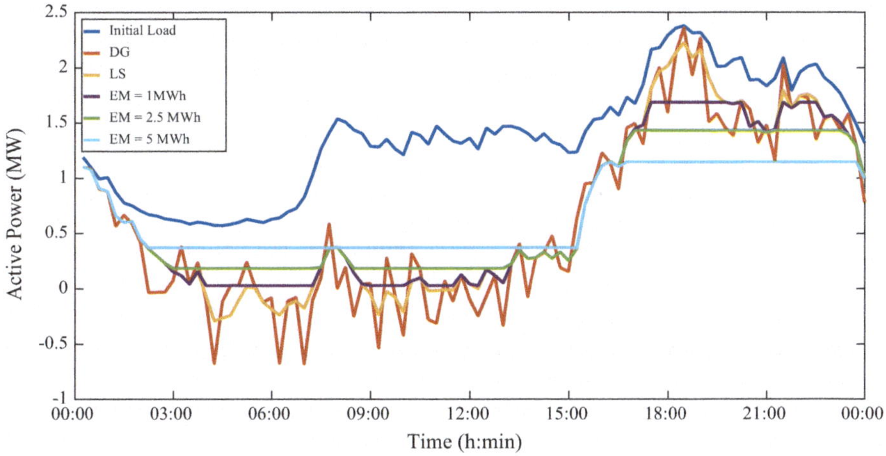

Fig. 8.23 Load curves for integrating DGs along ESS in LS and EM mode

and for EM near the Slack Bus, since it is theorized that it will be installed by the DNO. In Table 8.13 the energy losses are presented and in Fig. 8.23 the load curves for a time period of one day are illustrated. As can be seen, in this configuration the impact in energy losses of ESS is limited, regardless of mode or size, although great benefits have been provided for the DNO, from both modes, regarding the load curves. It should be emphasized though, that OESSP might prove promising in the field of energy loss minimization over a more extended and elaborate analysis. For instance, both sizing and siting could be examined concurrently, and ESSs systems capable of LS and EM operation, or even Frequency Regulation, as well.

References

1. T. Ackermann, V. Knyazkin, Interaction between distributed generation and the distribution network: operation aspects, in *IEEE/PES Transmission and Distribution Conference and Exhibition* (2002), pp. 1357–1362
2. N. Mohandas, R. Balamurugan, L. Lakshminarasimman, Optimal location and sizing of real power DG units to improve the voltage stability in the distribution system using ABC algorithm united with chaos. Int. J. Electr. Power Energy Syst. **66**, 41–52 (2015)
3. V.H. MendezQuezada, J. RivierAbbad, T. GomezSanRoman, Assessment of energy distribution losses for increasing penetration of distributed generation. IEEE Trans. Power Syst. **21**(2), 533–540 (2006)
4. K.O. Oureilidis, E.A. Bakirtzis, C.S. Demoulias, Frequency-based control of islanded microgrid with renewable energy sources and energy storage. J. Mod. Power Syst. Clean Energy **4**(1), 54–62 (2016)
5. P.A. Gkaidatzis, D.I. Doukas, A.S. Bouhouras, K.I. Sgouras, D.P. Labridis, Impact of penetration schemes to optimal DG placement forlossminimisation. Int. J. Sustain. Energy **36** (5), 473–488 (2017)

6. A.S. Bouhouras, K.I. Sgouras, P.A. Gkaidatzis, D.P. Labridis, Optimal active and reactive nodal power requirements towards loss minimization under reverse power flow constraint defining DG type. Int. J. Electr. Power Energy Syst. **78**, 445–454 (2016)
7. M. Esmaili, Placement of minimum distributed generation units observing power losses and voltage stability with network constraints. IET Gener. Transm. Distrib. **7**(8), 813–821 (2013)
8. S. Ge, L. Xu, H. Liu, J. Fang, Low-carbon benefit analysis on DG penetration distribution system. J. Mod. Power Syst. Clean Energy **3**(1), 139–148 (2015)
9. A. Soroudi, M. Ehsan, R. Caire, N. Hadjsaid, Hybrid immune-genetic algorithm method for benefit maximisation of distribution network operators and distributed generation owners in a deregulated environment. IET Gener. Transm. Distrib. **5**(9), 961 (2011)
10. Y. del Valle, G.K. Venayagamoorthy, S. Mohagheghi, J.-C. Hernandez, R.G. Harley, Particle swarm optimization: basic concepts, variants and applications in power systems. IEEE Trans. Evol. Comput. **12**(2), 171–195 (2008)
11. K.E. Parsopoulos, M.N. Vrahatis, *Particle Swarm Optimization and Intelligence: Advances and Applications* (IGI Global, Hershey, 2010)
12. U. Leeton, D. Uthitsunthorn, U. Kwannetr, N. Sinsuphun, T. Kulworawanichpong, Power loss minimization using optimal power flow based on particle swarm optimization, in *2010 IEEE International Conference on Electrical Engineering/Electronics Computer Telecommunications and Information Technology (ECTI-CON)* (2010), pp. 440–444
13. P.S. Georgilakis, N.D. Hatziargyriou, A review of power distribution planning in the modern power systems era: models, methods and future research. Electr. Power Syst. Res. **121**, 89–100 (2015)
14. D.Q. Hung, N. Mithulananthan, R.C. Bansal, Analytical expressions for DG allocation in primary distribution networks. IEEE Trans. Energy Convers. **25**(3), 814–820 (2010)
15. D.Q. Hung, N. Mithulananthan, Multiple distributed generator placement in primary distribution networks for loss reduction. IEEE Trans. Ind. Electron. **60**(4), 1700–1708 (2013)
16. D.Q. Hung, N. Mithulananthan, Loss reduction and loadability enhancement with DG: a dual-index analytical approach. Appl. Energy **115**, 233–241 (2014)
17. P. Prakash, D.K. Khatod, An analytical approach for optimal sizing and placement of distributed generation in radial distribution systems, in *1st IEEE International Conference on Power Electronics. Intelligent Control and Energy Systems (ICPEICES-2016)* (2016), pp. 1–5
18. T. Kumar, T. Thakur, Comparative analysis of particle swarm optimization variants on distributed generation allocation for network loss minimization, in *2014 First International Conference on Networks & Soft Computing (ICNSC2014)* (2014), pp. 167–171
19. A.A. Abou El-Ela, S.M. Allam, M.M. Shatla, Maximal optimal benefits of distributed generation using genetic algorithms. Electr. Power Syst. Res. **80**(7), 869–877 (2010)
20. K.-H. Kim, Y.-J. Lee, S.-B. Rhee, S.-K. Lee, S.-K. You, Dispersed generator placement using fuzzy-GA in distribution systems, in *IEEE Power Engineering Society Summer Meeting*, vol. 3 (2002), pp. 1148–1153
21. F.S. Abu-Mouti, M.E. El-Hawary, Optimal distributed generation allocation and sizing in distribution systems via artificial bee colony algorithm. IEEE Trans. Power Deliv. **26**(4), 2090–2101 (2011)
22. A.A. Seker, M.H. Hocaoglu, Artificial Bee Colony algorithm for optimal placement and sizing of distributed generation, in *2013 8th International Conference on Electrical and Electronics Engineering (ELECO)* (2013), pp. 127–131
23. N. Taher, I.T. Seyed, A. Jamshid, T. Sajad, N. Majid, A modified honey bee mating optimization algorithm for multiobjective placement of renewable energy resources. Appl. Energy **88**(12), 4817–4830 (2011)
24. W.S. Tan, M.Y. Hassan, M.S. Majid, H.A. Rahman, Allocation and sizing of DG using Cuckoo search algorithm, in *2012 IEEE International Conference on Power and Energy (PECon)* (2012), pp. 133–138
25. M. Zahra, A. Amir, A novel approach based on cuckoo search for {DG} allocation in distribution network. Int. J. Electr. Power Energy Syst. **44**(1), 672–679 (2013)

26. W. Buaklee, K. Hongesombut, Optimal DG allocation in a smart distribution grid using Cuckoo search algorithm, in *2013 10th International Conference on Electrical Engineering/ Electronics, Computer, Telecommunications and Information Technology (ECTI-CON)* (2013), pp. 1–6
27. S. Roy, S. Sultana, P.K. Roy, Oppositional cuckoo optimization algorithm to solve DG allocation problem of radial distribution system, in *2015 International Conference on Recent Developments in Control, Automation and Power Engineering (RDCAPE)* (2015), pp. 44–49
28. A.Y. Abdelaziz, R.A. Osama, S.M. Elkhodary, Using the harmony search algorithm for reconfiguration of power distribution networks with distributed generation units. J. Bioinform. Intell. Control **2**(3), 237–242 (2013)
29. S.I. Kumar, N.P. Kumar, A novel approach to identify optimal access point and capacity of multiple DGs in a small, medium and large scale radial distribution systems. Int. J. Electr. Power Energy Syst. **45**(1), 142–151 (2013)
30. R.S. Rao, K. Ravindra, K. Satish, S.V.L. Narasimham, Power loss minimization in distribution system using network reconfiguration in the presence of distributed generation. IEEE Trans. Power Syst. **28**(1), 317–325 (2013)
31. A. Mohamed Imran, M. Kowsalya, Optimal size and siting of multiple distributed generators in distribution system using bacterial foraging optimization. Swarm Evol. Comput. **15**, 58–65 (2014)
32. M.J. Hadidian-Moghaddam, S. Arabi-Nowdeh, M. Bigdeli, D. Azizian, A multi-objective optimal sizing and siting of distributed generation using ant lion optimization technique. Ain Shams Eng. J. 1–9 (2017)
33. A. Sobieh, M. Mandour, E.M. Saied, M.M. Salama, Optimal number size and location of distributed generation units in radial distribution systems using Grey Wolf optimizer. Int. Electr. Eng. J. **7**(9), 2367–2376 (2017)
34. M.H. Moradi, M. Abedini, A combination of genetic algorithm and particle swarm optimization for optimal DG location and sizing in distribution systems. Int. J. Electr. Power Energy Syst. **34**(1), 66–74 (2012)
35. A.J.G. Mena, J.A.M. Garcia, An efficient approach for the siting and sizing problem of distributed generation. Int. J. Electr. Power Energy Syst. **69**, 167–172 (2015)
36. R. Viral, D.K. Khatod, An analytical approach for sizing and siting of DGs in balanced radial distribution networks for loss minimization. Int. J. Electr. Power Energy Syst. **67**, 191–201 (2015)
37. R. Eberhart, J. Kennedy, A new optimizer using particle swarm theory, in *Proceedings of the Sixth International Symposium on Micro Machine and Human Science MHS'95* (1995), pp. 39–43
38. A.P. Engelbrecht, *Computational Intelligence: An Introduction*, vol. 115, 2nd edn. (Wiley, Chichester, 2008), pp. 3–78
39. K.E. Parsopoulos, M.N. Vrahatis, Parameter selection and adaptation in unified particle swarm optimization. Math. Comput. Model. **46**(1–2), 198–213 (2007)
40. P.A. Gkaidatzis, A.S. Bouhouras, D.I. Doukas, K.I. Sgouras, D.P. Labridis, Application and evaluation of UPSO to ODGP in radial distribution networks, in *2016 13th International Conference on the European Energy Market (EEM)*, vol. 2016, July (2016), pp. 1–5
41. J.H. Holland, Genetic algorithms. Sci. Am. **267**(1), 66–72 (1992)
42. D. Karaboga, B. Basturk, Artificial bee colony (ABC) optimization algorithm for solving constrained optimization problems, ed. by P. Melin, O. Castillo, L.T. Aguilar, J. Kacprzyk, W. Pedrycz, in *Proceedings of the Foundations of Fuzzy Logic and Soft Computing: 12th International Fuzzy Systems Association World Congress (IFSA 2007)*, Cancun, Mexico, 18–21 June 2007 (Springer, Berlin, Heidelberg, 2007), pp. 789–798
43. X.S. Yang, S. Deb, Cuckoo search via levy flights, in *World Congress on Nature Biologically Inspired Computing (NaBIC 2009)* (2009), pp. 210–214
44. Z.W. Geem, J.H. Kim, G.V. Loganathan, A new heuristic optimization algorithm: harmony search. Simulation **76**(2), 60–68 (2001)

45. M. Kashem, V. Ganapathy, G. Jasmon, M. Buhari, A novel method for loss minimization in distribution networks, in *International Conference on Electric Utility Deregulation and Restructuring and Power Technologies (DRPT2000). Proceedings (Cat. No.00EX382)*, no. 603 (2000), pp. 251–256
46. S. Ghosh, S.P. Ghoshal, S. Ghosh, Optimal sizing and placement of distributed generation in a network system. Int. J. Electr. Power Energy Syst. **32**(8), 849–856 (2010)
47. C. Wang, M.H. Nehrir, Analytical approaches for optimal placement of distributed generation sources in power systems. IEEE Trans. Power Syst. **19**(4), 2068–2076 (2004)
48. M.F. Akorede, H. Hizam, I. Aris, M.Z.A. Ab Kadir, Effective method for optimal allocation of distributed generation units in meshed electric power systems. IET Gener. Transm. Distrib. **5**(2), 276 (2011)
49. R.K. Singh, S.K. Goswami, Optimum siting and sizing of distributed generations in radial and networked systems. Electr. Power Components Syst. **37**(2), 127–145 (2009)
50. D.I. Doukas, P.A. Gkaidatzis, A.S. Bouhouras, K.I. Sgouras, D.P. Labridis, On reverse power flow modelling in distribution grids, in *Mediterranean Conference on Power Generation, Transmission, Distribution and Energy Conversion (MedPower 2016)* (2016), p. 65 (6.)
51. R. Yokoyama, S.H. Bae, T. Morita, H. Sasaki, Multiobjective optimal generation dispatch based on probability security criteria. IEEE Trans. Power Syst. **3**(1), 317–324 (1987)
52. P. Kayal, C.K. Chanda, Placement of wind and solar based DGs in distribution system for power loss minimization and voltage stability improvement. Int. J. Electr. Power Energy Syst. **53**, 795–809 (2013)
53. V.R. Pandi, H.H. Zeineldin, W. Xiao, Determining optimal location and size of distributed generation resources considering harmonic and protection coordination limits. IEEE Trans. Power Syst. **28**(2), 1245–1254 (2013)
54. A. Keane, M. O'Malley, Optimal distributed generation plant mix with novel loss adjustment factors, in *2006 IEEE Power Engineering Society General Meeting* (2006), 6 pp
55. Y.M. Atwa, E.F. El-Saadany, M.M.A. Salama, R. Seethapathy, Optimal renewable resources mix for distribution system energy loss minimization. IEEE Trans. Power Syst. **25**(1), 360–370 (2010)
56. C. Yammani, S. Maheswarapu, S. Matam, Optimal placement of multi DGs in distribution system with considering the DG bus available limits. Energy and Power **2**(1), 18–23 (2012)
57. P.A. Gkaidatzis, A.S. Bouhouras, K.I. Sgouras, D.I. Doukas, D.P. Labridis, Optimal distributed generation placement problem for renewable and DG units: an innovative approach, in *Mediterranean Conference on Power Generation, Transmission, Distribution and Energy Conversion (MedPower 2016)* (2016), p. 66 (7.)
58. S. Soudi, Distribution system planning with distributed generations considering benefits and costs. Int. J. Mod. Educ. Comput. Sci. **5**(October), 45–52 (2013)
59. L.F. Ochoa, G.P. Harrison, Minimizing energy losses: optimal accommodation and smart operation of renewable distributed generation. IEEE Trans. Power Syst. **26**(1), 198–205 (2011)
60. L.F. Ochoa, A. Padilha-Feltrin, G.P. Harrison, Evaluating distributed time-varying generation through a multiobjective index. IEEE Trans. Power Deliv. **23**(2), 1132–1138 (2008)
61. G.N. Koutroumpezis, A.S. Safigianni, Optimum allocation of the maximum possible distributed generation penetration in a distribution network. Electr. Power Syst. Res. **80**(12), 1421–1427 (2010)
62. Y.M. Atwa, E.F. El-Saadany, Probabilistic approach for optimal allocation of wind-based distributed generation in distribution systems. IET Renew. Power Gener. **5**(1), 79 (2011)
63. F. Rotaru, G. Chicco, G. Grigoras, G. Cartina, Two-stage distributed generation optimal sizing with clustering-based node selection. Int. J. Electr. Power Energy Syst. **40**(1), 120–129 (2012)
64. M.F. Shaaban, Y.M. Atwa, E.F. El-Saadany, DG allocation for benefit maximization in distribution networks. IEEE Trans. Power Syst. **28**(2), 939–949 (2013)
65. P. Subcommittee, IEEE reliability test system. IEEE Trans. Power Appar. Syst. **PAS-98**(6), 2047–2054 (1979)

66. A.S. Bouhouras, C. Parisses, P.A. Gkaidatzis, K.I. Sgouras, D.I. Doukas, D.P. Labridis, Energy loss reduction in distribution networks via ODGP, in *International Conference on the European Energy Market (EEM)*, vol. 2016–July (2016)
67. B. Pawar, S. Kaur, G.B. Kumbhar, An integrated approach for power loss reduction in primary distribution system, in *2016 IEEE 6th International Conference on Power Systems (ICPS)* (2016), pp. 1–6
68. W.M. Dahalan, H. Mokhlis, Network reconfiguration for loss reduction with distributed generations using PSO, in *2012 IEEE International Conference on Power and Energy (PECon)* (2012), pp. 823–828
69. W. Mohd Dahalan, H. Mokhlis, R. Ahmad, A.H. Abu Bakar, I. Musirin, Simultaneous network reconfiguration and DG using EP method. Int. Trans. Electr. Energy Syst. **25**(11), 2577–2594 (2015)
70. A.S. Bouhouras, P.A. Gkaidatzis, D.P. Labridis, Optimal application order of network reconfiguration and ODGP for loss reduction in distribution networks, in *17 IEEE International Conference on Environment and Electrical Engineering (EEEIC 2017)* (2017), pp. 1–6
71. A.S. Bouhouras, G.T. Andreou, D.P. Labridis, A.G. Bakirtzis, Selective automation upgrade in distribution networks towards a smarter grid. IEEE Trans. Smart Grid **1**(3), 278–285 (2010)
72. A.S. Bouhouras, D.P. Labridis, Influence of load alterations to optimal network configuration for loss reduction. Electr. Power Syst. Res. **86**, 17–27 (2012)
73. N.D. Hatziargyriou, D. Skrlec, T. Capuder, P.S. Georgilakis, M. Zidar, Review of energy storage allocation in power distribution networks: applications, methods and future research. IET Gener. Transm. Distrib. **10**(3), 645–652 (2016)

Chapter 9
Optimal Planning of Grid Reinforcement with Demand Response Control

Alexandre M. F. Dias and Pedro M. S. Carvalho

Abstract This chapter presents a hybrid methodology based on a local search algorithm and a genetic algorithm, used to address the multi-objective and multi-stage optimal distribution expansion planning problem. The methodology is conceived to solve optimal network investment problems under the new possibilities enabled by the smart grid, namely the new observability and controllability investments that will be available to enable demand response in the future. The multi-objective methodology is applied to an existing low-voltage electric distribution network under a congestion scenario to yield a Pareto-optimal set of solutions. The solutions are then projected onto the two investment possibilities considered: demand control investments and traditional network asset investments. The projected surface is then analyzed to discuss the merit of demand control with respect to postponing traditional asset investments.

Keywords Demand response · Distribution planning · Information and communications technology · Network optimization

9.1 Introduction

Distributed Generation (DG) and Electric Vehicles (EV) bring new challenges to the operation of electric distribution networks. New challenges involve dealing not only with peak load conditions, but also with potential reverse flows due to DG and

A. M. F. Dias (✉) · P. M. S. Carvalho
Department of Electrical and Computer Engineering, Instituto Superior Técnico,
University of Lisbon, Lisbon, Portugal
e-mail: alexandre.f.dias@tecnico.ulisboa.pt

P. M. S. Carvalho
e-mail: pcarvalho@tecnico.ulisboa.pt

A. M. F. Dias · P. M. S. Carvalho
Instituto de Engenharia de Sistemas e Computadores - Investigação
e Desenvolvimento (INESC-ID), Lisbon, Portugal

F. Shahnia et al. (eds.), *Electric Distribution Network Planning*,
Power Systems, https://doi.org/10.1007/978-981-10-7056-3_9

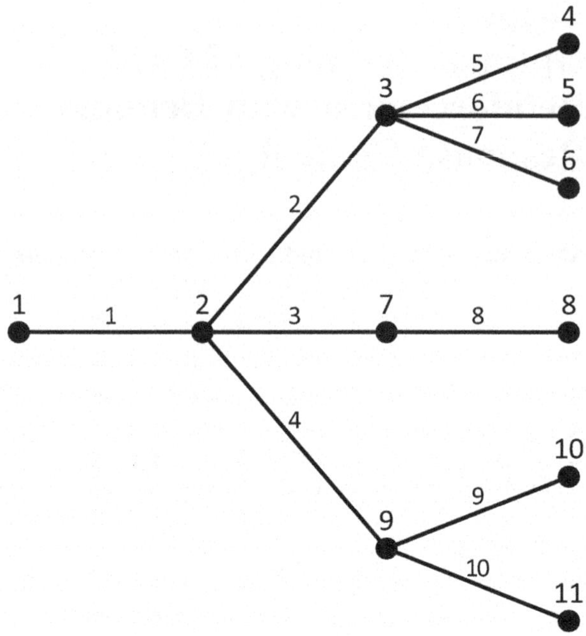

Fig. 9.1 Graph representation of an electric distribution network: circles represent network nodes (graph vertices), while lines connecting circles represent network branches (graph edges) that connect the network nodes

with the new active management possibilities enabled by the Information and Communications Technologies (ICT) such as demand response (DR) control.

Traditionally, distribution planning does not consider the investment and operational benefits of control as enabled by ICT. The benefits of additional controllability are currently not clearly identified and the impacts of ICT investments are not well understood.

A multi-objective optimization methodology is used to search for traditional network investments together with DR control investments, so that a set of investment projects and their corresponding time periods lead to a minimum overall value of a set of objective functions. A multi-objective formulation is quite valuable in this context since it allows projecting the Pareto surface onto the two investment dimensions of (i) investment on traditional network assets and (ii) investment on control equipment. Such a projection allows trading-off these two kinds of investments without speculating on the costs of ICT and control equipment, whose costs are currently difficult to predict.

Let us represent an electric distribution network by a graph G, where the vertices represent the network nodes, while the edges represent the existing lines and transformers (see Fig. 9.1). ICT (vertex) investments can be considered in order to reduce demand impact through demand and DER control, whereas line/transformer (edge) reinforcement investments can be considered in order to relieve overloads and avoid voltage drop or rise beyond predefined acceptable levels. Under such a solution space, the planning solution is a schedule of projects, both vertex and edge type projects, that aim to minimize investment and operational costs while respecting key planning criteria like adequate voltage levels under normal and contingency situations and loading limits for lines and transformers.

An overview of the existing literature shows that distribution planning used to rely on a set of methods to decide the location and type of reinforcements needed in order to cope with the traditional sources of uncertainty, such as expected load forecast, at minimum cost. Several approaches have been taken in the past to solve the problem [1–8].

Some models deal with a fixed horizon year and single network solution topology [9–16] and are thus known as single stage models. Other models deal with the dynamic nature of demand through time as well as with a sequence of network solution topologies (one per stage) and are hence known as multistage models [2, 3, 5, 8, 15].

In either single or multistage models, optimization techniques were used to solve the problem. These involve genetic algorithms [16], Benders' decomposition [17], simulated annealing [18], tabu search [19], greedy randomized adaptive search procedure (GRASP) [20] and game theory [21]. The common output of all these former approaches is the conception of a plan, i.e., a set of projects where the system reinforcements and equipment additions are scheduled.

A hybrid optimization strategy is presented in the following sections of this chapter, combining a local search algorithm with moderate search effort and a metaheuristic method to broaden the search space and to guide towards a close-to-optimum solution.

Finally, a realistic case study, where ICT reinforcements are traded off with conventional grid reinforcements, is analyzed through the application of the developed strategy.

9.2 Distribution Planning Methodology

9.2.1 Formulation of the Problem

Consider a vector of possible investment projects $P = [p_1, p_2, \ldots, p_N]$. A decision schedule can be represented by a vector of timings, $\bar{t}$, that index the projects of P, where $\bar{t} = [t_1, t_2, \ldots, t_N]$.

The optimal planning problem can then be formulated as the problem of finding the optimal timing for each investment project. As a multi-objective problem, the problem may be formulated as

$$\min_{(\bar{t})} \{f_1(\bar{t}), f_2(\bar{t}), \ldots, f_j(\bar{t})\} \tag{9.1}$$

$$\begin{aligned} s.t. \quad & \bar{t} = [t_1, t_2, \ldots, t_N] \\ & t_i \in \{1, 2, \ldots, T+1\}, i = 1, 2, \ldots, N \end{aligned}$$

where $f_j(\bar{t})$ represents the objective function j to be minimized for the planning horizon T, and the timing $T + 1$ is interpreted as the timing to be assigned to the projects that will not be undertaken within such an horizon.

9.2.2 Solution Approach

As decisions are multi-stage, the decision space is large-scale and the decision schedules are computationally expensive to evaluate, making the possible effective solution approaches very confined. Note that the scheduling problem alone is an NP-Hard problem [22] and therefore the search cannot guarantee the global optimum to be found. The solution approach must therefore be able to provide close-to-optimal plans involving short-term and longer-term investment decisions that need to be evaluated thoroughly.

Project schedules can be found by a classical local search algorithm (Gaussian-like search) with moderate search effort. However, such Gaussian algorithm is a local optimization approach, and being the stated problem a non-convex one [23], it does not guarantee close-to-optimum solutions [24]—solutions typically get trapped in local optima. The Gaussian Search (GS) is sensitive with respect to the order by which the different investment projects are analyzed [25]. Therefore, to find close-to-optimum schedules a specific Genetic Algorithm (GA) is presented to learn about the best order for analysis by the GS. The overall solution approach can then be formulated as a hybrid algorithm.

Thus, the GA is used to find the best order for the GS optimization, while GS is used to find the best project schedule given an order to analyze the projects. Within the GS evaluation, objective function values are updated based on the results of a DR optimized power flow, whose implementation is described in this chapter. The presented hybrid approach has proven to yield robust solutions in several planning contexts [25] and has been successfully implemented for a Medium Voltage feeder [26].

Within the application of this hybrid solution, if a multi-objective function is to be addressed explicitly, one may define GS and GA selection criteria to address search as multi-objective and return a set of Pareto non-dominated schedules.

The architecture of the overall algorithm is depicted in Fig. 9.2. The blocks represent the main methods, while the text and direction of the arrows represent, respectively, what information and the direction in which such information is exchanged between those methods. The methods of the hybrid solution approach are described in the following.

The execution of the algorithm starts with the initialization of a set of possible orders, O, for project evaluation by the GS. Within the context of this hybrid approach, a possible order is an order of N projects where each investment project $p_i \in P, i = 1, 2, \ldots, N$ appears exactly once in the order. For example, for $N = 6$

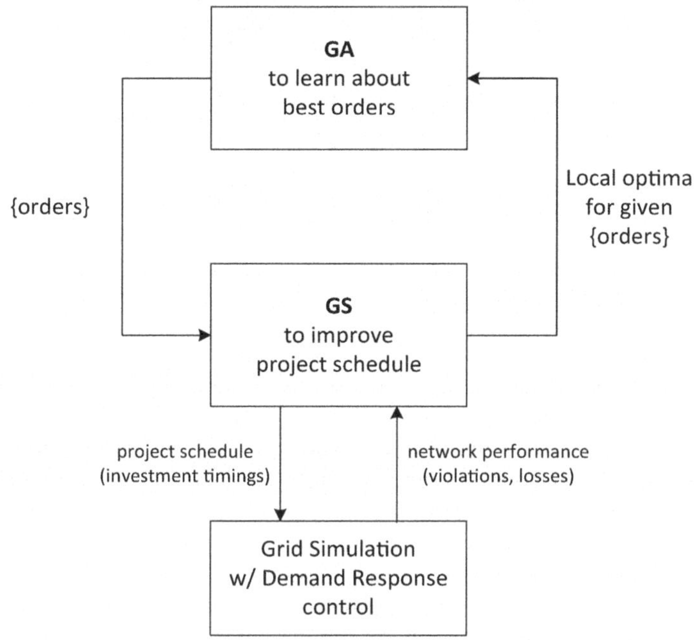

Fig. 9.2 Overview of the hybrid optimization methodology (GA and GS combination)

investment projects and an initial set of 4[1] orders, the initial population of orders could be the one illustrated in Fig. 9.3.

The population is then subjected to evaluation by the Gaussian Search, with each order o_k of O being evaluated separately. Throughout such evaluation, the optimal timings (stages) for allocation of the investment projects are continuously and iteratively updated so that a set of objective functions is minimized. When the set of optimal timings $\bar{t}$ of an order remains unchanged from the previous iteration, the process (GS) stops for that order. A possibility for the initial investment timings for the projects is $T + 1$ for all projects, meaning that at the beginning none of the projects has been allocated to the planning horizon.

More specifically, for a given order o_k of O, one iteration of the GS evaluation of such order corresponds to the update of the optimal investment timings for the $N = 6$ projects according to the relative positions of the projects in that order. As a convention, it is assumed that relative positions are established from left to right (referring to Fig. 9.3).

In the case of order o_1, this means that the GS starts by analyzing project p_3 and finding the optimal investment timing for such a project (t_3); then, it proceeds to the analysis of project p_5 and finding its optimal timing t_5. Similarly, the remaining

[1]Note that this value is relatively small and was chosen for demonstration purposes. The typical size of a genetic algorithm population is several tens or even hundreds of individuals.

o_1	p_3	p_5	p_2	p_4	p_1	p_6
o_2	p_1	p_5	p_6	p_3	p_2	p_4
o_3	p_2	p_4	p_3	p_5	p_6	p_1
o_4	p_4	p_1	p_2	p_3	p_6	p_5

Fig. 9.3 Example of an initial population of orders. The six projects of each order are highlighted in different colors for ease of recognition. Each project appears exactly once in each order

projects p_2, p_4, p_1 and p_6 are analyzed and the end of the order is reached. A certain set of investment timings $\bar{t}$ is obtained in this first iteration of the GS. The GS would then proceed to again analyzing all the projects of order o_1 in the aforementioned sequence, taking into account the optimal timings $\bar{t}$ of the previous iteration. When the end of order o_1 is reached again, the investment timings set is compared to that of the previous iteration. The GS procedure is repeated while the sets of investment timings of the last two iterations compare differently.

The GS determines the optimal investment timing of a project by making use of an Optimal Power Flow method, which considers, for all the other projects, those that have been added to the network graph G so far (investment timings between 1 and T) and, for the project being analyzed, all the possibilities of investment timings (1 to $T + 1$). Therefore, for the investment timings of all the other projects (unchanged) and the possibilities for the project under analysis ($T + 1$), $T + 1$ sets of objective function values are obtained.

From these objective function value sets (solutions), the (Pareto) non-dominated solutions are selected. A solution $a = \{f_1(\bar{t}_a), f_2(\bar{t}_a), \ldots, f_j(\bar{t}_a)\}$ is said to be non-dominated when there is no other solution $b = \{f_1(\bar{t}_b), f_2(\bar{t}_b), \ldots, f_j(\bar{t}_b)\}$ that satisfies $b \leq a$, i.e., the value of at least one objective function value in b is less than the value of the same objective function in a, while the values of the other objective functions in b are less or equal than the corresponding objective function values in a. In the latter case, b is said to dominate a and, conversely, a is said to be dominated by b.

Assume an electric distribution network, with a radial configuration and six branches (lines or cables) connecting the network nodes. The reinforcement of each network branch is considered ($N = 6$), and two objective functions to be minimized as well: network reinforcement and network losses. For the sake of simplicity, the cost of reinforcement is 1 unit (year zero), being discounted in the planning horizon stages with certain inflation and discount rates. Let us also suppose that, during the GS, some projects had already been analyzed and some of the optimal timings were between 1 and T (within the planning horizon), meaning that there were investment projects added to the network graph. When a certain reinforcement project was analyzed (suppose p_4), the results of the Optimal Power Flow for the $T + 1$ investment timing possibilities were the following ($T = 3$):

t_4 ($\bar{t} = [t_1, t_2, t_3, t_4, t_5, t_6]$)	$f_1(\bar{t})$ Present reinforcement cost	$f_2(\bar{t})$ Network losses
1	3.57	5
2	3.51	5
3	3.46	10
4	2.63	15

The impact of this reinforcement project on losses is the same whether it is allocated to stages 1 or 2. But, since its present cost decreases with time, the total reinforcement cost also decreases with time (the investment timings of the other projects, t_1, t_2, t_3, t_5 and t_6 remain unchanged). Therefore, the solution for $t_4 = 2$ dominates the solution for $t_4 = 1$. Thus, allocating project p_4 to stages 2, 3 or 4 leads to non-dominated solutions, since that postponing project p_4 means a decrease in the reinforcement cost but also a corresponding increase in the network losses.

We are then left with three possibilities for the optimal investment timing t_4. The criterion or criteria to select a value can vary. For example, one might want to invest as late as possible or simply choose a non-dominated solution randomly (the latter is used in the case study of this chapter).

The local optimization by the GS is heavily dependent on the Optimal Power Flow (OPF) method. The latter gives an indication to the GS of network performance (objective function values) considering the investments allocated so far within the planning horizon, and when the GS needs to decide on the investment timing of the project being analyzed, it does so based on the results of the OPF method. How investments affect network performance depends on how they impact network operation. More specifically, reinforcing a line or cable is different from ICT infrastructure investment intended to shed consumer load during congestion periods. The impacts and actions of network investments are simulated by the selected OPF method.

The OPF method presented in this chapter aims at simulating the actions facilitated by DR control investment, while still considering the impact of traditional reinforcement investments. If violations of network line or cable current ratings or of node voltage levels are detected, it tries to solve them by shifting (postponing) consumer load demand power. Priority is given to current rating violations over voltage level violations, since the former are more critical to network operation.

The OPF method ensures that the minimum number of controllable consumers is affected when trying to solve violations. On the one hand, it tries to solve current (and later voltage) violations by first looking at problematic branches (nodes) at the most downstream locations in the network topology and only affecting controllable consumers downstream such branches (nodes). Only when violations are solved or when no further reduction in demand power of downstream controllable consumers is possible it moves upstream the network to solve current (voltage) violations. On the other hand, in a given time period only the minimum number of consumers is affected as the demand power of controllable consumers that contribute most to violations (highest demand power) is reduced first, and only then controllable consumers with lower demand power may have their demand power reduced in order to solve violations.

After all the orders of population O are evaluated by the GS, a set of investment timings $\bar{t}$ for each order o_k is determined, as well as the corresponding objective function values. Following GS evaluation, individuals (orders) of the population are selected and then manipulated through genetic operators in order to create a new population. The process of selection and manipulation is described next.

The orders are selected considering how interior the Pareto fronts to which their solutions (objective function values) belong is. From a Pareto-optimal point of view, the more interior the solution front is the better—corresponding to lower overall values of the objective functions, in the case of a problem formulated as a minimization. To determine the Pareto fronts to which each solution belongs, the non-dominated sorting of the NSGA-II [27] is used.

Essentially, the NSGA-II sorting works in the following way: initially, considering all the solutions obtained, the subset of solutions that are not dominated by any other solution belongs to the most interior Pareto front of solutions (with rank equal to one); discarding this first subset of solutions, the remaining solutions that belong to the next most interior Pareto front (not dominated by any other solution) are determined, resulting in a second subset of solutions (with rank equal to two). This process of ranking solutions according to the Pareto front they belong to is repeated until there are no more solutions to be ranked. There can be as much Pareto fronts (ranks) as solutions (in the case of one solution per Pareto front).

Following rank determination, the orders are selected for genetic manipulation using binary tournaments. For each order of the population, an opponent order is selected. Each of these order-opponent order pairs corresponds to one tournament round. For each tournament round, it is decided if the existing population order is kept or is replaced by the opponent order: if the orders solutions have different ranks, the order with the lowest ranked solution wins; if the solution ranks are equal, then one order is randomly chosen as winner. In the case the opponent order wins, it replaces the existing population order. This way of selecting individuals for genetic manipulation ensures that the global search by the GA is guided towards a more Pareto-optimal population.

Genetic manipulation follows order selection. The genetic operators to be applied to the selected orders are order recombination and mutation. The goal of the recombination process is that information regarding relative project positions in the orders that lead to more interior Pareto solutions is exchanged between population orders (note that the selection that precedes manipulation favors lower ranked, more Pareto-optimal solutions). In turn, the mutation process aims at introducing randomness that might contribute to improve the Pareto-wise quality of the solutions (i.e., more interior solutions, namely non-dominated solutions).

The recombination works in the following way: the population of orders is divided into pairs, with each order only being present in one pair; for each pair, recombination will be undertaken (or not) according to a given probability; if a pair is to be recombined, a crossover point is chosen; finally, the positions of the projects

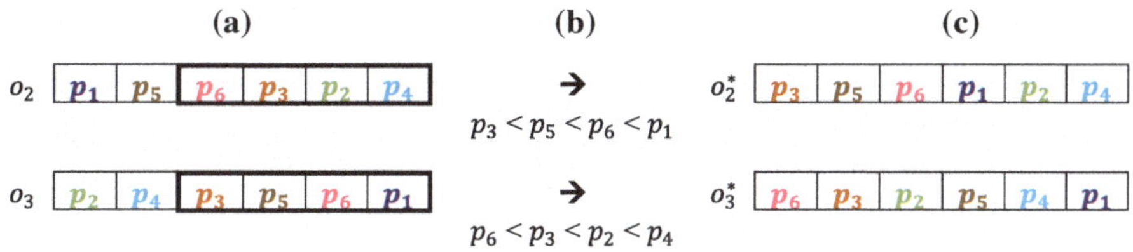

Fig. 9.4 Example of order recombination, starting from the third project position (**a**): relative project positions to be respected in each order (**b**) and offspring orders that result from recombination (**c**)

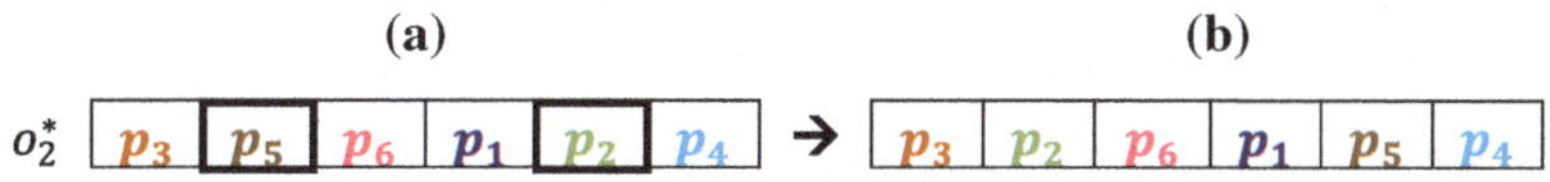

Fig. 9.5 Example of order mutation: chosen project positions (**a**) and resulting order (**b**)

in one order of the pair are altered in order to satisfy the relative project positions imposed by the other order of the pair, starting from the crossover point of the latter.

As an example, suppose that the orders of Fig. 9.3 have been divided into the pairs (o_1, o_4) and (o_2, o_3). According to the recombination probability, it was decided that the orders of the pair (o_2, o_3) were to be recombined. Also, a crossover point was randomly chosen[2] so that recombination takes place starting from the third project position. The relative project positions in one order of the pair, starting from the crossover point, impose relative project positions to be respected in the other order (Figs. 9.4a, b). The orders would have then been changed, with the resulting project positions being the ones in the offspring orders (indicated in Fig. 9.4c).

The next step is the mutation of the orders. This process occurs for each order of the population with a given probability. If it is decided that an order is to be mutated, then two distinct project positions need to be chosen (e.g. randomly). The projects in such positions are then swapped. An example of order mutation is shown in Fig. 9.5.

After genetic manipulation of the orders, a new genetic population is created (population of the next generation). This population will then be subjected to evaluation by the GS and to selection and manipulation in order to create another population. The whole process of population evaluation, selection and manipulation is repeated until some stop or convergence criteria is met.

The steps of the various methods described (Hybrid Genetic Algorithm, Gaussian Search, Optimal Power Flow, Binary tournaments, Recombination and Mutation) are presented below.

[2]For *N* projects, meaningful values of the crossover point are in the range of 2 to *N*-1.

(M1) Hybrid GA search

Step 1 Initialize a set O of possible orders for project evaluation (population of project orders).
Step 2 Evaluate the population of orders by running GS (method M2) on each order k of O (o_k).
Step 3 Determine the rank of each order of O according to the non-dominated sorting described in [27].
Step 4 Select the best orders from the population using binary tournaments (sub-method S1).
Step 5 Subject the best orders to genetic manipulation: recombination (sub-method S2) and mutation (sub-method S3).
Step 6 Go back to Step 2 until convergence is achieved.

(M2) Gaussian Search

Step 1 Initialize i ($i = 1$).
Step 2 Take the ith project of project order o_k and choose the optimal timing for its implementation, t_i, according to an OPF method considering DR control (method M3);
Step 3 If the optimal timing t_i is not $T + 1$, add the project to the network graph G.
Step 4 Increment i ($i \leftarrow i + 1$); go back to Step 2 until the Nth project of o_k is reached.
Step 5 If the end of the project order is reached, go back to Step 1 until the project timing array $\bar{t} = [t_1, t_2, \ldots, t_N]$ remains unchanged from the previous GS iteration (Steps 1–4).

(M3) DR optimized power flow

Step 1 Run power flow for unconstrained network loading.
Step 2 If the current rating of a network branch (line or cable) is violated, go to Step 3, otherwise go to Step 7.
Step 3 Select the most downstream branch (or one of the most downstream branches) with current rating violations.
Step 4 For all consumers downstream the selected branch having a control enabling device, decrease demand power of the consumers with highest demand power according to a predefined rate; the shed demand power is postponed (load shifting); if no (further) decrease of demand power is possible, go to Step 6.
Step 5 Run power flow. If the current rating of the branch is still violated, go back to Step 4, otherwise continue.
Step 6 If there is a branch with current rating violations at the same network topology level of the branch being considered that has yet not been selected, select that branch; otherwise, if there is at least one branch with current rating violations at a level upstream the branch that was being

considered, select a branch at the closest level upstream that branch. If a branch is selected, go back to Step 4.

Step 7 If the voltage level at a node is violated, go to Step 8, otherwise stop.

Step 8 Select the most downstream node (or one of the most downstream nodes) with voltage level violations.

Step 9 For all consumers downstream the selected node having a control enabling device, decrease demand power of the consumers with highest demand power according to a predefined rate; the shed demand power is postponed (load shifting); if no (further) decrease of demand power is possible, go to Step 11.

Step 10 Run power flow. If the voltage level of the node is still violated, go back to Step 9, otherwise continue.

Step 11 If there is a node with voltage violations at the same network topology level of the node being considered that has yet not been selected, select that node; otherwise, if there is at least one node with voltage level violations at a level upstream the node that was being considered, select a node at the closest level upstream that node. If a node is selected, go back to Step 9, otherwise stop.

(S1) Binary tournaments

Step 1 For each individual of the population (order o_k), randomly select an opponent order from the same population.

Step 2 For each individual of the population and its opponent order, decide the winner of the tournament round: if the orders have different ranks, the order with the lowest rank is the winner, otherwise the winning order is one of the two orders chosen at random.

Step 3 For each individual of the population (order o_k), if the winning order of the corresponding tournament round is the opponent order, it replaces the existing order.

(S2) Recombination

Step 1 Randomly divide the population in pairs, with each individual of the population (order o_k) only being present in one single pair.

Step 2 Select the first pair.

Step 3 Decide if the pair is to be recombined with a probability of p_{recomb}; if the pair is to be recombined, go to Step 4, otherwise go to Step 6.

Step 4 Randomly determine the crossover point of the pair.

Step 5 Change the positions of the projects in each order of the pair so that the exchanged relative positions of projects, beginning from the crossover point position to the last project position, are respected.

Step 6 Select the next pair and go back to Step 3 until there are no more pairs left to recombine.

(S3) Mutation

Step 1 Select the first order of the population (order o_1).
Step 2 Decide if the order is to be mutated with a probability of p_{mut}; if the order is to be mutated, go to Step 3, otherwise go to Step 5.
Step 3 Select two distinct project positions at random.
Step 4 Swap the projects in the selected positions.
Step 5 Select the next order of the population and go back to Step 2 until there are no more orders left to mutate.

9.3 Case Study

A Low-Voltage (LV) electric distribution network is considered as an investment case study. The network has a nominal voltage of 400 V and is comprised of thirty six nodes and thirty five cables (underground network with radial topology, see Figs. 9.6 and 9.7). Several network reinforcement and ICT investments are considered as possible projects:

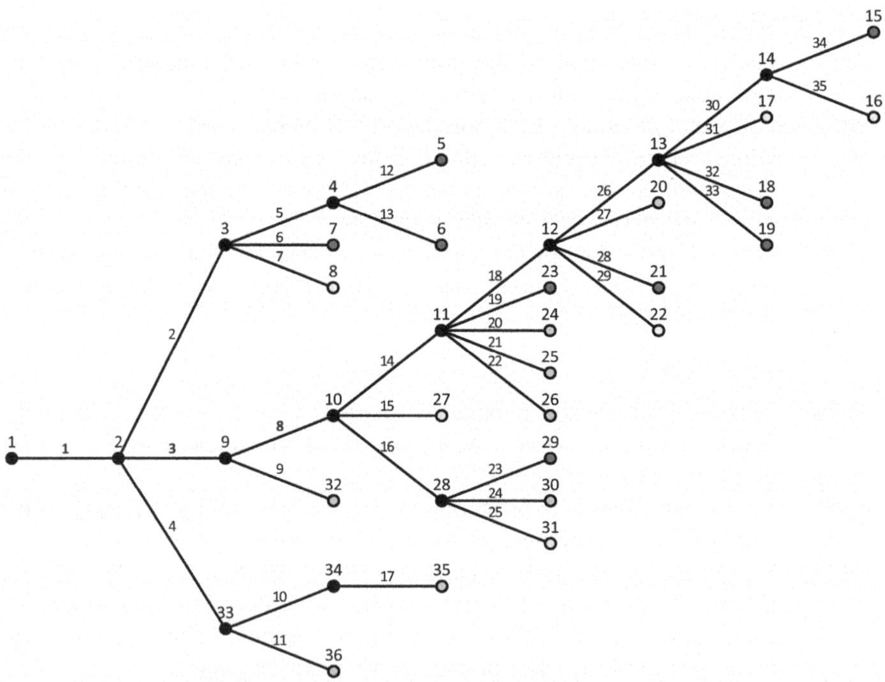

Fig. 9.6 Schematic representation of the network nodes, cables and consumers. Current rating violations are observed for cables 1 (feeder), 3 and 8, hence their numbers are highlighted in color. Nodes with connected consumers are colored according to the combined contracted power at the node: less or equal to 20 kVA (green), greater than 20 kVA and less or equal to 40 kVA (yellow), greater than 40 kVA (orange)

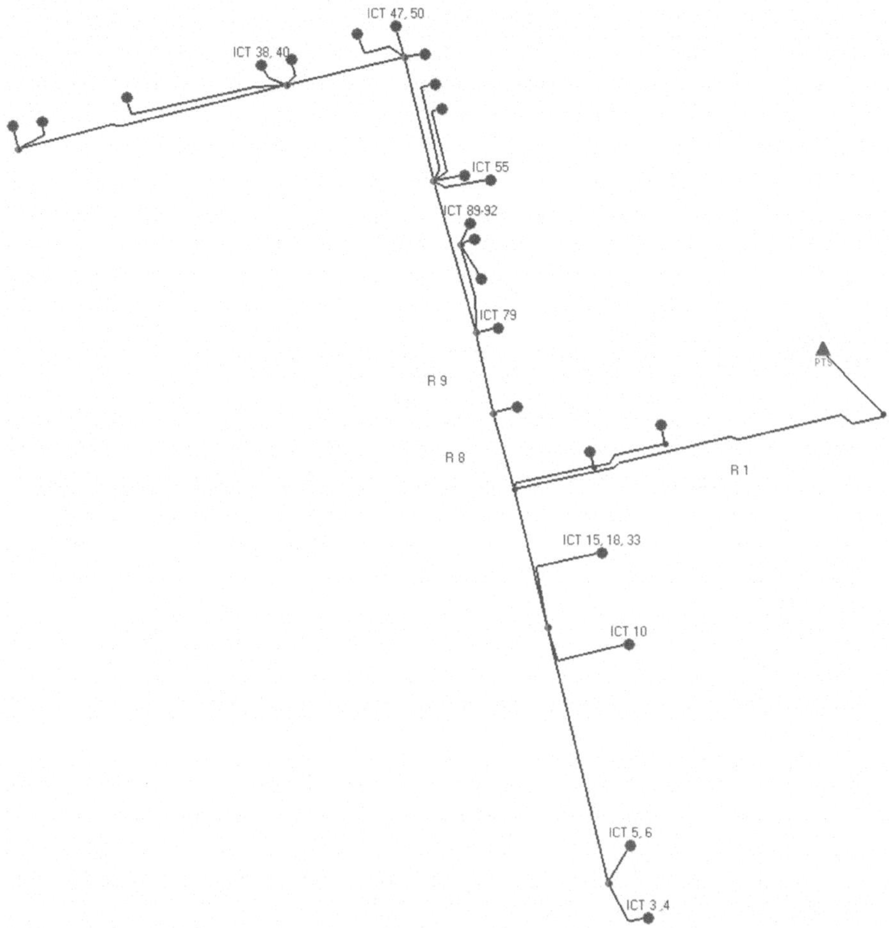

Fig. 9.7 Geographic view of the LV network of the case study, in which possibilities for investment in grid reinforcement (R) and DR reinforcements (ICT) are marked. The triangle refers to the secondary MV/LV substation

- The reinforcement projects are the replacement of three existing weak cables by 185 mm^2 Aluminum conductors (reinforcements denoted by R's in Fig. 9.7).
- The ICT projects are smart meter investments at the premises of every consumer, capable of enabling DR control of such consumers (smart meters denoted by ICT's in Fig. 9.7).

ICT investment is considered as a separate and normalized investment. Being considered as a separate normalized investment (and not minimized together with traditional investment) allows us to trade-off the two types of investment without establishing a cost for ICT, which would be very difficult as the use of ICT for DR control is not yet widespread as a mature technology.

The network has both single- and three-phase consumers (70 single-phase and 30 three-phase consumers) with contracted power values ranging from 1.15 to 20.7 kVA. The most common contracted power values for single-phase consumers are 3.45 and 6.9 kVA (28% and 22% of all the consumers), while for three-phase consumers these values are 10.35 and 17.25 kVA (12% and 8% of all the consumers).

Consumer load demand power is calculated for each consumer using historical smart meter data of a consumer with the same characteristics (identical number of connection phases and contracted power). Both single- and three-phase consumer demand power is evenly distributed among the three phases (for balanced load flow). Unity power factor is assumed. Consumer locations at network nodes are taken from real network data.

The operational conditions in which the case study is run are

- DR simulation starts at peak load time; whenever there is ICT equipment (smart meters) installed at the consumers' premises, the allowed demand power of control enabled consumers can be decreased in steps of 1.15 kVA, as much as needed in order to minimize violations of cable current ratings or node voltage levels.
- DR is simulated for a whole day in each stage of the planning horizon. Such days are assumed to be representative of the corresponding years (stages).
- The MV/LV substation secondary voltage is set to 1.05 pu.
- Monitoring equipment is assumed to be installed at secondary substation level so that cable currents can be measured and communicated to the ICT control system. The costs associated with the installation and operation of monitoring equipment are not considered.

A ten year planning horizon is considered with 11 possible investment stages ($T = 10$), being the investment stages evenly distributed in time (one stage per planning horizon year, plus one stage outside the planning horizon meaning that the project is not to be undertaken within the horizon). Two objective functions are optimized (minimized) together:

- Network reinforcement investment
- ICT (smart meter) investment

The inflation and discount rates considered are 3 and 10%, respectively.

The optimization algorithm is run for an initial population of fifty random project orders and five population generations, and for recombination and mutation probabilities of 80 and 10%, respectively.

From the obtained set of solutions (one optimized project schedule per project order), the non-dominated solutions were selected, and those with current or voltage violations were excluded, as they were considered unfeasible. The remaining non-dominated solutions were projected onto the objectives of ICT investment and reinforcement investment. Four zones corresponding to four project allocation modes were identified. The projected non-dominated solutions and the identified zones are shown in Fig. 9.8.

The solution with highest network reinforcement cost (2400 €) and no ICT investment corresponds to the reinforcement of all the three cables where current

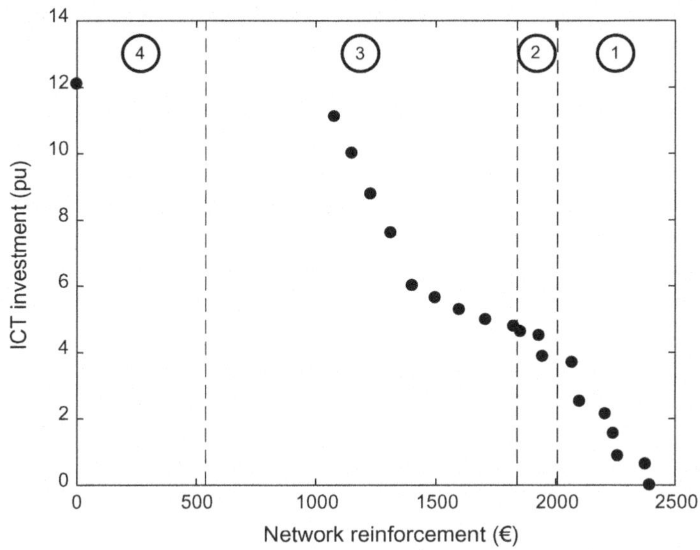

Fig. 9.8 Projection of the obtained non-dominated solutions onto the objectives of ICT investment and reinforcement investment for the LV electric distribution network

rating violations occur (cables 1, 3 and 8, referring to Fig. 9.6). The reinforcement of such cables is enough to ensure violation-free operation of the network within the planning horizon.

As ICT projects are allocated, DR control is possible and the reinforcement projects are postponed, making the reinforcement cost lower (considering inflation and discount rates, the present cost is lowered). The transition from zone 1 to 2 occurs when the allocated ICT projects are enough to avoid one reinforcement project (reinforcement of cable number 8).

Cables number 3 and 8 are almost identical (in length and cross section), and therefore when the OPF method reduces demand power downstream cable number 8, solving its current rating violations and thus avoiding its reinforcement, it also removes most of contributions to the violations of the current rating of cable number 3. As more ICT projects are allocated and more consumer demand power is shifted by the OPF, the reinforcement of cable number 3 is also avoided and transition from zone 2 to 3 occurs.

With the reinforcement of cables 3 and 8 avoided, what is left is the reinforcement of cable number 1 (feeder cable). Similarly to what happens in zones 1 and 2, allocating more ICT projects and shifting consumer demand allows postponing the reinforcement of cable 1. Since there are only ten stages (years) in the considered planning horizon, the reinforcement of cable 1 can only be postponed up to the last stage of the planning horizon.

Having enough ICT projects allocated and enough consumer demand power to be shifted makes the transition from zone 3 to 4 possible—the reinforcement of the feeder cable (and of the other two cables) is avoided. The optimal solution where all cable reinforcements are avoided corresponds to about 12 pu of investment in ICT projects (present cost) and 18 consumers (18% of all consumers) with DR enabling devices.

In Figs. 9.9 and 9.10, the distribution of the reinforcement and ICT investments according to their optimal timings for the two extreme solutions is illustrated. In Fig. 9.9, the allocation of the three reinforcement projects matches the stages when violations would first occur in each of the cables. On the other hand, it can be observed from Fig. 9.10 that the allocation of ICT projects is gradual and not abrupt, varying between 0.9 and 2.0 pu per stage of the planning horizon. This can also be verified through the cumulative ICT investment cost, whose evolution is very well fitted to a first-degree polynomial curve.

Figures 9.11, 9.12 and 9.13 correspond to the solutions preceding the transitions from zone 1 to 2 (Fig. 9.11), from zone 2 to 3 (Fig. 9.12) and from zone 3 to 4 (Fig. 9.13). From such figures, one can observe which reinforcement projects have been postponed or avoided so far, the limits for postponing reinforcement projects in the solutions found by the described hybrid approach and, similarly to Fig. 9.10, that ICT project allocation is gradual.

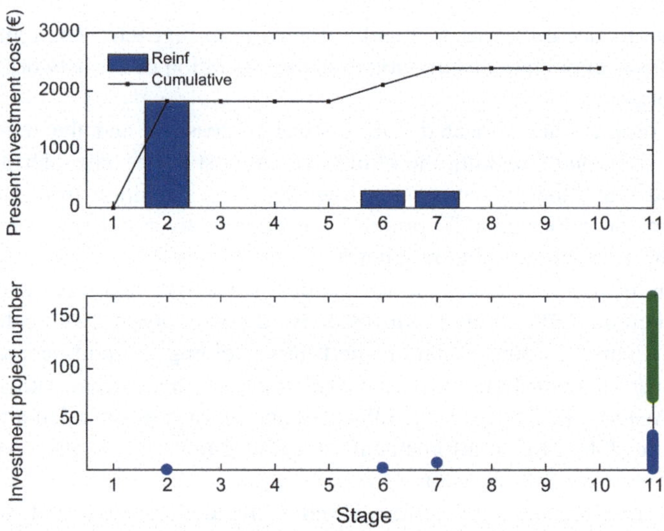

Fig. 9.9 Optimal investment stages and investment costs for reinforcement projects (blue dots and bars) of the solution where all the necessary reinforcement projects and no ICT projects are allocated

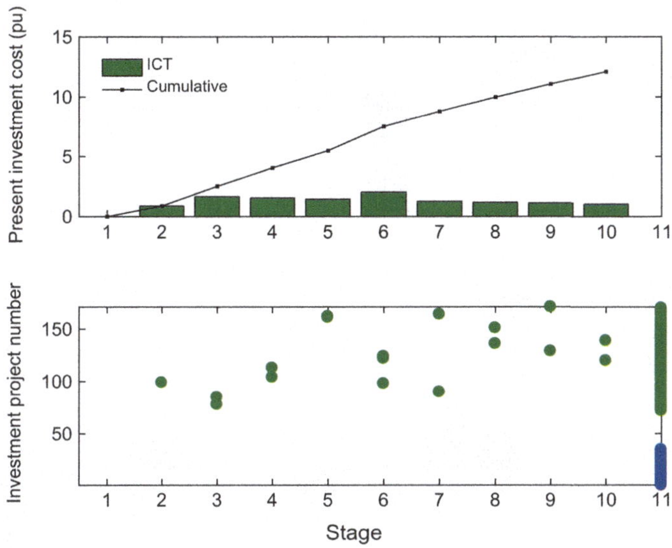

Fig. 9.10 Optimal investment stages and investment costs for ICT projects (green dots and bars) of the solution where the allocation of enough ICT projects avoids all cable reinforcements

The projection of the solutions in Fig. 9.8 can be looked at from another point of view: by doing a translation of the horizontal axis, the solutions can be seen from the perspective of avoided network reinforcement (due to the investment in ICT equipment) instead of required network reinforcement. The result of this translation is shown in Fig. 9.14. In this figure, the dashed lines connect the axes origin to the projection of certain solutions (marked within circles). These are the four solutions where ICT is most valuable in terms of avoided network reinforcement per installed ICT unit (€/pu of ICT)—the more horizontal the line connecting the solution projection to the axes origin, the higher the average ICT unit value.

For the extreme solution where enough ICT investment leads to complete avoidance of traditional network reinforcement, the current profile of the cables whose reinforcement is avoided, before and after investment optimization with DR, is shown in Figs. 9.15 and 9.16. It can be observed from these figures that DR control affects load demand power (and thus cable current) in two essential periods: from 14 to 16 h, when the current rating of the feeder cable would be violated, and from 20 to 24 h, when the ratings of the three cables would also be exceeded (blue lines). The excess demand power is shifted to later periods, decreasing cable loading in those two periods, which is offset by a slight increase in cable loading in the period of 17 to 19 h and a significant increase of such loading in the period of 1 to 5 h.

Changes occur in the load demand profiles of the consumers with DR enabling devices. Two groups of profile changes are identified: consumers whose profiles are affected by demand power shed in the period of 20 to 1 h and shifted to the period

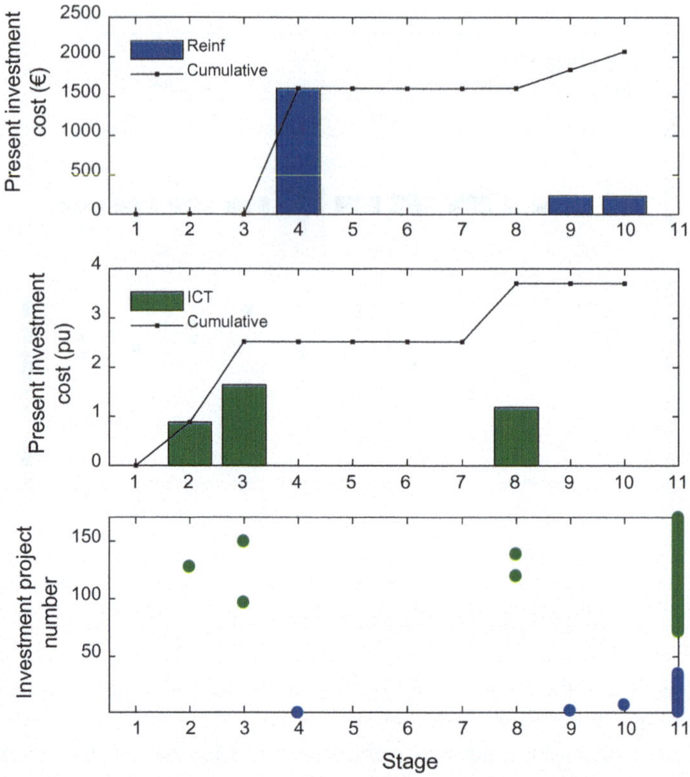

Fig. 9.11 Optimal investment stages for reinforcement and ICT projects (blue and green dots, respectively) and investment costs for reinforcement and ICT projects (blue and green bars, respectively) of the solution preceding the transition from zone 1 to 2 in Fig. 9.8. This solution corresponds to the limit found for postponing the reinforcement of cable number 8 (stage 10), while also reinforcing cables number 1 and 3

of 1 to 5 h (named DR pattern number 1) and consumers whose profiles, besides being affected by the same demand power shifting, are also affected by the postponing of demand power in the period of 14 to 17 h to the period of 16 to 19 h (named DR pattern number 2).

In Table 9.1, for the consumers affected by DR control (eighteen out of the total of one hundred), their network connection node, contracted power and DR pattern number are summarized. Figure 9.19 is similar to Fig. 9.6, with the differences being that the colored network nodes represent the nodes where consumers affected by DR patterns number 1 (blue) and number 2 (yellow) are located. In Figs. 9.17 and 9.18, two examples of optimized consumer demand profiles due to DR control (one example per DR pattern) are shown.

By comparing Figs. 9.6 and 9.19, it can be observed that there is a correlation between nodes with higher combined contracted power and the corresponding

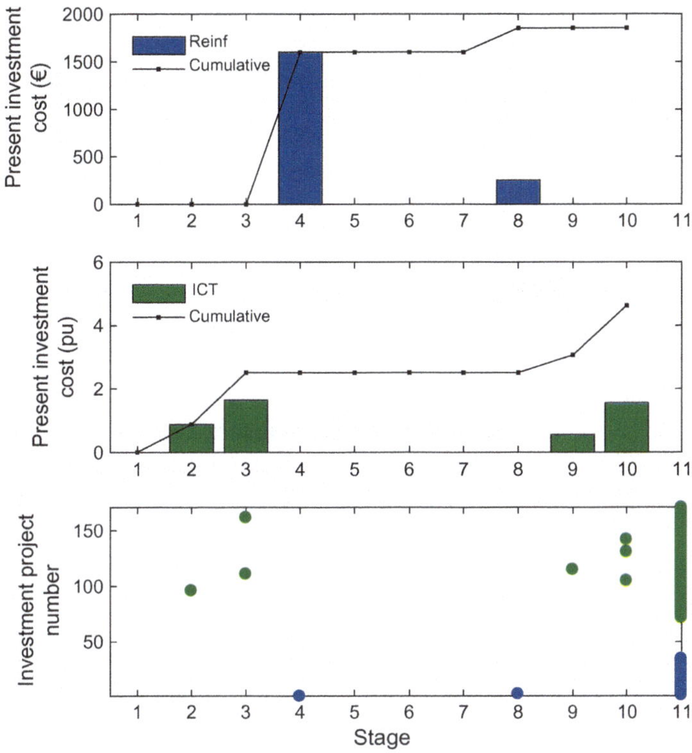

Fig. 9.12 Optimal investment stages for reinforcement and ICT projects (blue and green dots, respectively) and investment costs for reinforcement and ICT projects (blue and green bars, respectively) of the solution preceding the transition from zone 2 to 3 in Fig. 9.8. This solution corresponds to the limit found for postponing the reinforcement of cable number 3 (stage 8), while reinforcing cable number 1 and avoiding the reinforcement of cable number 8

consumers having DR enabling equipment—most of the nodes concerning consumers with DR equipment (blue and yellow colored nodes in Fig. 9.19) correspond to nodes with a combined contracted power greater than 20 kVA (yellow and orange colored nodes in Fig. 9.6). This is expected, as more consumers connected to the same node (higher combined contracted power) means higher aggregate loading demand power, thus enabling DR control for consumers at such nodes will decrease upstream cable loading when violations would occur.

It should be noted that these results were obtained for given reference costs of cable reinforcement and a given load density (respectively, around 40k € per kilometer of cable replacement and a contracted power density of 1900 kVA per kilometer of feeder length). For different cable reference costs and contracted power densities, the valuation of ICT investment would also be different.

If the same load density is considered with varying reference reinforcement costs, the ICT valuation would be directly proportional to such costs. If not, if the

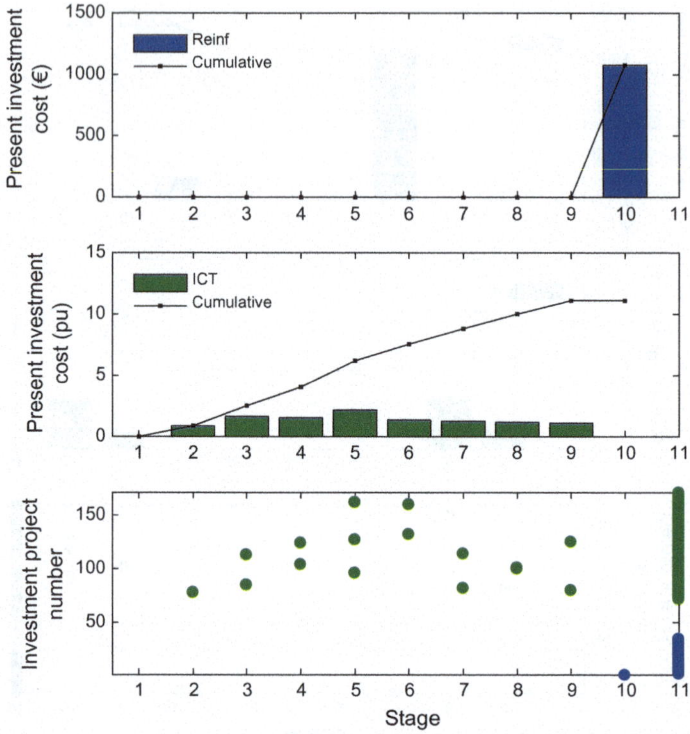

Fig. 9.13 Optimal investment stages for reinforcement and ICT projects (blue and green dots, respectively) and investment costs for reinforcement and ICT projects (blue and green bars, respectively) of the solution preceding the transition from zone 3 to 4 in Fig. 9.8. This solution corresponds to the limit found for postponing the reinforcement of cable number 1 (stage 10), while avoiding the reinforcement of cables number 3 and 8

network has the same reference costs but different load density, the ICT valuation would be inversely proportional to the load density. For instance, with a higher load density (same aggregate contracted power and lower network length) the installation of ICT infrastructure enables the management of the same load demand power as in a lower load density situation (higher network length), while the reinforcement expected to cope with that load demand power would be lower in length (lower in cost), and in that way the ICT valuation would be lower.

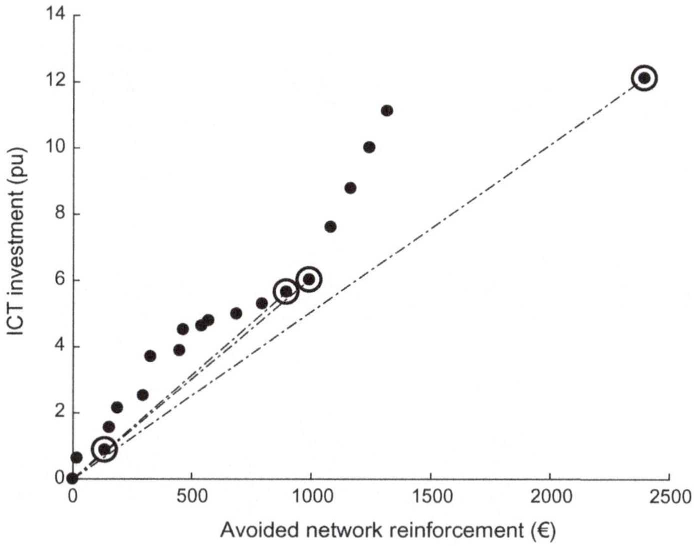

Fig. 9.14 Projection of the obtained non-dominated solutions onto the objectives of ICT investment and reinforcement investment for the LV electric distribution network

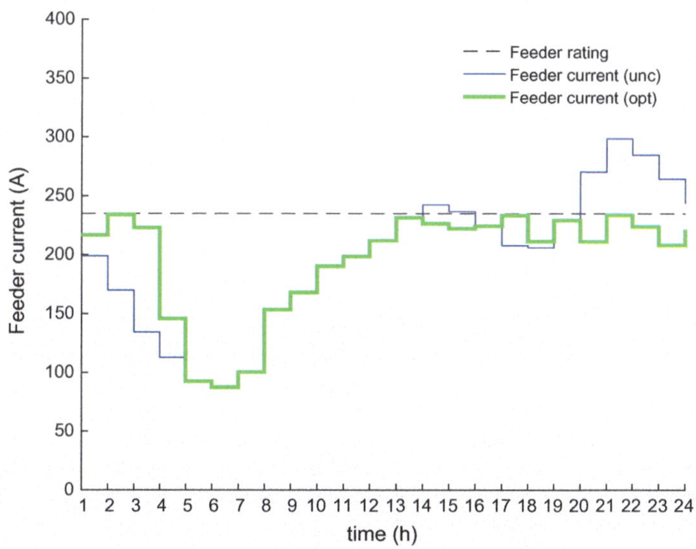

Fig. 9.15 Current in the feeder cable (year 10); due to the combination of network monitoring with DR control as enabled by ICT (peak shifting), feeder cable current does not exceed the cable current rating (dashed line), allowing its reinforcement to be avoided

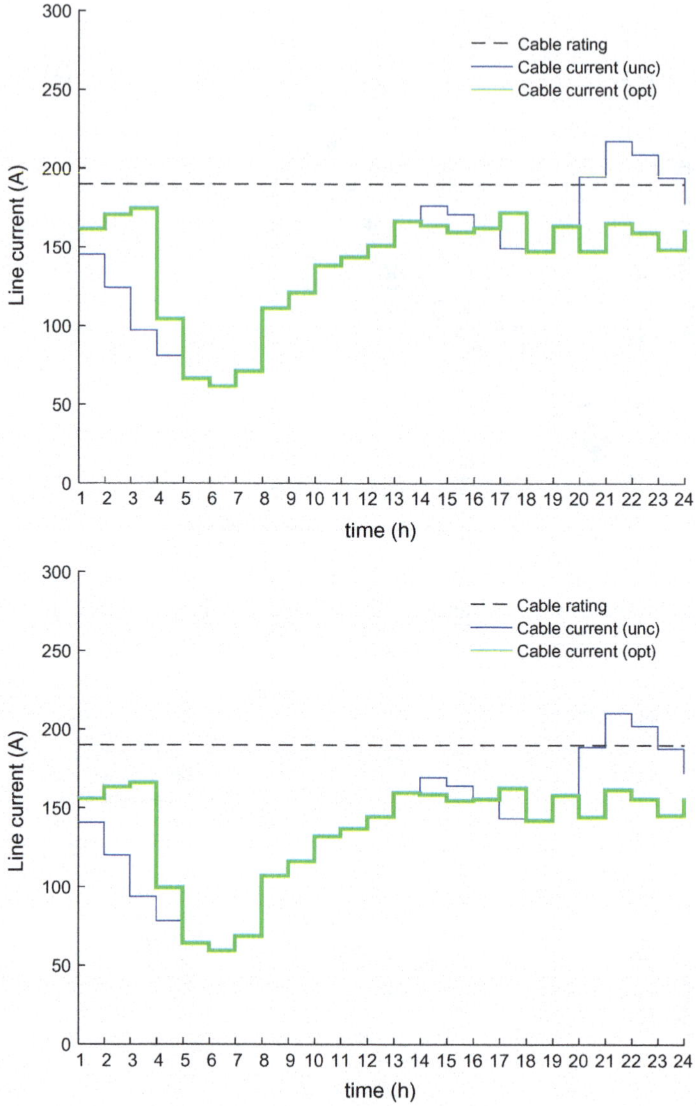

Fig. 9.16 Current in cables number 3 (top) and 8 (bottom), in the last stage of the planning horizon (year 10); due to the combination of network monitoring with DR control as enabled by ICT (peak shifting), cable current does not exceed cable current rating (dashed line), allowing the corresponding reinforcements to be avoided

Table 9.1 Summary of the network connection nodes, contracted power and DR pattern number of the consumers affected by DR control (eighteen consumers)

Consumer	Connection node	Contracted power (kVA)	DR control pattern
7	6	20.70	2
14	7	13.80	1
19	15	6.90	2
27	16	6.90	2
28	16	6.90	2
33	17	10.35	1
42	18	10.35	1
49	21	17.25	1
51	22	5.75	1
53	22	10.35	1
58	23	17.25	1
65	25	4.60	1
68	25	4.60	1
80	27	6.90	2
90	31	6.90	2
91	31	6.90	2
93	32	6.90	2
100	36	4.60	1

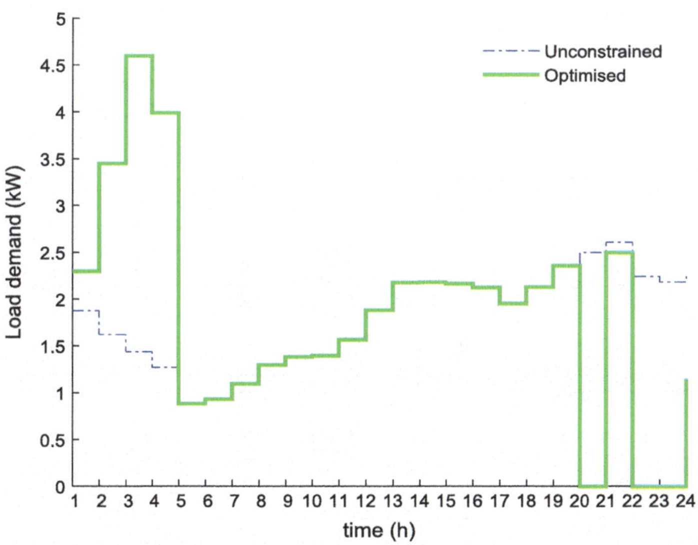

Fig. 9.17 Consumer load demand profile optimization (year-10) as a result of DR as enabled by ICT (smart meters): consumer 58 as an example of DR pattern number 1

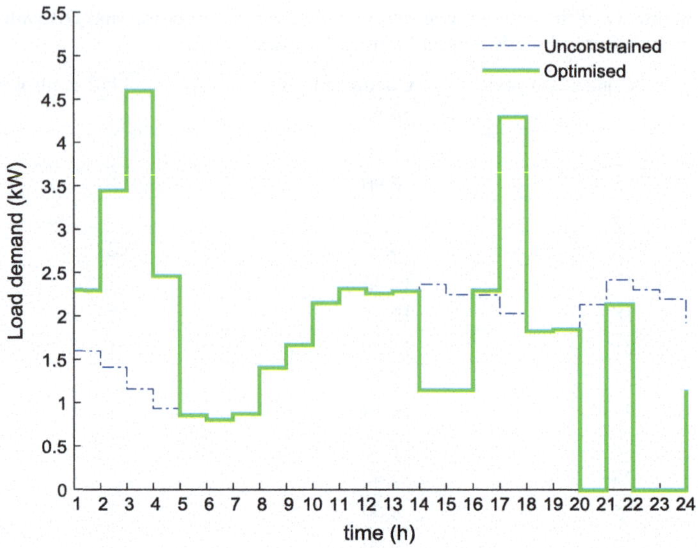

Fig. 9.18 Consumer load demand profile optimization (year-10) as a result of DR as enabled by ICT (smart meters): consumer 80 as an example of DR pattern number 2

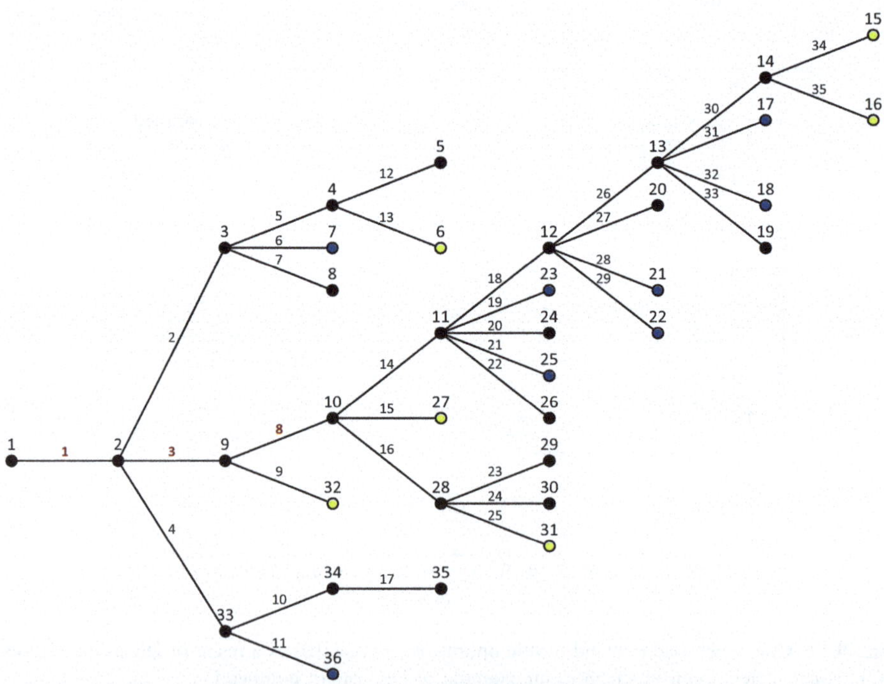

Fig. 9.19 Schematic representation of the network nodes, cables and consumers affected by DR control. Current rating violations are observed for cables 1 (feeder), 3 and 8, hence their numbers are highlighted in color. Nodes with DR enabled consumers are colored according to the DR control pattern: pattern number 1 (blue) and pattern number 2 (yellow)

Appendix

The notations used throughout this chapter are listed below:

f_j Objective function j
P Set of investment projects
O Set of orders for project analysis (population)
p_i Project i of P
o_k Order k of O (individual of the population)
t_i Timing of project p_i, $t_i \in \{1, 2, \ldots, T+1\}$
$\bar{t}$ Decision schedule: indexed array of timings t_i for projects p_i
N Number of projects
T Number of stages of the planning horizon
G Graph of the electric distribution network

References

1. M.V.F. Pereira, L.M.V.G. Pinto, S.H. Cunha, G.C. Oliveira, A decomposition approach to automated generation transmission expansion planning. IEEE Trans. Power Syst. **PAS-104** (11), 3074–3083 (1985)
2. R. Romero, A. Monticelli, A Hierarchical decomposition approach for transmission network expansion planning. IEEE Trans. Power Syst. **9**(1), 373–380 (1994)
3. R. Romero, A. Monticelli, A zero-one implicit enumeration method for optimizing investments in transmission expansion planning. IEEE Trans. Power Syst. **9**(3), 1385–1391 (1994)
4. G.C. Oliveira, A.P.C. Costa, S. Binato, Large scale transmission network planning using optimization and heuristic techniques. IEEE Trans. Power Syst. **10**(4), 1828–1833 (1995)
5. R. Romero, R.A. Gallego, A. Monticelli, Transmission expansion planning by simulated annealing. IEEE Trans. Power Syst. **11**(1), 364–369 (1996)
6. H. Rudnick, R. Palma, E. Cura, C. Silva, Economically adapted transmission systems in open access schemes—application of genetic algorithms. IEEE Trans. Power Syst. **11**(3), 1427–1440 (1996)
7. R.A. Gallego, A. Monticelli, R. Romero, Comparative studies on non-convex optimization methods for transmission network expansion planning. IEEE Trans. Power Syst. **13**(3), 822–828 (1998)
8. X. Wang, Y. Mao, Improved genetic algorithm for optimal multistage transmission system planning. IEEE (2001)
9. L.L. Garver, Transmission network estimation using linear programming. IEEE Trans. Power Syst. **PAS-89**(1), 1688–1697 (1970)
10. A. Monticelli, A. Santos, M.V.F. Pereira, S.H. Cunha, B.J. Parker, J.C.G. Praça, Interactive transmission network planning using a least-effort criterion. IEEE Trans. Power App. Syst. **PAS-101**(10), 3919–3925 (1982)
11. M.V.F. Pereira, L.M.V.G. Pinto, S.H. Cunha, G.C. Oliveira, A decomposition approach to automated generation/transmission expansion planning. IEEE Trans. Power Syst. **PAS-104** (11), 3074–3083 (1985)
12. R. Romero, A. Monticelli, A hierarchical decomposition approach for transmission network expansion planning. IEEE Trans. Power Syst. **9**(1), 373–380 (1994)

13. R. Romero, A. Monticelli, A zero-one implicit enumeration method for optimizing investments in transmission expansion planning. IEEE Trans. Power Syst. **9**(3), 1385–1391 (1994)
14. G.C. Oliveira, A.P.C. Costa, S. Binato, Large scale transmission network planning using optimization and heuristic techniques. IEEE Trans. Power Syst. **10**(4), 1828–1833 (1995)
15. A. Escobar, R.A. Gallego, R. Romero, Multistage and coordinated planning of the expansion of transmission systems. IEEE Trans. Power Syst. **19**(2), 735–744 (2004)
16. H. Rudnick, R. Palma, E. Cura, C. Silva, Economically adapted transmission systems in open access schemes—application of genetic algorithms. IEEE Trans. Power Syst. **11**(3), 1427–1440 (1996)
17. S. Binato, M.V. Pereira, S. Granville, A new benders decomposition approach to solve power transmission design problems. IEEE Trans. Power Syst. **16**(2), 235–240 (2001)
18. R. Romero, R.A. Gallego, A. Monticelli, Transmission expansion planning by simulated annealing. IEEE Trans. Power Syst. **11**(1), 364–369 (1996)
19. E.L. Silva, J.M.A. Ortiz, G.C. Oliveira, S. Binato, Transmission network expansion planning under a tabu search approach. IEEE Trans. Power Syst. **16**(1), 62–1440 (2001)
20. S. Binato, G.C. Oliveira, J.L. Araújo, A greedy randomized adaptive search procedure for transmission expansion planning. IEEE Trans. Power Syst. **16**(2), 247–253 (2001)
21. J. Contreras, F.F. Wu, A kernel-oriented algorithm for transmission expansion planning. IEEE Trans. Power Syst. **15**(4), 1434–1440 (2000)
22. M. Pinedo, *Scheduling—Theory, Algorithms, and Systems* (Prentice Hall, 1995). ISBN 0-13-706757-7
23. F.S. Reis, M. Pinto, P.M.S. Carvalho, L.A.F.M. Ferreira, *Short-Term Investment Scheduling in Transmission Power Systems by Evolutionary Computation—DRPT2000* (London, April 2000)
24. F.S. Reis, P.M.S. Carvalho, L.A.F.M. Ferreira, Combining gauss and genetic algorithms for multi-objective transmission expansion planning. WSEAS Trans. Syst. **3**(1), 206–209 (2004)
25. F.S. Reis, P.M.S. Carvalho, L.A.F.M. Ferreira, Reinforcement scheduling convergence in power systems transmission planning. IEEE Trans. Power Syst. **20**(2), 1151–1157 (2005)
26. A. Dias, P.M.S. Carvalho, P. Almeida, S. Rapoport, Multi-objective distribution planning approach for optimal network investment with EV charging control, in *Presented at PowerTech 2015* (June 2015) [Online], Available: http://ieeexplore.ieee.org/stamp/stamp.jsp?arnumber=7232674
27. K. Deb, A. Pratap, S. Agarwal, T. Meyarivan, A Fast and elitist multiobjective genetic algorithm: NSGA-II. IEEE Trans. Evol. Comput. **6**(2), 182–197 (2002)

Chapter 10
Simultaneous Network Reconfiguration and Sizing of Distributed Generation

Wardiah Mohd Dahalan and Hazlie Mokhlis

Abstract This chapter introduces simultaneous optimization concept of Network Reconfiguration and Distributed Generation sizing. The main objective of the introduced concept is to reduce the real power loss and improve the overall voltage profile in the electric distribution network through optimal network reconfiguration and Distributed Generation sizing, while at the same time satisfy the system operating constraints. The meta-heuristic methods have been applied in the optimization process due to its excellent capability for searching optimal solution in a complex problem. The applied meta-heuristics methods are Genetic Algorithm, Evolutionary Programming, Particle Swarm Optimization, Artificial Bee Colony and their respective modified types. A detail performance analysis is carried out on IEEE 33-bus systems to demonstrate the effectiveness of the proposed concept. Through simultaneous optimization, it was found that power loss reduction is more as compared to conducting reconfiguration or DG sizing approach alone. The test result also indicated that Evolutionary Particle Swarm Optimization produced better result in terms of power loss and voltage profile than other methods.

Keywords Distributed generation · Optimization techniques · Power loss reduction · Reconfiguration · Meta-Heuristic method

W. M. Dahalan
Department of Electrical Engineering, Universiti Kuala Lumpur,
Kuala Lumpur, Malaysia
e-mail: wardiah@unikl.edu.my

W. M. Dahalan
Malaysian Institute of Marine Engineering Technology, Perak, Malaysia

H. Mokhlis (✉)
Department of Electrical Engineering, University of Malaya,
Kuala Lumpur, Malaysia
e-mail: hazli@um.edu.my

F. Shahnia et al. (eds.), *Electric Distribution Network Planning*,
Power Systems, https://doi.org/10.1007/978-981-10-7056-3_10

10.1 Introduction

The major issues nowadays that most power utilities are trying to achieve in generation, transmission and distribution systems are to ensure high service quality, reliability and efficiency of the overall power system. Distribution system is the final stage in the process of power delivering power from the generation to the individual customer. It has contributed the greatest amount of power loss in which resulted to poor voltage magnitude. The performance of electric distribution networks becomes inefficient due to the increase in power loss and reduction in voltage magnitude especially in the heavily loaded network. The studies in [1, 2] report that 70% of the total losses are occurring in the electric distribution networks while transmission and sub transmission lines account for only 30% of the total losses is as shown in Fig. 10.1.

One of the well-known techniques to minimize power loss is through network reconfiguration [3]. This technique can reduce the power loss and improve the overall voltage profile provided that the most optimum configuration can be determined. The application of network reconfiguration in the electric distribution networks can be divided into two categories; planning and operation. In planning, network reconfiguration is needed to identify the best configuration via changing the on/off sectionalizing and tie-switches in the network. By doing that, the load will transfer from heavily loaded feeders to relatively less heavily loaded feeders, so that the power loss is minimized. Meanwhile, the network reconfiguration during operation plays an important role in the process of rerouting power supply in the network due to a fault. In this context, reconfiguration is required to restore power supply automatically and quickly to un-faulted sections of the system to improve the system reliability. Instantaneous response to the damage system prevents it from propagating allowing as many loads as possible to function. Network reconfiguration requires optimization technique to determine the best combination set of switches to be open. The execution of the process of selection should fulfill the

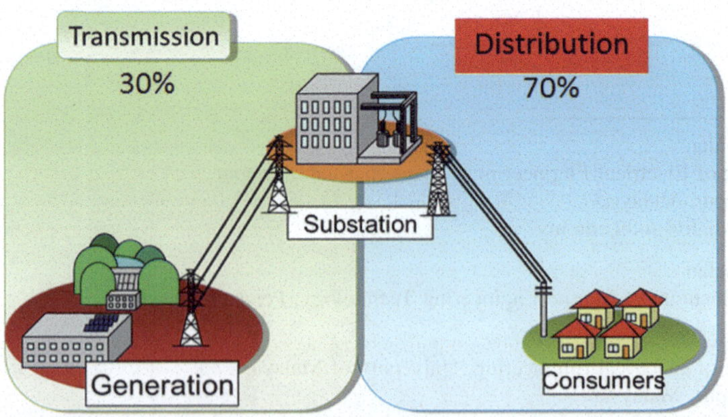

Fig. 10.1 Illustration of a power system delivering power to customers

requirement of optimization (minimize power loss) and satisfying the operating constraints. The application of reconfiguration network is much simpler and cost efficient compared to other techniques. In general, reconfiguration have two primary aim which are to provide the maximum amount of electrical supply to the end customers and reconfigure the network automatically as soon as there are problems arises such as fault.

Another technique that able to reduce power loss is by interconnecting local power supply in electric distribution networks. By having local supply, electrical power can be delivered to the load in a short distance, which able to reduce power loss. Local power supply from renewable energy sources such as mini-hydro, wind, solar and Bio-fuel are nowadays is connected in networks to generate electrical power. In view of this matter, a new identity appeared in the electric distribution networks known as "Distributed Generations" (DG). DG is related to the use of small generating units installed at strategic points on the networks and mainly close to the load centers. It can be used in an isolated way, supplying the consumer's local demand, or in an integrated way, supplying energy to the remaining system [4]. In general, DG is the generation of electricity by facilities smaller than the central plants, usually 10 MW or less [5].

From studies, DG penetration is predicted will surpass more than 25% of the total generation in the foreseeable future [6]. Studies also revealed that the usage of Renewable Energy DG could reduce 60% of the carbonic pollution from conventional power generation by 2050 [7]. With this regard, changing the environment of power systems design and operation has caused the need to consider active distribution network. The integration of distribution system would lead to the improvement of the voltage profile, load balancing, reliability such as service restoration and increase energy efficiency. Therefore, it is very crucial to ensure that the DG size is at the optimal value to maximize its benefits. An inappropriate size of DG will cause of power loss in the system to be higher than the initial configuration.

Many researchers have employed various methods to overcome the problem of optimal reconfiguration and DG sizing in the electric distribution networks [8–16]. However, the existing methods have some limitations and drawbacks in their solutions such as the solution might trap in the locally optimal solution. This is due to the process of finding the optimum solution used sequential approach (e.g.: find the optimal DG first and followed by optimal reconfiguration or vice versa). Moreover, the reconfiguration is a complicated combinatorial and non-differentiable constrained optimization problem. It involves with many candidate-switching combinations. These obstacles really put the reconfiguration process in difficulties to achieve the comprehensive optimal solution and take too long time to reach the convergence point. Apart from that, the combination switches obtained from their results sometimes is not in radial which is important characteristic in assist finding optimal power loss.

Although there are various methods for network reconfiguration, the DG effect in the network reconfiguration has not been considered widely by researchers. There are very few researchers who considered network reconfiguration with DG [17–19]. Most of them have already fixed the size of DG or solve sequentially and the impact of DG on the distribution system has not been discussed seriously. None of them tried to solve the reconfiguration and DG problem simultaneously.

The installation of DGs in the distribution system indeed can improve energy efficiency and voltage profile, and at the same time minimize power interruption power. However, in order to ensure the effectiveness of DG in the distribution system, selecting the optimal size of DG plays an important role in giving the greatest impact on the operations and control of the electric distribution networks. Thus, the correct size (dispatch value) of DG becomes a vital point for the network system in order to produce a lower amount of power waste. In literature, the term DG size is commonly applied to represents the power dispatch value of the DG. Thus, throughout this chapter, DG size refers to the DG dispatch value.

Considering the existing limitations, this chapter introduces simultaneous optimization concept in finding optimal network reconfiguration and DGs size. The meta-heuristics methods used in chapter are Genetic Algorithm (GA), Evolutionary Programming (EP), Particle Swarm Optimization (PSO) and Artificial Bee Colony (ABC). Meanwhile, the modified versions of these methods which contribute greatly to this chapter are Modified Genetic Algorithm (MGA), Evolutionary Particle Swarm Optimization (EPSO), Modified Particle Swarm Optimization (MPSO) and Simplified Artificial Bee Colony (SABC).

10.2 Optimal Network Reconfiguration and Distributed Generation Sizing

Electric distribution network reconfiguration can be seen as a combinatorial optimization problem, comprising distribution system planning, loss minimization and energy restoration. Generally, the network reconfiguration is defined as altering the topological structure of distribution feeders by changing the opened or closed state of sectionalization and tie switches (to transfer load from heavily loaded feeders to relatively less heavily loaded feeders) so that the power loss is minimized and at the same time constraints are met. These two types of switches are designed for both protection and configuration management. It is normally being configured radially for effective coordination of their protective systems. Network is reconfigured to reduce the system power loss (network reconfiguration for power loss reduction), improve the voltage profiles and relieve overloads in the network (network reconfiguration for load balancing) and finally increase energy efficiency of the system. This operation transfers load from one feeder to another, which will significantly improve the operating condition of the overall system. In order to deal with these problems, several methods such as GA [18], PSO [19], EP [20, 21] and ABC [14] have been applied in network reconfiguration.

The existing of DG in the system will allow the network to contribute in supplying the most optimum power to the load. However, selecting the optimal size of DG plays an important role to avoid any drawback to the network. The connection

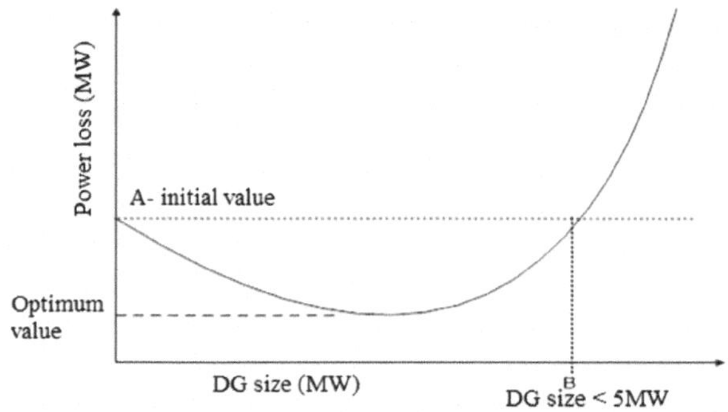

Fig. 10.2 Power loss dependence on DG size [22]

of high capacity and excess number of DG units to electrical power system will lead to very high power loss [22]. When DG is accessed to the distribution network, it can be simplified into 3 different scenarios: in scenario-1, the loads at each bus are all greater than power generation of each DG. In scenario-2, the total loads are greater than the total power generation of DG, while in scenario-3, the total loads are less than the total power generation of DG. For scenario-1, DG's access can reduce the power loss of all lines. However, in scenario-2, DGs access may increase the power loss of some lines, but the total power loss reduces. Meanwhile, in scenario-3, if the total power generation is less than two times of the total loading, the influence is the same as the scenario-2, or DGs access will increase the power loss. However, if the total power generation is in a high proportion of the system, it will bring down the power quality.

There is a tendency for losses to follow the U-shape trajectory as shown in Fig. 10.2 [22]. Specifically, losses begin to decrease when connecting small amounts of DG size until they achieve their minimum level. If the DG increases then losses begin to rise. Thus, it is worth pointing out that at high DG sizes, losses can become larger than those without DG connected. In this chapter, the DG size varies from 0 to 5 MW. According to the U-Shape when the DG size is larger than the B point value, the power loss in the system has become larger than A, which is the initial value. This factor makes the optimal size of DG become an important consideration for the network to have lower power loss value. Thus, it can be seen that DGs access may reduce or increase the power loss depends on the size of DG and the network structure. The use of the reconfiguration method in cooperating with the DG units with appropriate size can help the system to have a much lower power loss in the distribution system.

10.3 Problem Formulation

Reconfiguration techniques in the distribution network will change the direction of power flow throughout the network. In this chapter, the main objective for doing the reconfiguration is to obtain the minimum active power loss in the system based on active current formulation. Therefore, the objective function is:

$$\text{Minimise}\left\{P_{losses}=\sum_{l=1}^{n}I_l|^2k_lR_l\right\} \tag{10.1}$$

where:

I Number of lines in the system.
I_l Line real active current.
R_l Line resistance.
k_l is the variable that represents the topology status of the branches (1 = close, 0 = open).

The technical constraints that must be considered for the reconfiguration optimization are:

(a) Distributed Generator operation:

$$p_i^{\min}\leq p_{dg,i}\leq p_i^{\max} \tag{10.2}$$

where P_i^{min} and P_i^{max} are the lower and upper bound of DG output and all DG units shall function within the acceptable limit.

(b) Power injection:

$$\sum_{i=1}^{k}P_{DG}<(P_{Load}+P_{Losses}),\quad k=no.\ of\ DG \tag{10.3}$$

In order to avoid problem in protection setting, extra power from DG units are not allowed to be injected into the main grid (Substation). At all time the total power output from DG units should be less than the total load demand in the electric distribution network. Thus, there will be a power supply from the main grid to the network at all time.

(c) Power balance:

$$\sum_{i=1}^{k}P_{DG}+P_{Substation}=P_{Load}+P_{Losses} \tag{10.4}$$

The sum of power generated from DG units and power from substation must be equal to the summation of power load and power loss. This is to comply with the principle of equilibrium in power generation and load demand concept.

(d) Voltage bus:

$$V_{\min} \leq V_{bus} \leq V_{\max} \tag{10.5}$$

The voltage for each bus should operate within the acceptable limit which is in between 1.05 and 0.95 (±5% of rated value).

(e) Radial configuration:
The radiality of the network should be maintained throughout the reconfiguration process. In order to ensure radial network is maintained, a set of rules has been adopted for selections of switches [23].

a. All switches that do not belong to any loop are to be closed state.
b. All switches connected to the sources are to be closed state.
c. All switches contributed to a meshed network need to be closed state.

For the implementation of the optimization methods, the variable used for tie switches represented by *S* and as for *DG* size is represented by P_{Dg}. The proposed chromosome or particle can be written as

$$X_{im} = \left[S_1, S_2, \ldots, S_N, p_{Dg,1}, p_{Dg,2}, \ldots, p_{Dg,k}\right] \tag{10.6}$$

where i = 1,2,3…m. The variable m indicates the population size from a set of random distributions. N = number of tie switches and k = numbers of DG. If the method only to find the optimum value of DG that can minimize the power loss, the chromosome or particle can be written as

$$X_{im} = \left[p_{Dg,1}, p_{Dg,2}, \ldots, p_{Dg,k}\right] \tag{10.7}$$

10.4 Description of Modified Meta-Heuristic Methods

In this chapter, besides applying conventional GA [18], EP [20], PSO [23] and ABC [24] methods, its modified version have been applied as well. The modification of each method is summarized in Table 10.1. Details description on the modification can be found in the respective reference.

Table 10.1 Modification of meta-heuristic methods

Method	Modification from the conventional method
Modified Genetic Algorithm (MGA) [18]	Basically, the steps involved in MGA are mostly similar to GA steps except a slight difference in the mutation process. The chromosomes which consist of tie-switches and DG size are represented in real coded compared to binary coded as represented in conventional method in order to increase the efficiency and reduce the computational time. Too long string or chromosome will increase the time consuming in the searching space for the optimum especially when the system operates in the larger and more complex system. The advantage of MGA is the acceleration in the searching speed because the encoding and decoding process is not needed as required in binary-coded. Furthermore, it is a simple design tool to treat complex constraints because the method is close to problem spaces
Evolutionary Particle Swarm Optimization (EPSO) [19]	EPSO is developed based on merging two methods PSO and EP. EPSO is proposed to improve and enhance the convergence speed of conventional PSO. The proposed EPSO undergoes the similar steps as the traditional PSO accept selection process part where EP employs a selection through the tournament scheme to choose the survivals for the next generation. Three steps involved are as – Combination old and new position – sort the population based on fitness value – select the best element from the survival particle (lower value) With these the particles can move quickly to the optimal point compared to the conventional PSO
Modified Particle Swarm Optimization (MPSO) [23]	Generally, the steps involved in MPSO are almost similar to conventional PSO. However, the quality and efficiency of the current PSO has been slightly modified. A new parameter (bold) is inserted into the original PSO Equation as shown below; $V_j^{k+1} = \omega \times V_j^k + C_1 \times rand_1 \times \left(P_{bestj}^k - X_j^k\right) + C_2 \times rand_2 \times \left(G_{best}^k - X_j^k\right) + \boldsymbol{C_3 rand_3} \times \left(\boldsymbol{B_{best}^k} - \boldsymbol{X_j^k}\right)$ The purpose of the additional new parameter is to avoid the fitness value being trapped in local optima and increasing the exploration capability of particles in the search space. Therefore, the exploration and exploitation capability of MPSO is improved and provide the best solution quality and consistent results near to the global optimum
Simplified Artificial Bee Colony (SABC) [24]	The operation of the SABC is nearly similar to the original ABC. A slight modification in term of the searching for new food sources procedure has been implied on this simplified mode. A new and better concept of changing information between the bees in the population is applied. The following equations show a new searching area that is used by Employed and Onlooker Bees in the SABC algorithm $B_{i,rand(D)} = SW_{rand(N),rand(D)}$ where B_i is presenting the new searching location by *it* bees in the current iteration and SW is the switches. Therefore, the new switches that need to be opened in next iteration are: $SW^{new} = f\left(SW^{old}\left(B_{i,rand(D)}\right)\right)$ The implementation of B_i in the SABC will avoid the unacceptable switch number appeared during the reconfiguration process compared to original ABC

10.5 Implementation of the Proposed Concept

From the base system, five different scenarios are formed to analyze the robustness and efficiency of the proposed concept based on different meta-heuristic methods (GA, EP, PSO, ABC, MGA, EPSO, MPSO and SABC). The lists of each scenario are as

- scenario-1: Original network as a base scenario
- scenario-2: Optimal network reconfiguration
- scenario-3: Optimal DG sizing
- scenario-4: Optimal reconfiguration and DG sizing based on Sequential approach
- scenario-5: Optimal reconfiguration and DG sizing based on Simultaneous approach.

The parameter used in the simulation of each algorithm depends on the characteristics of the method used. However, the basic data of the network such as bus data and line data shall be the same for all the methods. The initialization population is determined by selecting tie switches from the set of the original tie switches as well as the DG size. Those variables are generated randomly by the program and it is utilized to compute the power loss in the next step. The number of population is 50 and the maximum number iteration is set 100 which is applied and used in all the proposed algorithms. The minimum and maximum voltages are set between 0.95 and 1.05 pu respectively. The simulation process will stop once the results achieve maximum number of iteration or convergence level. All the tests and simulations developed in this work are conducted on a Personal Computer with processor Intel Core Duo CPU 3.07 GHz.

The flow chart of the reconfiguration process is illustrated in the Fig. 10.3. The range of the DG size varies between 0 and 5 MW which is based on the literature review [25] where most researchers use similar networks in their work.

The following steps can be used to develop programming codes to apply the proposed simultaneously optimal network reconfiguration and DG sizing.

Step 1 Randomize N number of switches and DG output. Check either the random number fulfils all the constraints. If *yes* then save the opened switches and DG output. *Else* delete and re-randomize the new output.

Step 2 Evaluate the fitness functions $(f\ (x))$ for successful population: $f(x)$ = Power loss (Eq. (10.1)).

Step 3 Change the tie switches and sectionalizing switches through simulation process according to the proposed methods respectively (GA, EP, PSO, ABC, MGA, EPSO, MPSO and SABC).

Step 4 Evaluate the new fitness function (load flow analysis) and check the radiality of the output through Graph Theory.

Step 5 Check the stopping criteria, if the iteration number > iter max or all population give the similar values and then stops. Otherwise go to *Step 3*.

Step 6 Show the optimal results. End.

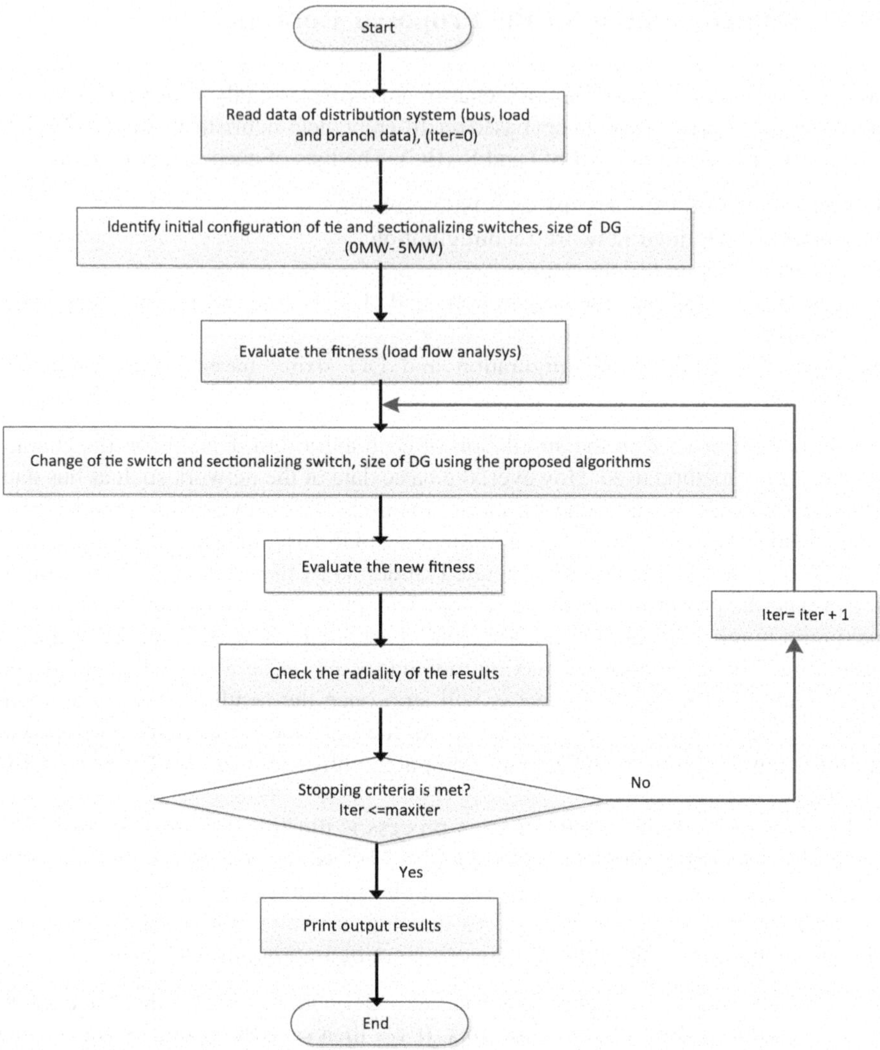

Fig. 10.3 Flowchart of network reconfiguration process

10.6 Test Results of 33-Bus System

The initial test network of the 33-bus system is shown in Fig. 10.4. The overall information of 33-bus distribution network is given in Appendix A.1.2. The network consists of 33 buses, 38 lines, 5 tie switches represented by dotted lines and 3 branches (excluding the main branch). The total load of the system is 3715 kW and 2300 kVar. In scenario-3, scenario-4 and scenario-5, three DGs unit have been installed and placed at bus number 6, 16 and 25 [26, 27] respectively.

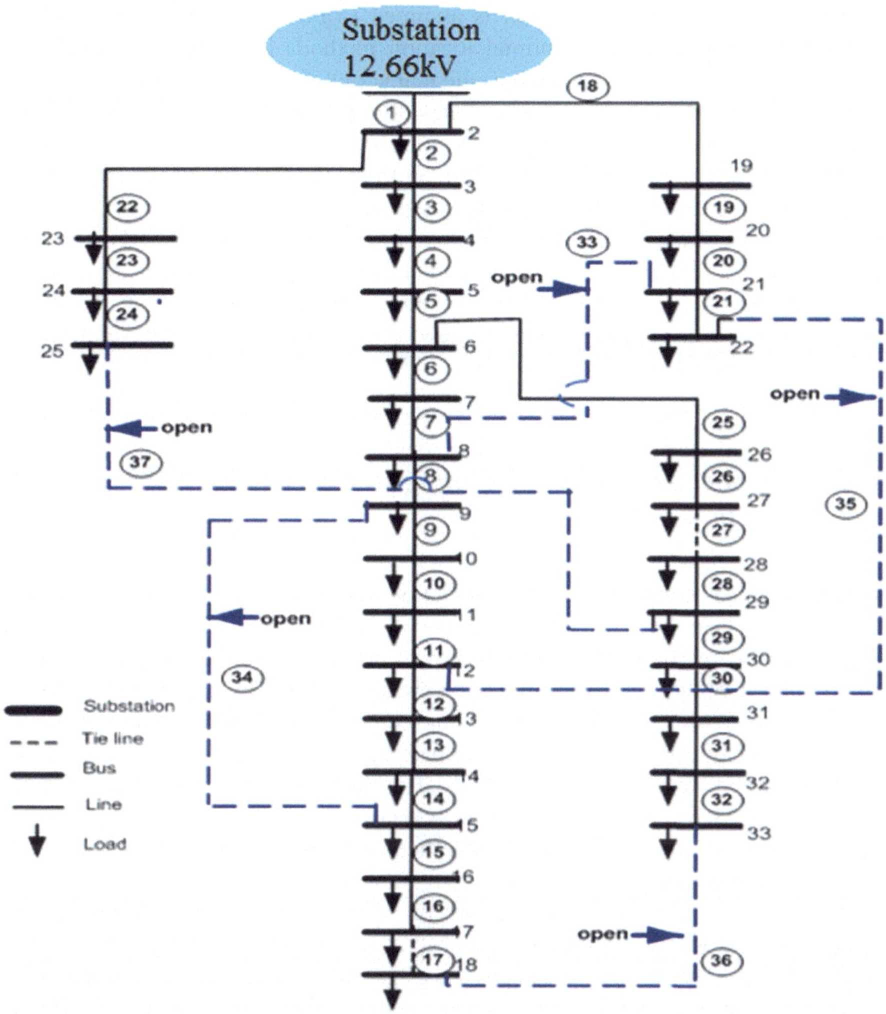

Fig. 10.4 Initial configuration of the 33-bus radial distribution system

As mentioned in the previous section, the analysis of reconfiguration involves the network with and without DG units. In the scenario of a network with DG, the optimal size of DG units is obtained from the simulation in which both parameters DG size and the opened switches are adjusted during simulation simultaneously. The size of each DG is set within the limits of the DG capacity. For example, in this chapter, the range is set between 0 and 5 MW. The capacity depends on the type of DG such as medium distribution generation 5 < 50 MW and large distributed generation 50 < 300 MW [28].

In this chapter, the DG location is assumed to be based on its suitability of geographical location or any optimal location methods [20, 29]. Tie switch and sectionalizing switch are considered as the main control variables. Since the process of randomization at each iteration produces different results, there is a need to do simulation repeated for several times in order to get the best results. Thus in this scenario, the simulation is conducted for 30 times. The value which appears the same for many times shall be chosen and assumed to be the best results. The result obtained consist of the opened switches, total power loss and optimal DG sizing are shown as in Eq. (10.6). However, the number of elements in particle or chromosome depends on how many tie-switch and DG used in the system. Then, the network after reconfiguration shows the new opened switches which have produced the lowest power loss.

10.6.1 Impact of Network Reconfiguration and DG Sizing on Power Losses

From the analysis conducted on the simulation, the results are discussed in details. At the initial stage (scenario-1), the network of 33-bus system is run without the presence of reconfiguration and DG. The network has given the initial total power loss of 202.3 kW through five initial open switches of 33, 34, 35, 36 and 37 for all methods used. With regard to Table 10.2 in scenario-2, reconfiguration is employed in the network of 33-bus system. The impact of reconfiguration of the power loss reduction can be observed for all methods. The total power loss has been improved by 34.5% for MGA, MPSO, and EPSO whenever the network reconfiguration is applied. However, there is a slight difference in SABC method where the power loss are improved about 32.77%.

Meanwhile, in scenario-3 the network of 33-bus system is operated using DG which is placed on bus number 6, 16 and 25. The impacts of the DG presence are then analyzed. The results obtained show greater power loss as compared to scenario-2. The total power loss has been reduced between 61.9 and 66.5%. In scenario-4, operation conducted involving both reconfiguration and DG has taken place. However, reconfiguration process is only done upon obtaining the right size of DG. In other word, both techniques are run sequentially. The results obtained reveal even greater power loss reduction as compared to scenario-2 and scenario-3. Thus, the presence of DG in the reconfiguration process has indeed caused the reduction of power loss.

The network condition of scenario-5 is almost identical to scenario-4 except that this time both reconfiguration and DG are being applied simultaneously. In other words, the switches that will open and the size of DG are determined

Table 10.2 The overall performance of simulation results of the 33-bus distribution system

Scenario	Method	Opened switches	Power loss (MW)	Loss reduction (%)	DG size (MW)		
					6	16	25
scenario-2	GA	6, 10, 14, 17, 28	136.5	32.53	–	–	–
	PSO	7, 10, 28, 14, 32	136.4	32.58	–	–	–
	EP	16, 5, 10, 25, 13	135.2	33.17	–	–	–
	ABC	7, 9, 14, 32, 37	139.5	31.04	–	–	–
	MGA	6, 9, 13, 17, 25	132.5	34.50	–	–	–
	EPSO	7, 10, 13, 16, 25	130.5	35.49	–	–	–
	MPSO	7, 10, 28, 14, 34	132.43	34.54	–	–	–
	SABC	6, 9, 14, 31, 37	136.0	32.77	–	–	–
scenario-3	GA	33, 34, 35, 36, 37	110.6	45.33	1.4107	0.902	0.5061
	PSO	33, 34, 35, 36, 37	109.6	45.82	1.0038	0.9004	0.5167
	EP	33, 34, 35, 36, 37	106	47.60	0.7315	0.7224	1.0270
	ABC	33, 34, 35, 36, 37	110.5	45.38	0.7540	0.5300	1.5004
	MGA	33, 34, 35, 36, 37	104.0	48.59	1.0190	0.9120	0.5061
	EPSO	33, 34, 35, 36, 37	102.5	49.33	0.7310	0.6564	1.1560
	MPSO	33, 34, 35, 36, 37	109.2	46.02	1.1488	0.9023	0.5167
	SABC	33, 34, 35, 36, 37	109.5	45.87	0.7740	0.5310	1.5004
scenario-4	GA	7, 8, 10, 16, 28	112.0	44.64	1.041	0.905	0.7001
	PSO	7, 10, 14, 28, 32	93.5	53.78	1.0439	0.9061	0.7012
	EP	7, 9, 34, 36, 37	99.4	50.87	1.0499	0.9098	0.7099
	ABC	11, 20, 24, 32, 34	129.7	35.89	1.3004	0.53	0.7054
	MGA	7, 9, 28, 36, 37	99.5	50.82	1.048	0.907	0.7001
	EPSO	7, 9, 33, 36, 37	94.2	53.44	1.1127	0.918	0.729
	MPSO	7, 10,14, 28, 32	97.1	52.00	1.0489	0.9118	0.7312
	SABC	11, 20, 24, 32, 34	101.6	49.78	1.2604	0.531	0.774

(continued)

Table 10.2 (continued)

Scenario	Method	Opened switches	Power loss (MW)	Loss reduction (%)	DG size (MW)		
					6	16	25
scenario-5	GA	7, 10, 14, 28, 30	100.9	50.12	1.1490	0.9427	0.6332
	PSO	7, 9, 14, 28, 32	92.3	54.37	1.1523	0.9545	0.6312
	EP	7, 10, 12, 16, 28	94.1	53.48	1.1519	0.9378	0.6680
	ABC	11, 20, 31, 34, 37	103.9	48.64	1.133	0.9510	0.6220
	MGA	7, 10, 12, 16, 28	96.88	52.11	1.1519	0.9335	0.6678
	EPSO	6, 10, 13, 16, 28	89.4	55.81	1.1590	0.9747	0.6632
	MPSO	7, 9, 14, 28, 32	92.46	54.30	1.1733	0.9651	0.6363
	SABC	11, 20, 31, 34, 37	97.5	51.80	1.1019	0.7575	0.7780

simultaneously during simulation. The results for the network of 33-bus system show the greatest improvement on power loss reduction between 31.4 and 51.8%. The performance of the test of power loss for scenario-5 of the proposed methods is depicted in Fig. 10.5. From the results obtained in scenario-5, the SABC is observed to generate highest improvement which is 6.19% as compared to the original method. This is then followed by EPSO (4.89%), MGA (3.9%) and MPSO (2.97%). However, in overall results, EPSO still maintain as the method which produces the lowest power loss due to the changing of switch during the simulation process and injection of the active power by the DG simultaneously.

After reconfiguration, the optimal sizes of DG of each method have also been altered as illustrated in Table 10.3. With reference to scenario-5, three DG are installed at different locations which have been fixed earlier. Once the program is run, the sizes of DG will vary automatically between the predetermined ranges until it reaches the optimal values. This can be seen from the Table 10.2 which shows that the optimum DG size is different of each method. SABC produced the smallest size of DG which is 2.6374 MW followed by MGA 2.7532 MW. While for MPSO and EPSO are 2.7747 and 2.7969 MW respectively. If the total size of DG is in a high proportion of the total load system, it will bring down the power quality. The analysis indicates that the maximum energy saving is achieved when the DG size is placed at bus no. 6, 16 and 25 as shown on the diagram. However, the total size of DG of each method is still within the range.

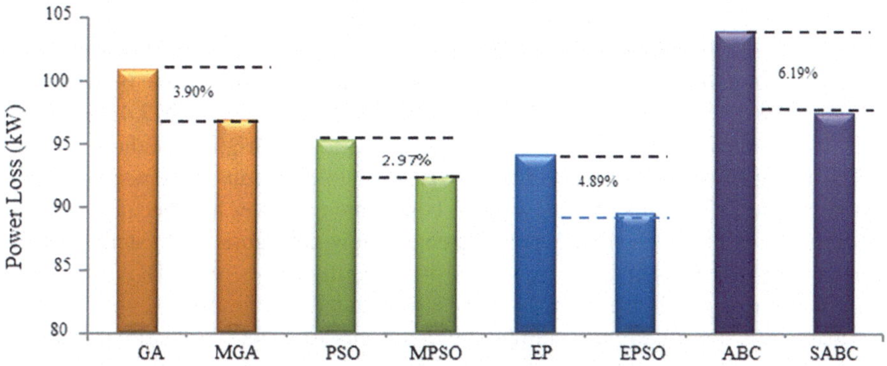

Fig. 10.5 Power losses improvement using the proposed algorithms (scenario-5)

Table 10.3 The performance of the proposed method based on optimal DG sizing

Method	DG size (MW)—scenario-5			Total size of DG (MW)
	6	16	25	
MGA	1.1519	0.9335	0.6678	2.7532
EPSO	1.1590	0.9747	0.6632	2.7969
MPSO	1.1733	0.9651	0.6363	2.7747
SABC	1.1019	0.7575	0.7780	2.6374

Table 10.4 Comparisons of performance of the proposed methods (scenario-5)

30 run times	Power loss (kW)			
	MGA	MPSO	EPSO	SABC
Min solution	96.8	92.4	89.4	97.5
Max solution	98.9	99.8	96.1	99.7
Average	97.4	94.37	92.24	98.3
Standard deviation	0.0009	0.00297	0.00211	0.00107
No. of iteration	35	21	13	30
CPU time (s)	29.4	16.1	12.8	21.5
Original method (s)	GA—60	PSO—28.1	EP—16.8	ABC—44.5

Meanwhile, Table 10.4 shows the comparisons of the performance of the proposed method after 30 runs of the 33-bus distribution system. Every repeated process (run times) was initialized with random new combination switch and DG size, thus the robustness and efficiency of the algorithm in finding the minimum power loss can be evaluated.

From the analysis of the results, the minimum solution or the best output achieved by EPSO is 89.494 kW. However, the maximum solution which indicates the unfavorable value is produced by MPSO and SABC which are 99.8 and 99.7 kW respectively. MPSO gives the highest standard deviation with 0.00297 among others. Nevertheless, the percentage of differences between the 'min solution' and 'max solution' of each method is 2.12% until 7.42%, which gives the smallest standard deviation as compared to the original method.

The analysis of this scenario has proven the simultaneous presence of reconfiguration and DG together yields a much better rate of power loss because the new set of switches has been rearranged to create a new configuration system plus the optimal size of DG. Thus, the new network reconfiguration with DGs of 33-bus system operated on simultaneously basis is illustrated in Fig. 10.6.

Meanwhile, the maximum number of iterations to reach the optimal value is 35 iterations for MGA, 30 iterations for SABC. Meanwhile MPSO need 21 iterations and only 13 iterations of the EPSO to reach the optimal point. EPSO method takes only 13 iterations which need 12.8 s to converge while MGA method shows the longest computing time of 29.4 s compared to other methods. This is due to the reason that MGA requires more steps before converging. It means the highest number of iterations, the longer the computing time

The convergence curve summarizes the capability and efficiency of each method and the speed of the algorithm in reaching the optimal point. Figure 10.7 shows the convergence characteristics of the proposed method of 33-bus system. With the updated technique, the value of power loss is improved until the best solution is reached. From the observation, the EPSO is the fastest (13 iterations) algorithm to reach the optimal solution followed by MPSO, MGA and SABC.

10.6.2 Impact of Network Reconfiguration and DG Sizing on Voltage Profile

The impact on the voltage profile for scenario-5 using the proposed method is depicted in Fig. 10.8. By observing the results, it can be concluded that the voltage profile has been improved and the average bus voltages have reached 0.950 pu as compared to the base scenario which is 0.9131 pu In scenario 5, the system which is operated with reconfiguration and DG simultaneously, the voltage profile of the system has been improved considerably with a minimum node voltage of 0.977692 pu at point 33 of MPSO method and 0.97698 pu at point 33 of SABC method. The improvement is about 7.07%. While the minimum node voltage for MGA and EPSO occurs at the same point which is 0.985983 pu

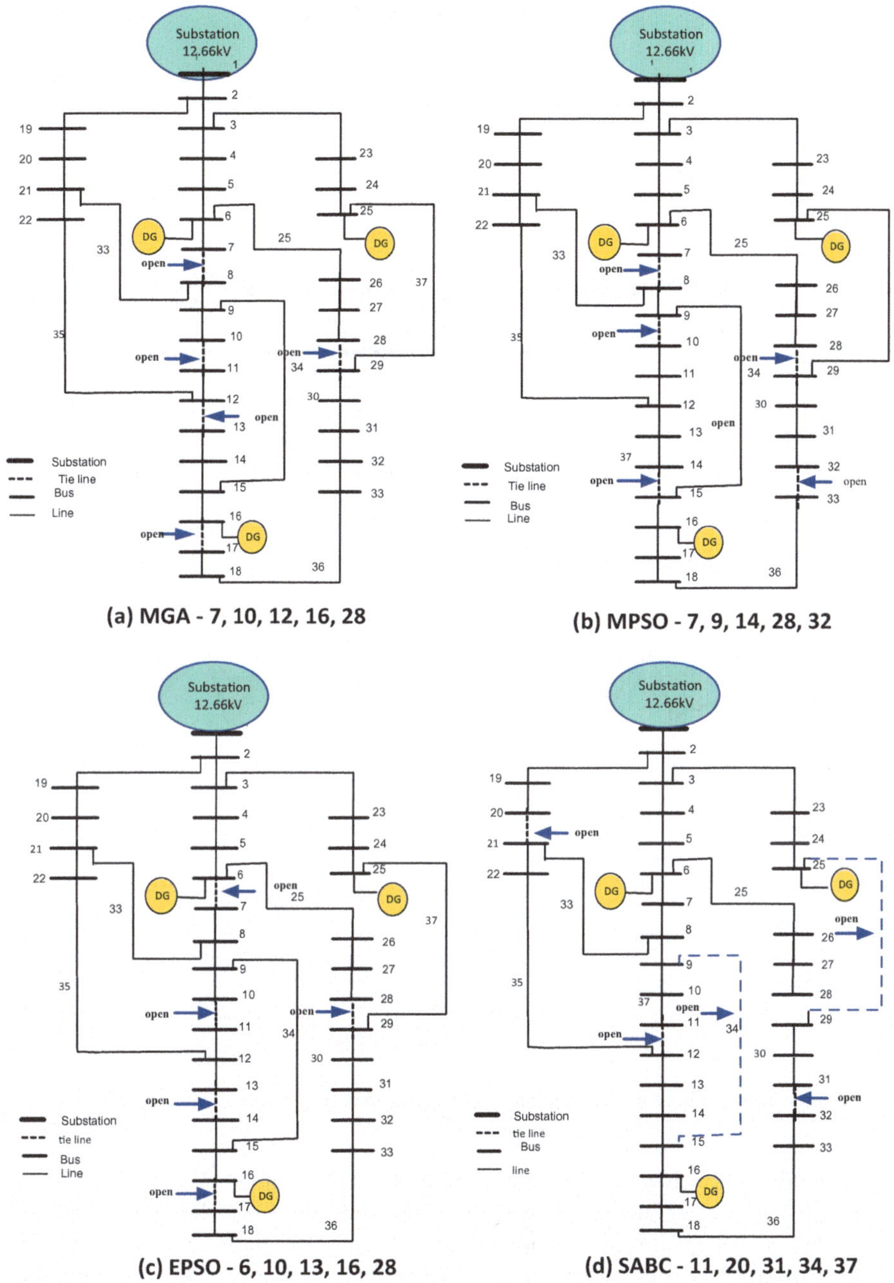

Fig. 10.6 Network Reconfiguration and DG size simultaneously using **a** MGA, **b** MPSO, **c** EPSO and **d** SABC (scenario-5)

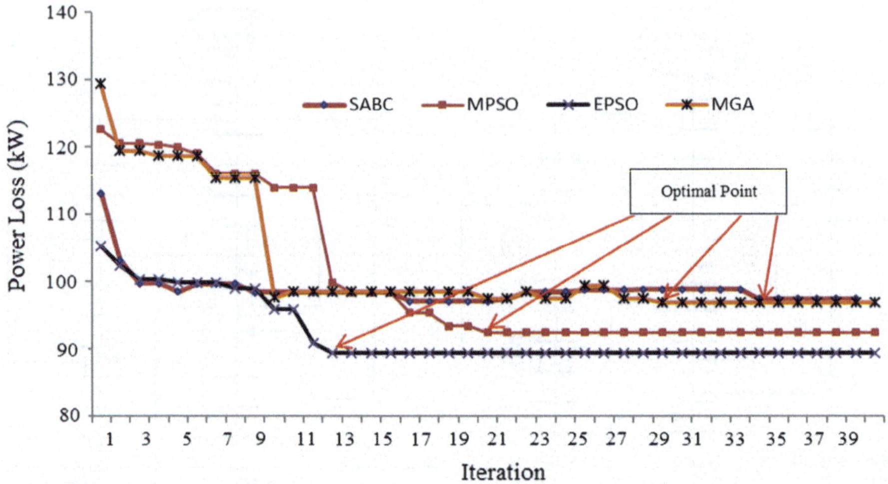

Fig. 10.7 Convergence characteristics of the SABC, MPSO, MGA and EPSO algorithms

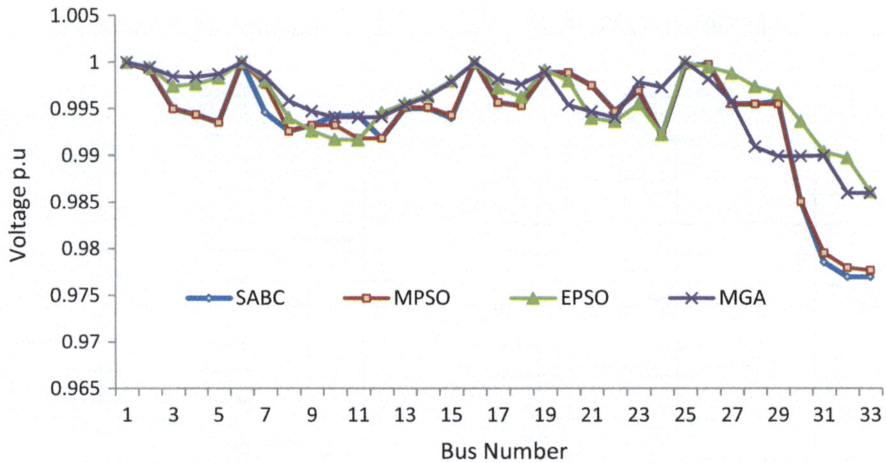

Fig. 10.8 Performance of the voltage profile of the proposed methods

Generally, the voltage profile shows slight differences among the methods accept at point 33. Therefore, the implementation of reconfiguration technique and DG has given better voltage profile compared to without reconfiguration and DG. The voltage profile has been improved more effectively whenever the reconfiguration and DG in the system is operated simultaneously where all bus voltages satisfy the 0.95 pu voltage constraints and near to 1 pu

References

1. N. Suresh, Dr. T. Gowri Manohar, Optimal citing of custom power controller in distribution system for loss reduction, in National Conference on GSSSETW, Oct 2009, Mysore, Kartanaka
2. K. Prasad, R. Ranjan, N. Sahoo, A. Chaturvedi, Optimal reconfiguration of radial distribution systems using a fuzzy mutated genetic algorithm. Power Delivery, IEEE Trans. **20**(2), 1211–1213 (2005)
3. N. Gupta, A. Swarnkar, K. Niazi, Reconfiguration of distribution systems for real power loss minimization using adaptive particle swarm optimization. Electr. Power Compon. Syst. **39**(4), 317–330 (2011)
4. C.L. Borges, D.M. Falcão, Impact of distributed generation allocation and sizing on reliability, losses and voltage profile. Paper presented at the Power Tech Conference Proceedings, 2003 IEEE Bologna (2003)
5. T.S. Basso, R. DeBlasio, IEEE 1547 series of standards: interconnection issues. Power Electron IEEE Trans. **19**(5), 1159–1162 (2004)
6. O. Javanmardi, M. Nasri, I. Sadeghkhani, Investigation of distributed generation effects on the voltage profile and power losses in distribution systems. Adv. Electr. Eng. Syst. **1**(2), 74–77 (2012)
7. S. Sivanagaraju, Y. Srikanth, E.J. Babu, An efficient genetic algorithm for loss minimum distribution system reconfiguration. Electr. Power Compon. Syst. **34**(3), 249–258 (2006)
8. G. Wang, P. Wang, Y.-H. Song, A. Johns, Co-ordinated system of fuzzy logic and evolutionary programming based network reconfiguration for loss reduction in distribution systems. Paper presented at the Fuzzy Systems, 1996. Proceedings of the Fifth IEEE International Conference on (1996)
9. D. Shirmohammadi, H.W. Hong, Reconfiguration of electric distribution networks for resistive line losses reduction. Power Deliv. IEEE Trans. **4**(2), 1492–1498 (1989)
10. Y. Song, G. Wang, A. Johns, P. Wang, Distribution network reconfiguration for loss reduction using fuzzy controlled evolutionary programming. *Paper presented at the Generation, Transmission and Distribution, IEE Proceedings* (1997)
11. J.Z. Zhu, Optimal reconfiguration of electrical distribution network using the refined genetic algorithm. Electr. Power Syst. Res. **62**(1), 37–42 (2002)
12. L. Ganesan, P. Venkatesh, Distribution system reconfiguration for loss reduction using genetic algorithm. J. Electr. Syst. **2**(4), 198–207 (2006)
13. S. Sivanagaraju, J.V. Rao, P.S. Raju, Discrete particle swarm optimization to network reconfiguration for loss reduction and load balancing. Electr. Power Compon. Syst. **36**(5), 513–524 (2008)
14. C.-T. Su, C.-F. Chang, J.-P. Chiou, Distribution network reconfiguration for loss reduction by Ant Colony Search algorithm. Electr. Power Syst. Res. **75**(2), 190–199 (2005)
15. J. Chakravorty, Network reconfiguration of distribution system using fuzzy controlled evolutionary programming. Int. J. Eng. Sci. Adv. Technol. **2**(2), 176–182 (2012)
16. N. Gupta, A. Swarnkar, K. Niazi, Reconfiguration of distribution systems for real power loss minimization using adaptive particle swarm optimization. Electr. Power Compon. Syst. **39**(4), 317–330 (2011)
17. N. Rugthaicharoencheep, S. Sirisumrannukul, Feeder reconfiguration for loss reduction in distribution system with distributed generators by Tabu Search. GMSARN Int. J. **3**, 47–54 (2009)
18. Y.-K. Wu, C.-Y. Lee, L.-C. Liu, S.-H. Tsai, Study of reconfiguration for the distribution system with distributed generators. Power Delivery, IEEE Trans. **25**(3), 1678–1685 (2010)
19. J. Olamaei, T. Niknam, G. Gharehpetian, Application of particle swarm optimization for distribution feeder reconfiguration considering distributed generators. Appl. Math. Comput. **201**(1), 575–586 (2008)

20. Z. Bingda, Y. Liu, A novel algorithm for distribution network reconfiguration based on evolutionary programming. Paper presented at the Advanced Power System Automation and Protection (APAP), 2011 International Conference on (2011)
21. A.C. Nerves, J.C.K. Roncesvalles, Application of evolutionary programming to optimal siting and sizing and optimal scheduling of distributed generation. Paper presented at the TENCON 2009–2009 IEEE Region 10 Conference (2009)
22. M.P. Lalitha, V.V. Reddy, V. Usha, Optimal DG placement for minimum real power loss in radial distribution systems using PSO. J. Theoret. Appl. Inf. Technol. **13**(2), 107–116 (2010)
23. A. Arya, Y. Kumar, M. Dubey, Reconfiguration of electric distribution network using modified particle swarm optimization. Int. J. Comput. Appl. **34**(6) (2011)
24. M.P. Lalitha, N.S. Reddy, V.V. Reddy, Optimal DG placement for maximum loss reduction in radial distribution system using ABC Algorithm. J. Theoret. Appl. Inf. Technol. (2010)
25. K. Nara, A. Shiose, M. Kitagawa, T. Ishihara, Implementation of genetic algorithm for distribution systems loss minimum re-configuration. Power Syst. IEEE Trans. **7**(3), 1044–1051 (1992)
26. T. Ackermann, G. Andersson, L. Söder, Distributed generation: a definition. Electr. Power Syst. Chapter **57**(3), 195–204 (2001)
27. A. El-Zonkoly, Optimal placement of multi-distributed generation units including different load models using Particle Swarm Optimisation. IET Gener. Transm. Distrib. **5**(7), 760–771 (2011)
28. J. Mendoza, R. López, D. Morales, E. López, P. Dessante, R. Moraga, Minimal loss reconfiguration using genetic algorithms with restricted population and addressed operators: real application. Power Syst. IEEE Trans. **21**(2), 948–954 (2006)
29. Y. Del Valle, G.K. Venayagamoorthy, S. Mohagheghi, J.-C. Hernandez, R.G. Harley, Particle swarm optimization: basic concepts, variants and applications in power systems. Evol. Comput. IEEE Trans. **12**(2), 171–195 (2008)

Chapter 11
Optimal Incentive Plans for Plug-in Electric Vehicles

Mehdi Rahmani-Andebili, Mahmud Fotuhi Firuzabad and Moein Moeini-Aghtaie

Abstract This chapter investigates implementation of some parking lots for a plug-in electric vehicle (PEV) aggregator to participate in energy market. Herein, behaviors of the PEVs' drivers regarding their cooperation with the aggregator with respect to the introduced incentive (value of discount on charging fee of PEVs) are modeled. The considered incentive includes the value of discount on the charging fee of PEVs' batteries. In addition, the capability of parking lots for transacting electrical energy is modeled based on the hourly arrival/departure time of PEVs to/from the parking lots and the hourly state of charge (SOC) of PEVs' batteries. Also, the degradation of PEVs' batteries is modeled based on the effective ampere-hours throughput of the PEVs' batteries due to vehicle-to-grid (V2G). Moreover, the economic factors such as inflation and interest rates and the technical factors including the PEVs' batteries power limit, the depth of discharge (DOD) constraint of PEVs' batteries, the yearly maintenance of parking lot, and the yearly replacement rate of the conventional vehicles with the PEVs are taken into consideration in the problem over the definite planning horizon. Furthermore, due to variability and uncertainties involved with the energy market prices and the PEVs' drivers' behavior, the planning problem is solved stochastically.

Keywords Modeling capability of parking lot for energy transaction
Modeling behavior of PEVs' drivers · Modeling life loss of PEVs' batteries
Optimal incentive plans · Stochastic optimization

M. Rahmani-Andebili (✉) · M. F. Firuzabad
Department of Electrical Engineering, Sharif University of Technology, Tehran, Iran
e-mail: mehdi.rahmani.andebili@gmail.com

M. F. Firuzabad
e-mail: fotuhi@sharif.edu

M. Moeini-Aghtaie
Sharif Energy Research Institute, Sharif University of Technology, Tehran, Iran
e-mail: moeini@sharif.edu

F. Shahnia et al. (eds.), *Electric Distribution Network Planning*, Power Systems, https://doi.org/10.1007/978-981-10-7056-3_11

11.1 Introduction

Replacing internal combustion vehicles with plug-in electric vehicles (PEVs) is a promising strategy to calm the energy security and environmental issues, since PEVs can be charged by the electricity generated by renewables or clean energy resources [1]. Nowadays, governments around the world call for the deployment of PEVs and hybrid PEVs [2–5]. A recent study demonstrates that almost 27% of total energy consumption and 33% of greenhouse gas emissions in the world are related to the transportation sector [6]. Based on the studies presented in [7, 8], PEVs utilization is being increased rapidly in some developed countries due to the advancement of battery technology. Recently, the role of energy storage has become more important with development of smart grids [9]. An individual PEV has a trifle impact on an electric distribution network; however, aggregation of a large number of PEVs can noticeably affect the network performance [10, 11]. Through effective coordination and communication technologies, the PEVs can be considered as the mobile energy storage and play an important role in the smart grids [12]. Nonetheless, replacing conventional vehicles with electric ones may put the network at risk and bring about new issues such as system overload and spikes in energy market prices due to uncontrolled charging of the PEVs' batteries [13, 14]. Herein, a PEV aggregator can play an effective role to calm the above mentioned issues, since it can motivate the PEVs' drivers (by introducing a variety of incentives to them) to park their vehicles in the specific locations (parking lots) to manage and coordinate the charging time of the PEVs' batteries. By implementing this strategy, the aggregator can take part in different power markets and provide benefit for itself, for the PEVs' drivers, and also for the network.

It has been reported that private vehicles are parked at parking lots in idle state for more than 90% during a day [15]. Therefore, the PEVs as the energy storage units have a huge potential for doing energy transactions in power market. Since every individual PEV's driver is not able to participate in energy market and compete with other powerful market players, due to a low power capacity, a PEV aggregator is introduced to aggregate them [16]. A comprehensive literature overview regarding the economic and technical management of a PEV aggregator has been given in [17]. In [18], the methods for optimal charging management of PEVs have been reviewed. Moreover, advantages and disadvantages, and also economic and technical characteristics of V2G technology have been discussed in [10, 19–21].

In [22], the feasibility of utilizing Ontario's grid for charging PEVs has been analyzed applying a zonal model of Ontario's transmission network and base-load generation capacities for the period of 2009–2025. In [23, 24], real-time load management strategies for coordinating the charging time of PEVs for minimum energy losses and voltage control have been proposed. In [25], reliability assessment of network considering PEVs fleet has been studied. In [26], feeder reconfiguration has been used for coordinating V2G of PEVs in a stochastic framework.

In [27], energy management of one PEV connected to a smart home has been investigated.

Presence of the PEV aggregator in different power markets has been investigated in several studies [28–37]. In [28–30], PEVs have been utilized to support smart grids by offering ancillary service including frequency regulation. In [31, 32], participation of the PEVs in the spinning reserve market has been studied. In [33], the PEV charging scheduling by an aggregator in a day-ahead energy market applying mixed integer linear programming (MILP) has been investigated. In [34], an optimal bidding strategy of a PEV aggregator participating in day-ahead energy and regulation markets using stochastic optimization has been presented. The authors in [35] have presented a method to manage the PEVs charging in real-time for participation of the PEV aggregators in the energy market. In [36], solar parking lots have been sized and allocated in an electrical distribution system based on their optimal power factor applying quantum annealing.

In spite of the numerous studies in the literature about PEVs and their aggregator, the behavior of PEVs' drivers regarding their cooperation level with the aggregator with respect to the incentive plans has not been modeled. In this chapter, in addition to the PEVs' drivers' responsiveness level, capability of parking lots for energy transaction in energy market is modeled. Moreover, model of a battery life loss presented in [37] is applied for the PEVs' batteries in the problem simulation.

11.2 Modeling Capability of the Parking Lots for Energy Transaction

Figure 11.1 illustrates the schematic diagram of a PEV's battery indicating its capacity, state of charge (SOC) level, and the defined depth of discharge (DOD) limit. As can be seen, the value of available grid-to-vehicle (G2V) power can be determined based on the difference between the PEV's battery capacity and its SOC level. Also, the value of available vehicle-to-grid (V2G) power can be calculated based on the difference between the SOC level of the PEV's battery and the given DOD limit for the PEV's battery. Therefore, in a parking lot, at every

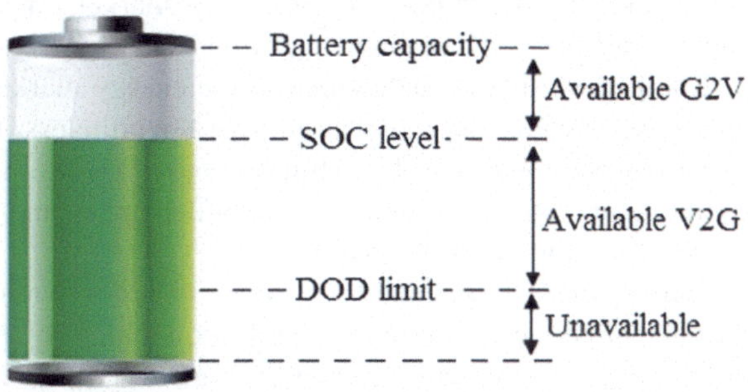

Fig. 11.1 Schematic diagram of a PEV's battery indicating its capacity, SOC level, and the defined DOD limit

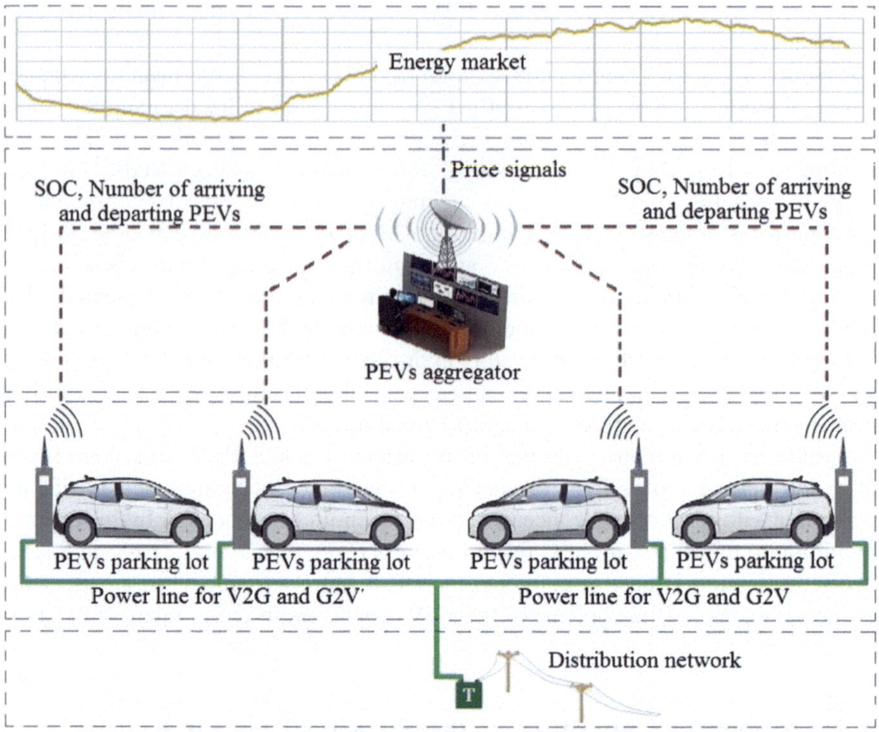

Fig. 11.2 The aggregator as an intermediate agent between the energy market and the PEVs connected to the distribution network

hour, the total value of available V2G and G2V powers can be computed by knowing the values of some parameters concerned with the PEVs' drivers' behavior. These parameters include the number of PEVs arriving to the parking lots, the number of PEVs departing from the parking lots, and the SOC level of the batteries of the PEVs arrived to the parking lots [31]. Figure 11.2 shows the role of aggregator as an intermediate agent between the energy market and the PEVs connected to the distribution system through the parking lots. Based on this, the aggregator can participate in the energy market transactions considering the price signals and the total available V2G and G2V powers of the parking lots.

The PEVs' drivers' behavior and the energy market price are uncertain and they may have a wide range of variability. Nevertheless, the variability range of every uncertain parameter can be estimated based on the hourly historical data gathered by the aggregator. Figure 11.3 shows the upper and lower bands for the possible value of an uncertain parameter at every hour of the day.

Herein, the data gathered for every uncertain parameter and for every hour of a day are fitted on a Gaussian distribution function as the most appropriate distribution function, as can be seen in Fig. 11.4 [31]. Then, in order to address the

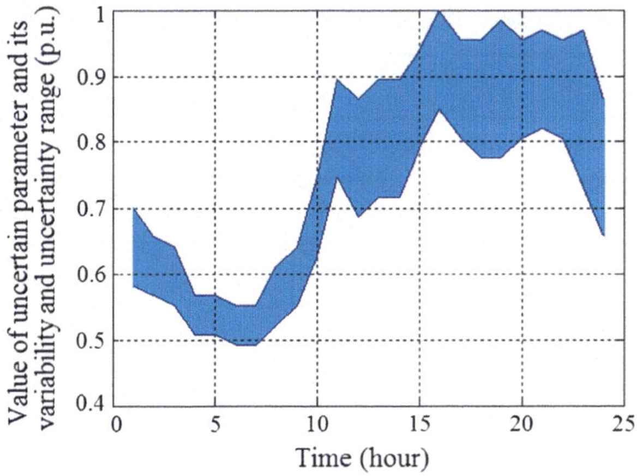

Fig. 11.3 The upper and lower bands for the possible value of an uncertain parameter at every hour of the day

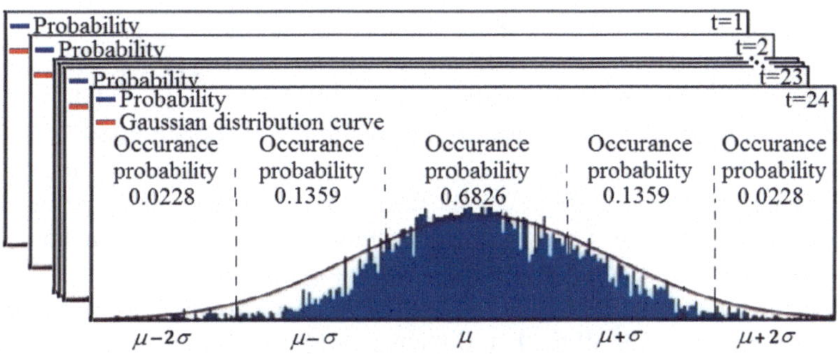

Fig. 11.4 Considering five distinct values for each uncertain parameter at every hour of the day

prediction uncertainty, five distinct values are considered for every uncertain parameter with the probabilities 0.0228, 0.1359, 0.6826, 0.1359, and 0.0228 according to the areas in the related Gaussian distribution function concerned with the $\mu - 2\sigma$, $\mu - \sigma$, μ, $\mu + \sigma$, $\mu + 2\sigma$. Figure 11.4 graphically illustrates the above mentioned approach. After that, in order to investigate the problem stochastically, 15 comprehensive and diverse scenarios are defined for the hourly value of each uncertain parameter throughout the day, as can be seen in Fig. 11.5. These scenarios have been defined arbitrarily; however, it has been tried to design the diverse and comprehensive scenarios to include the most probable scenarios and eliminate the similar ones.

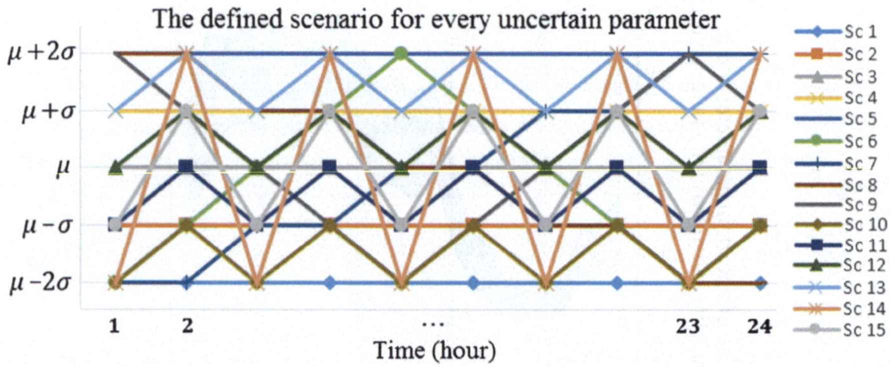

Fig. 11.5 The considered scenarios for the values of each uncertain parameter over the day

11.3 Modeling Cooperation Between PEVs' Drivers and Aggregator

In this part, cooperation between PEVs' drivers (ξ) and the aggregator with respect to the value of discount on charging fee (ψ) is modeled applying linear, power, logarithmic, and exponential functions. As can be seen in (11.1)–(11.4) in Table 11.1, these models have been designed for 100% cooperation of the PEVs' drivers under free charging and no cooperation under zero discount on the charging fee. The cooperation percentage curves related to the linear model, power model with exponents 0.1, 0.3, 1.5, and 3, logarithmic model, and exponential model respect to value of discount on charging fee for the range of (0%, 100%) are illustrated in Fig. 11.6. As can be seen, the considered models are very comprehensive, since they cover all the two dimensional space. Therefore, all the possible linear and nonlinear behaviors of the PEVs' drivers are taken into consideration.

Table 11.1 Models for cooperation percentage of the PEVs' drivers with the aggregator as the function of discount on charging fee

Model	Cooperation percentage of PEVs' drivers (%)	
Linear	$\xi_{Lin} = \psi$	(11.1)
Power	$\xi_{Pow} = 100 \times \left(\frac{\psi}{100}\right)^n, \quad n \in \mathbb{R}$	(11.2)
Exponential	$\xi_{Exp} = 100 \times e^{M \times \left(\frac{\psi}{100} - 1\right)}, \quad M \gg 1$	(11.3)
Logarithmic	$\xi_{Log} = 100 \times ln\left(\frac{\psi}{100} \times (\exp(1) - 1) + 1\right)$	(11.4)

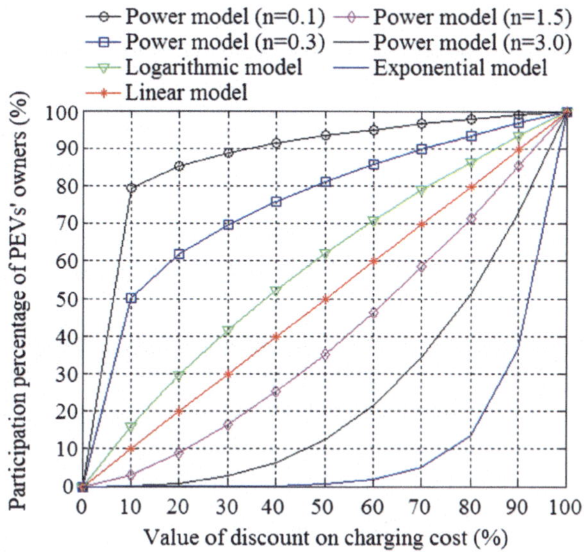

Fig. 11.6 Curves of cooperation percentage of the PEVs' drivers with the aggregator respect to value of the discount on charging fee assuming various linear and nonlinear models

11.4 Modeling PEV's Battery Life Loss Cost Due to V2G

Herein, the value of life loss cost or aging cost of a PEV's battery is modeled based on the effective ampere-hours throughput of the PEV's battery due to the V2G actions [37], as can be seen in (11.5). In (11.5), $Ah_{b,t}$ is the ampere-hours throughput of the PEV's battery due to V2G at hour t, Ah_b^{Tot} is the total cumulative ampere-hours throughput of the PEV's battery in its life cycle, $Price^{Bat}$ is the price of a PEVs' battery, and λ, as the effective weighting factor, is determined using the model introduced in [37]. As can be seen in Fig. 11.7, in the presented model, the value of the effective weighting factor has a nonlinear relationship with the SOC level of the PEV's battery. For instance, at SOC of 50%, removing 1 A h from the PEV's battery is equivalent to removing 1.3 A h from the total cumulative ampere-hours throughput of the PEV's battery. However, at SOC of 100%,

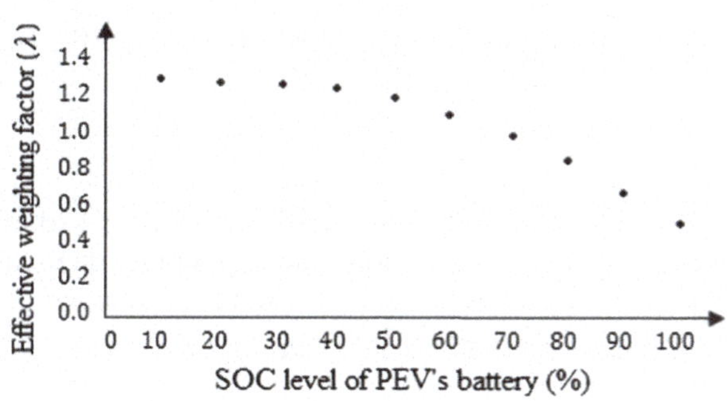

Fig. 11.7 Relationship between effective weighting factor and SOC level of a PEV's battery [37]

removing 1 A h results in only about 0.55 A h reduction. This reality indicates that the PEVs' batteries should be operated at high SOC levels to optimize their lifetime.

$$Cost^{LL}(\mathrm{t}) = \frac{\lambda \times Ah_b(t)}{Ah_b^{Tot}} \times Price^{Bat} \tag{11.5}$$

11.5 Planning Problem Formulation

In this chapter, the PEV aggregator builds and implements some parking lots in a residential area to participate in the energy market transactions to maximize its profit over the given planning horizon.

11.5.1 Objective Function

Different terms of the objective function include the income resulted from transactions in energy market, investment cost for structuring and implementing parking lots and equipping them, yearly maintenance cost of the parking lots, aging cost of the PEVs' batteries due to V2G, and cost of considering discount on the charging fee of the PEVs' batteries. Herein, some economic factors such as inflation and interest rates and several technical factors including the PEVs' batteries' power limit, DOD constraint of the PEVs' batteries, PEVs' batteries aging due to V2G, yearly maintenance of the parking lot, and yearly replacement rate of internal combustion engine based vehicles with electric ones are considered in the problem. Furthermore, due to variability and uncertainties involved with the energy market prices and the PEVs' drivers' behavior, the planning problem is solved stochastically considering several comprehensive scenarios for every uncertain parameter. The uncertain parameters include the energy market price, the number of PEVs arriving to the parking lot, the number of PEVs departing from the parking lot, and the SOC level of the PEVs' batteries arrived to the parking lots. The objective function of the planning problem is presented in (11.6).

$$\begin{aligned} OF_{pp} = Max\{&-Cost_{Tot}^{Inv} - PWV\left(Cost^{M}\right) + PWV\left(Income^{T}\right) - PWV\left(Cost^{BA}\right) \\ &-PWV\left(Cost^{Inc}\right)\} \end{aligned} \tag{11.6}$$

The first term of the objective function is related to the total investment cost for structuring the parking lots and equipping them and the second term of the objective function is concerned with the present worth value of maintenance cost of the parking lots over the planning period.

Table 11.2 The binary variables used to code the decisions of the aggregator

u^a	u^b	Decision
0	0	Idle
	1	
1	0	G2V
	1	V2G

The third term of the objective function is related to the present worth value of the aggregator's income over the planning period because of optimal transactions in the energy market by doing optimal V2G and G2V actions considering all the defined scenarios for the uncertain parameters. Equation (11.7) presents the deterministic value of aggregator's income per year. As can be seen in Table 11.2, u^a and u^b as the binary numbers are used to code the decision of the aggregator for being in idle state or performing one of the V2G or G2V actions. Equation (11.8) presents the number of PEVs' drivers who are willing to cooperate with the aggregator and connect their PEVs to the parking lots. Herein, Nev_{Tot} indicates the total number of PEVs in the area. Also, ξ_{Model}, as the cooperation percentage of the PEVs' drivers with the aggregator respect to value of the discount on charging fee (ψ), is determined using Table 11.1 or Fig. 11.6. Equations (11.9) and (11.10) present the stochastic value of aggregator's income per year and the present worth value of aggregator's income over the planning period, respectively.

$$Income_y^T = \sum_{d=1}^{365}\sum_{t=1}^{24} u^a(t) \times \left(u^b(t) \times \pi^E(t) \sum_{ev=1}^{Nev} V2G_{ev,y,d}(t) - \left(1 - u^b(t)\right) \times \pi^E(t) \sum_{ev=1}^{Nev} G2V_{ev,y,d}(t) \right) \tag{11.7}$$

$$Nev = Nev_{Tot} \times \xi_{Model} \tag{11.8}$$

$$Stoch\left(Income_y^T\right) = \sum_{i \in S^{\pi}} \sum_{j \in S^{Narr}} \sum_{k \in S^{Ndep}} \sum_{l \in S^{SOC}} \left\{Income_y^T\right\} \times Pr_i^{\pi} \times Pr_j^{Narr} \times Pr_k^{Ndep} \times Pr_l^{SOC} \tag{11.9}$$

$$PWV\left(Stoch\left(Income_y^T\right)\right) = \sum_{y=1}^{pp} Stoch\left(Income_y^T\right) \times \left(\frac{1 + IFR}{1 + ITR}\right)^y \tag{11.10}$$

The forth term of the objective function is related to the present worth value of aging cost of the PEVs' batteries over the planning period due to V2G actions considering all the defined scenarios for the uncertain parameters. Equations (11.11) and (11.13) give the deterministic value of yearly aging cost of the PEVs' batteries, the stochastic value of yearly aging cost of the PEVs' batteries, and the present worth value of aging cost of the PEVs' batteries over the planning period, respectively.

$$Cost_y^{BA} = \sum_{d=1}^{365} \sum_{t=1}^{24} u^a(t) \times u^b(t) \sum_{ev=1}^{Nev} Cost_{ev,y,d}^{LL}(t) \tag{11.11}$$

$$Stoch\left(Cost_y^{BA}\right) = \sum_{i \in S^{\pi}} \sum_{j \in S^{Narr}} \sum_{k \in S^{Ndep}} \sum_{l \in S^{SOC}} \left\{Cost_y^{BA}\right\} \times Pr_i^{\pi} \times Pr_j^{Narr} \times Pr_k^{Ndep} \times Pr_l^{SOC} \tag{11.12}$$

$$PWV\left(Cost^{BA}\right) = \sum_{y=1}^{pp} Stoch\left(Cost_y^{BA}\right) \times \left(\frac{1+IFR}{1+ITR}\right)^y \tag{11.13}$$

The fifth term of the objective function is related to the present worth value of discount on charging fee over the planning period considering all the scenarios defined for the uncertain parameters. Herein, it is assumed that the cooperative drivers' PEVs' batteries will be charged from the initial SOC level to the full charge considering discount on the charging fee. Equations (11.14) and (11.16) present the deterministic value of yearly incentive paid to the drivers, the stochastic value of yearly incentive paid to the drivers, and the present worth value of incentive paid to the drivers over the planning period, respectively.

$$Cost_y^{Inc} = \sum_{d=1}^{365} \sum_{t=1}^{24} \sum_{ev=1}^{Nev} \left(1 - \frac{SOC_{ev,y,d}^{arr}(t)}{100}\right) \times P_{ev} \times \frac{\psi}{100} \times \pi^{ch} \tag{11.14}$$

$$Stoch\left(Cost_y^{Inc}\right) = \sum_{i \in S^{\pi}} \sum_{j \in S^{Narr}} \sum_{k \in S^{Ndep}} \sum_{l \in S^{SOC}} \left\{Cost_y^{Inc}\right\} \times Pr_i^{\pi} \times Pr_j^{Narr} \times Pr_k^{Ndep} \times Pr_l^{SOC} \tag{11.15}$$

$$PWV\left(Cost^{Inc}\right) = \sum_{y=1}^{pp} Stoch\left(Cost_y^{Inc}\right) \times \left(\frac{1+IFR}{1+ITR}\right)^y \tag{11.16}$$

11.5.2 Constraints

The first constraint of the problem relates to supplying each PEV after daily V2G and G2V actions. As can be seen in (11.7), the daily energy demand of each PEV must be supplied considering the daily cumulative values of G2V and V2G done by the PEV.

$$\sum_{t=1}^{24} u^a(t) \times \left(1 - u^b(t)\right) \times G2V_{ev,y,d}(t) - \sum_{t=1}^{24} u^a(t) \times u^b(t) \times V2G_{ev,y,d}(t) = \sum_{t=1}^{24} \left(1 - \frac{SOC_{ev,y,d}^{arr}(t)}{100}\right) \times P_{ev} \tag{11.17}$$

$\forall i \in S^{\pi}, \forall j \in S^{Narr}, \forall k \in S^{Ndep}, \forall l \in S^{SOC}, \forall ev = 1, \ldots, Nev, \forall y = 1, \ldots, pp, \forall d = 1, \ldots, 365$

The second and third constraints concerned with the allowable injectable power of every PEV's battery into the grid and the allowable injectable power of the grid into every PEV's battery, respectively. These constraints must be regarded at every hour of the planning period and in every scenario.

$$V2G_{ev,y,d}(t) = P_{ev} \tag{11.18}$$

$$G2V_{ev,y,d}(t) = P_{ev} \tag{11.19}$$

$\forall i \in S^{\pi}, \forall j \in S^{Narr}, \forall k \in S^{Ndep}, \forall l \in S^{SOC}, \forall ev = 1, \ldots, Nev, \forall y = 1, \ldots, pp, \forall d = 1, \ldots, 365, \forall t = 1, \ldots, 24$

The forth constraint is related to the obligation of the aggregator respect to the PEVs' drivers. In order to prolong the lifetime of PEVs' batteries, at every hour of the planning period and in every defined scenario, the battery of every PEV must not be discharged more than the defined DOD limit. In addition, the SOC level cannot be considered more than 100%.

$$DOD^{limit} \leq SOC_{ev,y,d}(t) \leq 100 \tag{11.20}$$

$\forall i \in S^{\pi}, \forall j \in S^{Narr}, \forall k \in S^{Ndep}, \forall l \in S^{SOC}, \forall ev = 1, \ldots, Nev, \forall y = 1, \ldots, pp, \forall d = 1, \ldots, 365, \forall t = 1, \ldots, 24$

11.6 Proposed Optimization Technique

In this chapter, the problem is solved by applying genetic algorithm (GA) as the optimization methodology [38]. Other optimization algorithms could be used in this problem, however capability of GA for parallel optimization and its competence in complex and nonlinear environments are the main reasons for utilization of GA in this problem.

Variables of the optimization problem include u^a and u^b (the indicator of the aggregator's decision regarding being in idle state or performing one of the V2G or G2V actions) at every hour of a day. Based on this, every chromosome in the

Fig. 11.8 Structure of the defined chromosome

Transaction matrix

	u^a	u^b
Hour 1	0 or 1	0 or 1
Hour 2	0 or 1	0 or 1
⋮	⋮	⋮
Hour t	0 or 1	0 or 1
⋮	⋮	⋮
Hour 24	0 or 1	0 or 1

population is defined as the transaction matrix with 24×2 as its dimensions. Figure 11.8 illustrates the structure of the defined chromosome. Herein, the value of net profit of the aggregator over the planning period is defined as the value of fitness of the chromosome. Different steps for applying GA in the problem are presented and described below.

Step 1: Obtaining primary data

Parameters for applying GA: These parameters includes mutation probability of the genes $(P^{Mutation})$ and the size of population (N_{ch}).

Parameters of the problem: The values of all the problem parameters and the initial data are obtained. Moreover, the value of discount on charging fee and the cooperation model of the PEVs' drivers with the aggregator are determined.

Initial population: The chromosomes of the population are initialized with random binary values.

Step 2: Updating the population

Applying crossover operator: Two crossover points are randomly selected for every pair chromosomes, and then, crossover operator is applied on every two chromosomes of the population to reproduce two new chromosomes as the offspring, as can be seen in Fig. 11.9.

Applying mutation operator: This operator is applied on every gene of every chromosome of the population with the definite probability $P^{Mutation}$.

Step 3: Selecting new population

Evaluating fitness of the chromosomes: For every chromosome, the problem is run and if all the constraints are satisfied, the fitness of chromosome is measured.

Applying selection process: As can be seen in (11.21), new chromosomes are selected through the probabilistic fitness-based selection process, where the fitter chromosomes are more likely to be chosen. The value of selection probability of every chromosome is calculated using (11.22) which is proportional to the fitness of chromosome.

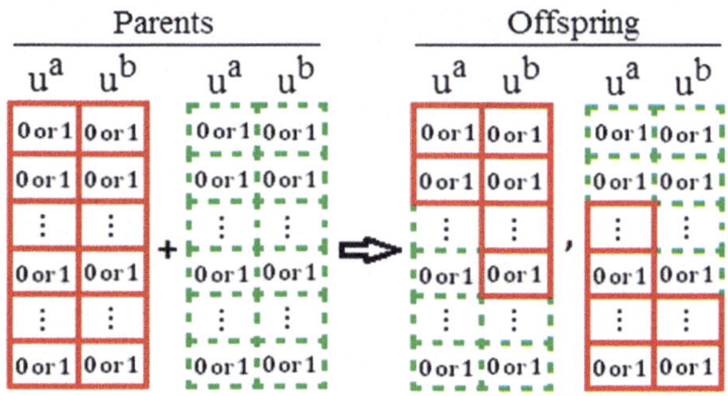

Fig. 11.9 Applying crossover operator on two chromosomes for reproducing new chromosomes

$$a_{ch} = \begin{cases} 1 & P_{ch}^{Selection} > r_{ch} \\ 0 & P_{ch}^{Selection} < r_{ch} \end{cases} \tag{11.21}$$

$$P_{ch}^{Selection} = \frac{f_{ch}}{Max(S^f)}, \quad S^f = \{f_1, \ldots, f_{ch}, \ldots, f_{Nch}\} \tag{11.22}$$

Step 4: Checking termination criterion

Herein, the convergence status of the optimization procedure is checked. Based on this, the values of improvements in fitness of the chromosomes of the old and new populations are measured and if there are no significant improvements (1% of the fitness of chromosome) in them, the optimization process is finished, otherwise, the algorithm is continued form Step 2.

Step 5: Introducing the outcomes

The consequences include the best fitted chromosome as the optimal transaction matrix.

This process is repeated for all possible values of discount on charging fee with a 10% step, and also for every cooperation model of the PEVs' drivers with the aggregator. After that, the optimal incentive, the optimal cooperation percentage of the PEVs' drivers with the aggregator, and the maximum net profit of the aggregator over the given time horizon are determined.

11.7 Numerical Studies

11.7.1 Primary Data

The initial data and the value of problem parameters are presented in Table 11.3. Figures 11.10, 11.11, 11.12 and 11.13 illustrate the variability range of the energy market price and the uncertain parameters of the PEVs' drivers' behavior including the number of PEVs arriving to the parking lot, the number of PEVs departing from the parking lot, and the SOC level of the PEVs' batteries arrived to the parking lots

Table 11.3 The initial data and parameters of the problem

Planning period (y)	20	Total cumulative ampere-hours throughput of a PEV's battery in its life cycle	700,000
Inflation rate (%/y)	10	Power of PEV's battery (kW)[a]	10
Interest rate (%/y)	15	Capacity of PEV's battery (kWh)[a]	50
Investment cost for a parking lot ($)	100,000	Charging/discharging voltage level (volt)[a]	480
Maintenance cost for a parking lot ($/y)	1000	DOD^{limit} based on the contract (%)	20
Size of a parking lot	200	Growth rate of PEVs (%/y)	1
Total number of parking lots	10	Charging fee ($/kWh)	0.043
PEVs' battery price ($)	10,000	Size of population in GA	100
Mutation probability of genes	0.05		

[a]TESLA, level 3 charging

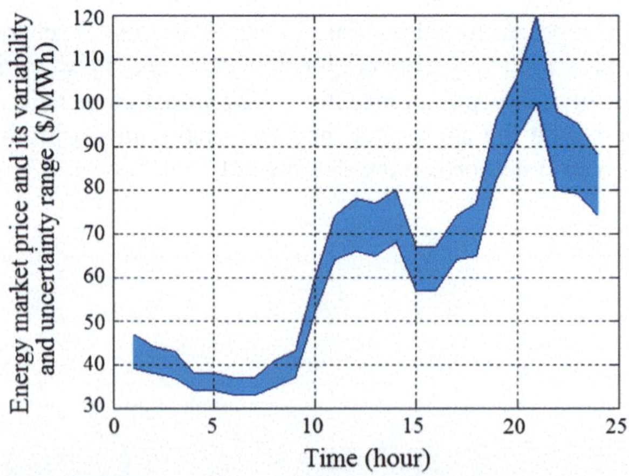

Fig. 11.10 The hourly upper and lower bands for the possible energy market price

Fig. 11.11 The hourly upper and lower bands for the possible percentage of PEVs arriving to the parking lots

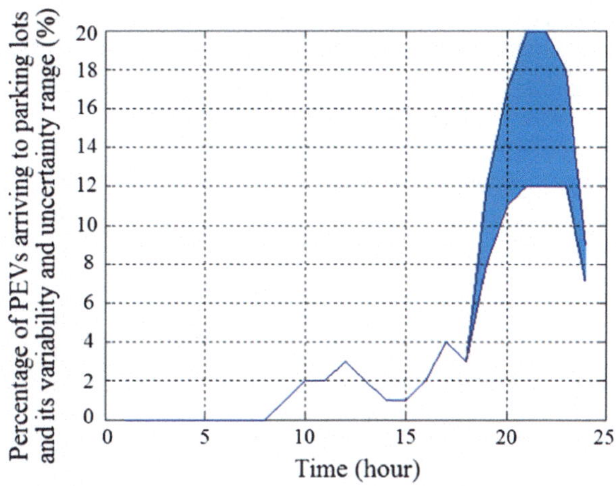

Fig. 11.12 The hourly upper and lower bands for the possible percentage of PEVs departing from the parking lots

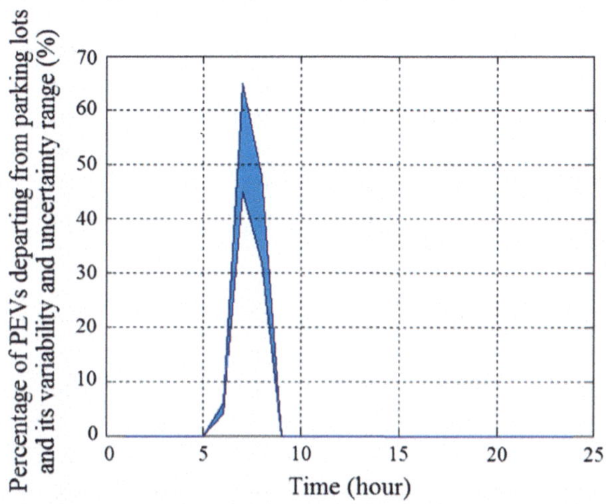

Fig. 11.13 The hourly upper and lower bands for the possible level of SOC of the PEVs batteries

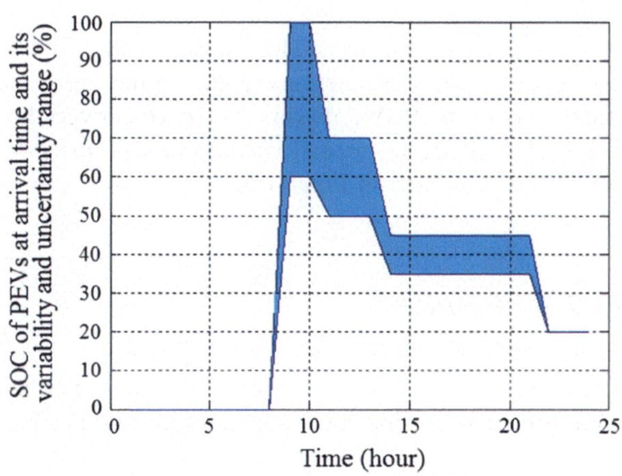

Table 11.4 The average value and the standard deviation of Gaussian distribution functions related to the uncertain parameters

Hour	Arriving PEVs (%)		Departing PEVs (%)		SOC (%)		Energy market price ($/MWh)	
	μ	σ	μ	σ	μ	σ	μ	σ
1	0	0	0	0	0	0	43	4
2	0	0	0	0	0	0	41	3
3	0	0	0	0	0	0	40	3
4	0	0	0	0	0	0	36	2
5	0	0	0	0	0	0	36	2
6	0	0	5	1	0	0	35	2
7	0	0	55	10	0	0	35	2
8	0	0	40	8	0	0	38	3
9	1	0	0	0	80	20	40	3
10	2	0	0	0	80	20	56	4
11	2	0	0	0	60	10	69	5
12	3	0	0	0	60	10	72	6
13	2	0	0	0	60	10	71	6
14	1	0	0	0	40	5	74	6
15	1	0	0	0	40	5	62	5
16	2	0	0	0	40	5	62	5
17	4	0	0	0	40	5	69	5
18	3	0	0	0	40	5	71	6
19	10	2	0	0	40	5	89	7
20	14	3	0	0	40	5	99	7
21	16	4	0	0	40	5	110	10
22	16	4	0	0	20	0	89	9
23	15	3	0	0	20	0	87	8
24	8	1	0	0	20	0	81	7

at every hour of the day. The average value and the standard deviation of the Gaussian distribution functions related to the energy market price and the uncertain parameters of the PEVs' drivers' behavior at every hour of the day are presented in Table 11.4. Also, the scenarios considered in the problem simulation are based on the details presented in Fig. 11.5.

11.7.2 Results

The curves related to the value of aggregator's net benefit over the planning period with respect to the discount value are illustrated in Fig. 11.14 for every cooperation

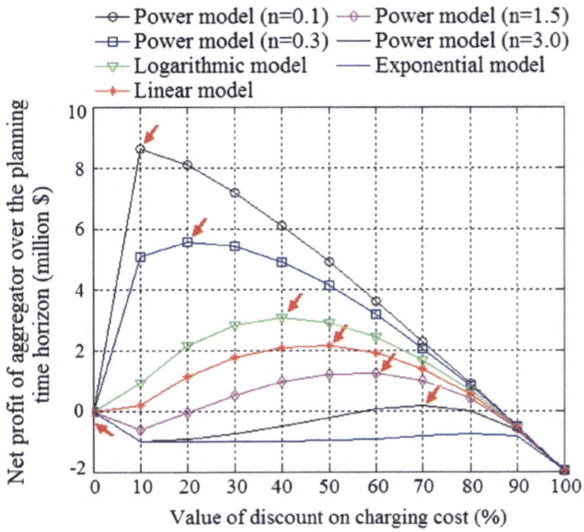

Fig. 11.14 Aggregator's net profit curves versus the value of discount on charging fee considering a variety of cooperation models (the arrows show the peak point of the curves)

model. As can be seen, by changing the value of discount on the charging fee, the profit of aggregator over the planning horizon is changed. However, increasing the value of incentive for raising motivation of the PEVs' drivers is not always effective, since the benefit curves do not have pure ascending trend. In other words, the curves are nonlinear and there is just one optimal value for the incentive in every model. Moreover, the optimal value of the incentive is different in every cooperation model. Thus, it can be concluded that assuming an incidental value of incentive would not lead to the maximum profit of aggregator and even it may result in detriment for the aggregator in some models.

The detailed results of the problem simulation including optimal value of incentive, cooperation percentage of the PEVs' drivers with the aggregator, the values of income and cost terms of the objective function, and the value of maximum benefit of the aggregator over the given horizon for every model are presented in Table 11.5. As can be seen, the cooperation between the aggregator and the PEVs' drivers with power behavioral model (n = 0.1) results in maximum benefit for the aggregator. In addition, no profit is achieved for the aggregator due to cooperation of the aggregator with the PEVs' drivers with exponential behavioral model. Therefore, this cooperation is not practical and beneficial.

Table 11.5 Detailed results of the problem simulation

	Optimal discount (%)	PEVs cooperation percentage (%)	Investment cost ($/pp)	Maintenance cost ($/pp)	Batteries aging cost ($/pp)	Incentive cost ($/pp)	Income of transaction ($/pp)	Net benefit ($/pp)
Power model with n = 0.1	10	79	1,000,000	10,000	7,008,300	1,154,200	17,794,000	8,621,500
Power model with n = 0.3	20	61	1,000,000	10,000	5,444,000	1,793,100	13,823,000	5,575,900
Logarithmic model	40	52	1,000,000	10,000	4,615,600	3,040,500	11,719,000	3,052,900
Linear model	50	50	1,000,000	10,000	4,411,400	3,632,500	11,201,000	2,147,100
Power model with n = 1.5	60	46	1,000,000	10,000	4,100,500	4,051,800	10,411,000	1,248,700
Power model with n = 3	70	34	1,000,000	10,000	3,026,200	3,488,700	7,683,800	158,900
Exponential model	0	0	1,000,000	10,000	0	0	0	0

Appendix

The notation used throughout the chapter is listed below:

Problem parameters and Variables

$Ah_b(.)$	Ampere-hours throughput of the PEV's battery due to V2G
Ah_b^{Tot}	Total cumulative ampere-hours throughput of the PEV's battery in its life cycle
$Cost^{LL}(.)$	Battery life loss cost due to V2G
$Cost_{Tot}^{Inv}$	Total investment cost for structuring the parking lots and equipping them
$Cost^{M}$	Maintenance cost of the parking lots
$Cost^{BA}$	Aging cost of the PEVs' batteries due to V2G
$Cost^{Inc}$	Cost of considering discount on the PEVs' batteries charging fee
$Income^{T}$	Income resulted from transactions in energy market
$G2V(.)$	Grid-to-vehicle
IFR, ITR	Inflation rate and interest rate
OF_{pp}	Objective function of the problem over the given planning period
P_{ev}	Nominal input or output power of the PEV
$Price^{Bat}$	Price of a PEV's battery
Pr_i^{π}	Occurrence probability of the *i*th scenario related to the energy market price
Pr_j^{Narr}	Occurrence probability of the *j*th scenario related to number of arriving PEVs to the parking lot
Pr_k^{Ndep}	Occurrence probability of the *k*th scenario related to number of departing PEVs from the parking lot
Pr_l^{SOC}	Occurrence probability of the *l*th scenario related to SOC level of the PEVs' batteries
$SOC(.)$	State of charge of the PEVs' batteries
$SOC^{arr}(.)$	State of charge of the PEVs' batteries arrived to the parking lots
DOD^{limit}	Depth of discharge limit based on the contract that must be respected by the aggregator
$u^{a}(.), u^{b}(.)$	Controlling parameters for indicating decision of the aggregator for being in idle state or doing one of the V2G and G2V actions
$V2G(.)$	Vehicle-to-grid
ψ	Value of discount on charging fee
ξ	Cooperation percentage of the PEVs' drivers with the aggregator
π^{E}	Energy market price
π^{ch}	PEV's battery charging fee
μ, σ	Mean and standard deviation of the uncertain parameter
λ	Effective weighting factor.

GA Parameters

$p^{Mutation}$ Mutation probability of the genes
N_{ch} Size of the population
a_{ch} Binary variable as the indicator for selection of the chromosome for the new population
r_{ch} Random number in the range of $(0, 1)$
$p_{ch}^{Selection}$ Value of selection probability of a chromosome
f_{ch} Value of fitness of a chromosome
S^f Set of chromosomes' fitness.

References

1. W. Kempton et al., *A Test of Vehicle-to-Grid (V2G) for Energy Storage and Frequency Regulation in the PJM System* (University of Delaware, Newark, DE, 2008)
2. D.W. Kurtz, R.R. Levin, EHV systems technology—a look at the principles and current status. IEEE Trans. Veh. Technol. **32**(1), 42–50 (Feb. 1983)
3. Y. Wu, H. Gao, Optimization of fuel cell and super capacitor for fuel-cell electric vehicles. IEEE Trans. Veh. Technol. **55**(6), 1748–1755 (2006)
4. S. Li, S. Sharkh, F. Walsh, C. Zhang, Energy and battery management of a plug-in series hybrid electric vehicle using fuzzy logic. IEEE Trans. Veh. Technol. **60**(8), 3571–3585 (2011)
5. B. Zhao, Y. Shi, X. Dong, Pricing and revenue maximization for battery charging services in PHEV markets. IEEE Trans. Veh. Technol. **63**(4), 1987–1993 (2014)
6. S.F. Tie, C.W. Tan, A review of energy sources and energy management system in electric vehicles. Renew. Sustain. Energy Rev. **20**, 82–102 (2013)
7. Canadian Automobile Association Electric Vehicles: What You Need to Know. http://electricvehicles.caa.ca/government-incentives/
8. According to BC Hydro's Draft Integrated Resource Plan, Appendix 2A–2011 Electric Load Forecast. https://www.bchydro.com/content/dam/hydro/medialib/internet/documents/environment/EVcharging_infrastructure_guidelines09.pdf
9. M.L. Di Silvestre, G. Graditi, E.R. Sanseverino, A generalized framework for optimal sizing of distributed energy resources in micro-grids using an indicator-based swarm approach. IEEE Trans. Ind. Informat. **10**(1), 152–162 (2014)
10. W. Su, H. Eichi, W. Zeng, M. Chow, A survey on the electrification of transportation in a smart grid environment. IEEE Trans. Ind. Informat. **8**(1), 1–10 (2012)
11. Z. Darabi, M. Ferdowsi, Anevent-based simulation framework to examine the response of power grid to the charging demand of plug-in hybrid electric vehicles. IEEE Trans. Ind. Informat. **10**(1), 313–322 (2014)
12. F. Kennel, D. Gorges, S. Liu, Energy management for smart grids with electric vehicles based on hierarchical MPC. IEEE Trans. Ind. Informat. **9**(3), 1528–1537 (2013)
13. A.S. Masoum, S. Deilami, P.S. Moses, M.A.S. Masoum, A. Abu-Siada, Smart load management of plug-in electric vehicles in distribution and residential networks with charging stations for peak shaving and loss minimization considering voltage regulation. IET Gener. Trans. Distrib. **5**, 877–888 (2011)

14. L.P. Fernandez, T.G. San Roman, R. Cossent, C.M. Domingo, P. Frias, Assessment of the impact of plug-in electric vehicles on distribution networks. IEEE Trans. Power Syst. **26**, 206–213 (2011)
15. W. Kempton, Vehicle to grid power, FERC (2007)
16. W. Kempton, J. Tomic, S. Letendre, A. Brooks, T. Lipman, Vehicle to grid power: battery, hybrid, and fuel cell vehicles as resources for distributed electric power in California. *University of California Davis Institute for Transportation Studies*, Rep. ECD-ITS-RR-01-03 (2001)
17. R.J. Bessa, M.A. Matos, Economic and technical management of an aggregation agent for electric vehicles: a literature survey. *Eur. Trans. Elect. Power* (2011) [Online]. Available: http://onlinelibrary.wiley.com/doi/10.1002/etep.565/abstract
18. S.G. Wirasingha, A. Emadi, Classification and review of control strategies for plug-in hybrid electric vehicles. IEEE Trans. Veh. Technol. **60**(1), 111–122 (2011)
19. J.C. Ferreira et al., Vehicle-to-anything application (V2Anything App) for electric vehicles. IEEE Trans. Ind. Informat. **10**(3), 1927–1937 (2014)
20. M. Bertoluzzo, G. Buja, Development of electric propulsion systems for light electric vehicles. IEEE Trans. Ind. Informat. **7**(3), 428–435 (2011)
21. J.H. Zhao, F. Wen, Z.Y. Dong, Y. Xue, K. Wong, Optimal dispatch of electric vehicles and wind power using enhanced particle swarm optimization. IEEE Trans. Ind. Informat. **8**(4), 889–899 (2012)
22. A. Hajimiragha, C.A. Caizares, M.W. Fowler, A. Elkamel, Optimal transition to plug-in hybrid electric vehicles in Ontario, Canada, considering the electricity-grid limitations. IEEE Trans. Indust. Electron. **57**, 690–701 (2010)
23. D.Q. Oliveira, A.C. Zambroni de Souza, L.F.N. Delboni, Optimal plug-in hybrid electric vehicles recharge in distribution power systems. Elect. Power Syst. Res. **98**, 77–85 (2013)
24. Z. Liu, F. Wen, G. Ledwich, Optimal planning of electric-vehicle charging stations in distribution systems. IEEE Trans. Power Del. **28**, 102–110 (2013)
25. C. Chen, S. Duan, Optimal integration of plug-in hybrid electric vehicles in microgrids. IEEE Trans. Ind. Informat. **10**(3), 1917–1926 (2014)
26. A. Kavousi-Fard, M.A. Rostami, T. Niknam, Reliability-oriented reconfiguration of vehicle-to-grid networks. IEEE Trans. Ind. Informat **11**(3), 682–691 (2015)
27. X. Wu, X. Hu, S. Moura, X. Yin, V. Pickert, Stochastic control of smart home energy management with plug-in electric vehicle battery energy storage and photovoltaic array. J. Power Sources **333**, 203–212 (2016)
28. S. Han, S. Han, K. Sezaki, Development of an optimal vehicle-to-grid aggregator for frequency regulation. *IEEE Trans. Smart Grid*, pp. 65–72 (2010)
29. E. Sortomme, M.A. El-Sharkawi, Optimal charging strategies for unidirectional vehicle-to-grid. IEEE Trans. Smart Grid **2**, 131–138 (2011)
30. J.R. Pillai, B. Bak-Jensen, Integration of vehicle-to-grid in the Western Danish power system. IEEE Trans. Sustain. Energy **2**, 12–19 (2011)
31. M. Rahmani-andebili, Spinning reserve supply with presence of plug-in electric vehicles aggregator considering compromise between cost and reliability. IET Gener. Trans. Distrib. **7**, 1442–1452 (2013)
32. R.J. Bessa, M.A. Matos, F.J. Soares, J.A. Peças Lopes, Optimized bidding of a EV aggregation agent in the electricity market. IEEE Trans. Smart Grid **3**(1), 443–452 (2012)
33. C. Jin, J. Tang, P. Ghosh, Optimizing electric vehicle charging with energy storage in the electricity market. IEEE Trans. Smart Grid **4**(1), 311–320 (2013)
34. S.I. Vagropoulos, A.G. Bakirtzis, Optimal bidding strategy for electric vehicle aggregators in electricity markets. IEEE Trans. Power Syst. **28**(4), 4031–4041 (2013)
35. F.J. Soares, P.M. Rocha Almeida, J.A. Pecas Lopes, Quasi-real-time management of Electric Vehicles charging. Elect. Power Syst. Res. **108**, 293–303 (2014)

36. M. Rahmani-andebili, Optimal power factor for optimally located and sized solar parking lots applying quantum annealing. IET Gener. Transm. Distrib. **10**, 2538–2547 (2016)
37. D.P. Jenkins J. Fletcher, D. Kane, Lifetime prediction and sizing of lead–acid batteries for micro generation storage applications. *IET Renew. Power Gener.* **2**(3), 191–200 (Sept. 2008)
38. L. Zhang, Z. Wang, X. Hu, F. Sun, D.G. Dorrell, A comparative study of equivalent circuit models of ultracapacitors for electric vehicles. J. Power Sources **274**, 899–906 (2015)

Chapter 12
Optimal Allocation of Compensators

Mohamed Ebeed, Salah Kamel, Shady H. E. Abdel Aleem and Almoataz Y. Abdelaziz

Abstract Electric distribution networks mainly deliver the electric power from the high-voltage transmission system to the consumers. In these networks, the R/X ratio is significantly high compared to transmission systems hence power loss is high (about 10–13% of the generated power). Moreover, poor quality of power including the voltage profile and voltage stability issues may arise. The inclusion of shunt capacitors and distributed Flexible ac transmission system (D-FACTS) devices can significantly enhance the performance of distribution networks by providing the required reactive power. D-FACTS include different members such as; distributed static compensator (DSTATCOM), Distribution Static Var Compensator (D-SVC) and unified power quality conditioner (UPQC). Optimal allocation of these controllers in the distribution networks is an important task for researchers for power loss minimizing, voltage profile improvement, voltage stability enhancement, reducing the overall system costs and maximizing the system load ability and reliability. Several analytical and optimization methods have been presented to find the optimal siting and sizing of capacitors and shunt compensators in electric distribution networks. This chapter presents a survey of new optimization techniques which are used to find the optimal sizes and locations of such devices. This chapter also presents an application of new optimization technique called

M. Ebeed · S. Kamel
Electrical Engineering Department, Aswan Faculty of Engineering,
Aswan University, Aswan, Egypt
e-mail: mohamedebeed11@gmail.com

S. Kamel
e-mail: skamel@aswu.edu.eg

S. H. E. Abdel Aleem
Mathematical, Physical and Engineering Sciences Department,
15th of May Higher Institute of Engineering, Cairo, Egypt
e-mail: engyshady@ieee.org

A. Y. Abdelaziz (✉)
Electric Power and Machines Department, Faculty of Engineering,
Ain Shams University, Cairo, Egypt
e-mail: almoatazabdelaziz@hotmail.com

F. Shahnia et al. (eds.), *Electric Distribution Network Planning*,
Power Systems, https://doi.org/10.1007/978-981-10-7056-3_12

Grasshopper Optimization Algorithm (GOA) to determine the optimal locations and sizes of capacitor banks and DSTATCOMs. The obtained results are compared with different algorithms such as; Grey Wolf Optimizer (GWO), Sine Cosine Algorithm (SCA).

Keywords D-FACTS · UPQC · Capacitor · DSTATCOM · Optimization

12.1 Introduction

Reactive power compensation can be used for enhancing system power quality, reducing power loss, improving voltage profile, increasing power factor and network capacity and reliability, reducing power flow in feeder lines, and enhancing the network's loadability and stability, as well as minimizing energy cost.

The most conventional devices that have been applied for reactive power compensation are capacitor banks which include the switched and fixed types, in addition to phase shifters and shunt reactor. D-FACTS devices have been incorporated in the distribution network for reactive power compensation. The main advantages of D-FACTS devices are fast response, fine controllable and continuous adjustment compared to conventional devices. Several types of D-FACTS devices have been presented for enhancing the performance of distribution networks such as DSTATCOM [1], UPQC [2] and Distribution Static Synchronous Series Compensator (DSSSC) [3].

Optimal allocation of such compensation devices is an important issue to maximize the benefits of these devices. Several techniques have been presented for solving the optimal allocation problem of compensation devices in distribution networks such as analytical techniques, numerical programming techniques, heuristic techniques and artificial intelligence techniques [4]. The analytical methods are based on calculus analytical approaches to determine the maximum of a certain objective function, and the shortage of these methods is the obtained capacitor sizes aren't matched with the standard sizes hence the solution is rounded up to standard capacitor sizes which may lead to overvoltage or less loss saving [5–7]. The numerical programming techniques are iterative optimization approach that can be applied to determine the optimal size and locations of compensation devices [8–11]. It should point out that the obtained results using these methods are more accurate compared to the analytical methods, but these techniques could be trapped in local optimal solution. Heuristic techniques are applied for minimizing the search space of optimization techniques where heuristic techniques are based on determining the most candidate nodes for reactive power compensation using sensitivity analysis [12]. Recently, artificial intelligence (AI) techniques are widely used for solving the allocation problem of compensation devices in distribution networks. Most of AI techniques are inspired from the natural phenomena behaviors. The AI methods can be applied to the nonlinear and complex problems.

This chapter introduces an application of Grasshopper Optimization Algorithm (GOA) for solving problem allocation of compensators in distribution network where GOA is employed to determine the optimal placement of shunt capacitor banks for minimizing the total cost (energy loss cost along with capacitor cost) moreover GOA is applied for assigning the optimal location and size of DSTATCOM for minimizing the total loss, improving the voltage profile and enhancing the voltage stability simultaneously.

12.2 Operation Principles of Distributed Compensators

The fixed and switched capacitor types are the most common devices that have been incorporated for reactive power compensation. Different FACTS devices are implemented for changing the parameters of network such as; transmission line impedance, the bus voltage, the active and reactive power through networks for enhancing the performance of electric systems [13, 14]. FACTS devices can be classified as: (a) series members such as Thyristor Controlled Series Capacitor (TCSC) and Static Synchronous Series Compensator (SSSC) (b) Shunt connected devices include Static VAR Compensator (SVC), Static Synchronous Compensator (STATCOM) and (c) Combined shunt-series controllers like Interline Power Flow Controller (IPFC) and Generalized Unified Power Flow Controller (GUPFC) [15–18].

12.2.1 Shunt Capacitor

The power flow equations of distribution system can be obtained from Fig. 12.1 as

$$P_{n+1} = P_n - P_{L,n+1} - R_n\left(\frac{P_n^2 + jQ_n^2}{|V_n|^2}\right) \tag{12.1}$$

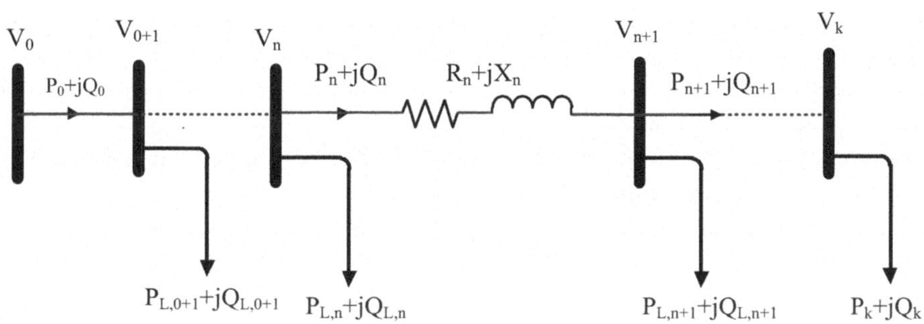

Fig. 12.1 Single line diagram of a radial distribution network

$$Q_{n+1} = Q_n - Q_{Ln+1} - X_n\left(\frac{P_n^2 + jQ_n^2}{|V_n|^2}\right) \tag{12.2}$$

$$V_{n+1}^2 = V_n^2 - 2(R_nP_n + X_nQ_n) + (R_n^2 + X_n^2)\left(\frac{P_n^2 + jQ_n^2}{|V_n|^2}\right) \tag{12.3}$$

where

P_n, Q_n Real and reactive power flows into the receiving end of branch $n + 1$ connecting bus n and node $n + 1$.
R_n, X_n Resistance and reactance of the line section between buses n and $n + 1$.
V_n The bus voltage magnitude at bus n

The active and reactive power loss of the nth line between buses n and $n + 1$ are given as

$$P_{loss(n,n+1)} = R_n\left(\frac{P_n^2 + jQ_n^2}{|V_n|^2}\right) \tag{12.4}$$

$$Q_{loss(n,n+1)} = X_n\left(\frac{P_n^2 + jQ_n^2}{|V_n|^2}\right) \tag{12.5}$$

The system security level can be realized using the voltage stability index [19] as

$$VSI_{(n+1)} = |V_n|^4 - 4(P_{n+1}X_n - Q_{n+1}R_n)^2 - 4(P_{n+1}X_n + Q_{n+1}R_n)|V_n|^2 \tag{12.6}$$

where $VSI_{(n+1)}$ is the voltage stability index at bus $n + 1$. Enhancing the voltage profile depends upon minimizing the voltage deviations as

$$VD = \sum_{n=1}^{k}(V_n - V_{ref})^2 \tag{12.7}$$

where k is a number of buses and V_{ref} is the reference voltage that commonly equals to 1 pu.

The capacitor banks are included in distribution systems for enhancing the power quality and minimizing the total cost by injecting reactive power into the systems. Figure 12.2 illustrates a shunt capacitor that is incorporated at bus $n + 1$ and the reactive power through the transmission line is given as

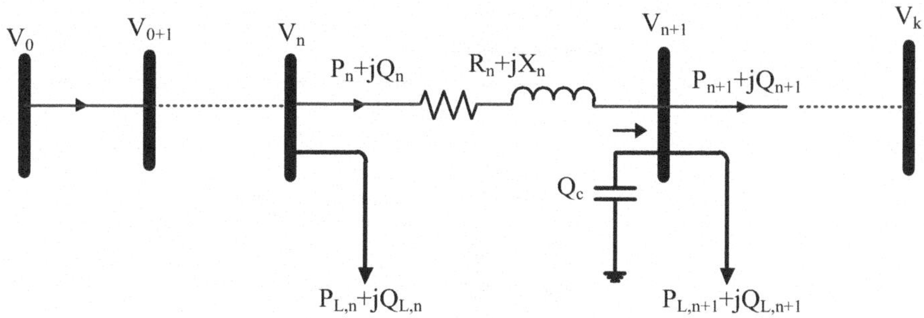

Fig. 12.2 Radial distribution system with a shunt capacitor

$$Q_{n+1} = Q_n - Q_{Ln+1} - X_n \left(\frac{P_n^2 + jQ_n^2}{|V_n|^2} \right) + Q_{C,n+1} \tag{12.8}$$

12.2.2 Distributed Static Compensator (DSTATCOM)

New members of FACTS controllers have been emerged due to progress of power electronic devices. DSTATCOM is a developed controller based on voltage source converter (VSC). DSTATCOM can inject or absorb both active and reactive power at a point of coupling connection (PCC) by injecting a variable magnitude and phase angle voltage at PCC. DSTATCOM is incorporated in electric systems for enhancing the power quality, improving the power factor, balancing the loading, mitigating the harmonic, reactive power compensation, reducing the power fluctuations of photovoltaic units minimizing the voltage sag, mitigating the flicker in the electric system and minimizing the power loss [20–23].

DSTATCOM consists of voltage source converter, dc bus capacitor, ripple filter and coupling transformer as shown in Fig. 12.3. VSC is constructed by using insulated gate bipolar transistors (IGBT) and MOSFET where the switching of component is based on pulse-width modulation (PWM) sequences. The coupling transformer is utilized for matching the inverter voltage with the bus voltage. The DSTATCOM topologies are categorized based on three-phase three-wire (3P3 W) and three-phase four-wire (3P4 W) as illustrated in [24].

DSTATCOM has an ability to exchange active and reactive current with the network. A steady state modeling DSTATCOM has been presented in [25].

Figure 12.4 shows DSTATCOM controller which included in the radial distribution system at bus $n + 1$ where DSTATCOM inject or absorb I_D at this bus. By applying KVL, the voltage at bus $n + 1$ can be obtained as

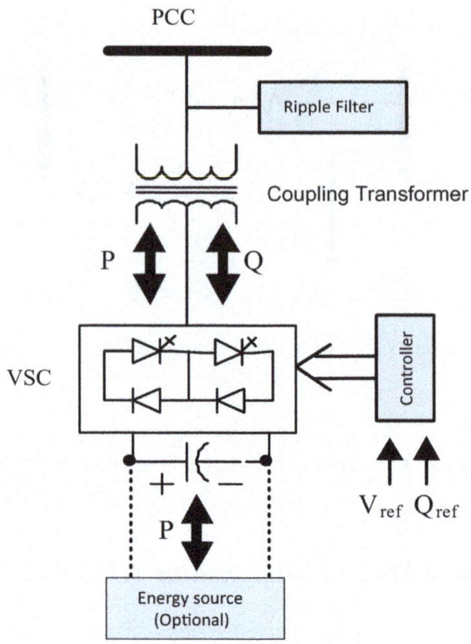

Fig. 12.3 Schematic diagram of DSTATCOM device

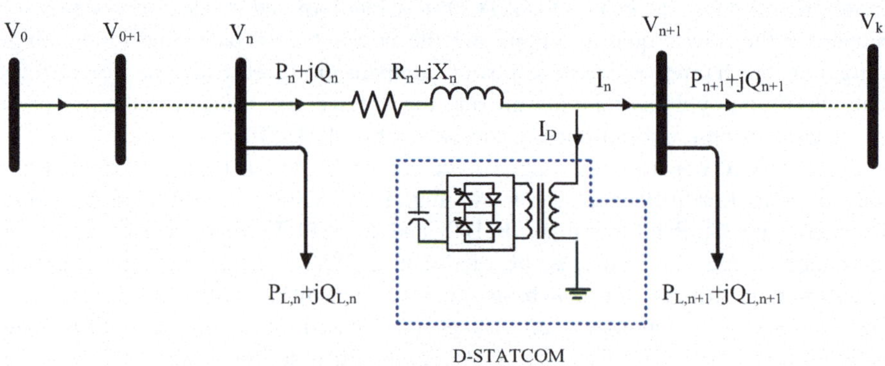

Fig. 12.4 Radial distribution system with DSTATCOM

$$V_{n+1}\angle\theta_{n+1} = V_n\angle\theta_n - (R_n + jX_n)\left(I_n\angle\delta + I_D\angle\left(\theta_{n+1} + \frac{\pi}{2}\right)\right) \tag{12.9}$$

where

$V_{n+1}\angle\theta_{n+1}$	Voltage of bus *n* + *1* after inclusion DSTATCOM.
I_D	The injected current by DSTATCOM.
I_n	The line current after inclusion of DSTATCOM

Equation (12.9) represents the essential idea for modeling DSTATCOM which can be solved by separating it to real and imaginary terms as

$$\begin{aligned} V_{n+1}\mathrm{Cos}(\theta_{n+1}) = {} & Re(V_n\angle\theta_n) - Re(I_n\angle\delta(R_n + jX_n)) + X_nI_D\mathrm{Sin}\left(\theta_{n+1} + \frac{\pi}{2}\right) \\ & - R_nI_D\mathrm{Cos}\left(\theta_{n+1} + \frac{\pi}{2}\right) \end{aligned} \tag{12.10}$$

$$\begin{aligned} V_{n+1}\mathrm{Sin}(\theta_{n+1}) = {} & Im(V_n\angle\theta_n) - Im(I_n\angle\delta(R_n + jX_n)) - X_nI_D\mathrm{Cos}\left(\theta_{n+1} + \frac{\pi}{2}\right) \\ & - R_nI_DSin\left(\theta_{n+1} + \frac{\pi}{2}\right) \end{aligned} \tag{12.11}$$

Equations (12.10) and (12.11) can be simplified as

$$a\mathrm{Cos}x_2 = k_1 - b_1x_1\mathrm{Sin}x_2 - b_2x_1\mathrm{Cos}x_2 \tag{12.12}$$

$$a\mathrm{Sin}x_2 = k_2 - b_2x_1\mathrm{Sin}x_2 + b_1x_1\mathrm{Cos}x_2 \tag{12.13}$$

where

$$k_1 = Re(V_n\angle\theta_n) - Re(I_n\angle\delta(R_n + jX_n))$$

$$k_2 = Im(V_n\angle\theta_n) - Im(I_n\angle\delta(R_n + jX_n))$$

$$a = V_{n+1}$$

$$b_1 = -R_n$$

$$b_2 = -X_n$$

$$x_1 = I_D$$

$$x_2 = \theta_{n+1}$$

Equations (12.12) and (12.13) can be rewritten as

$$x_1 = \frac{a\text{Cos}x_2 - k_1}{-b_1\text{Sin}x_2 - b_2\text{Cos}x_2} \tag{12.14}$$

$$x_1 = \frac{a\text{Sin}x_2 - k_2}{-b_2\text{Sin}x_2 + b_1\text{Cos}x_2} \tag{12.15}$$

Solving (12.14) and (12.15) yields

$$(k_1b_2 - k_2b_1)\text{Sin}x_2 + (-k_1b_1 - k_2b_2)\text{Cos}x_2 + ab_1 = 0 \tag{12.16}$$

The previous equation can be simplified as

$$(d_1^2 + d_2^2)x^2 + (2d_1ab_1)x + (a^2b_1^2 - d_2^2) = 0 \tag{12.17}$$

where

$$x = Sin(x_2)$$

$$d_1 = (k_1b_2 - k_2b_1)$$

$$d_2 = (-k_1b_1 - k_2b_2)$$

Hence, (12.17) can be solved as

$$x = \frac{-B \pm \sqrt{B^2 - 4AC}}{2A} \tag{12.18}$$

where

$$A = (d_1^2 + d_2^2)$$

$$B = (2d_1ab_1)$$

$$C = (a^2b_1^2 - d_2^2)$$

Hence

$$\theta_{n+1} = \text{Sin}^{-1}(x) \tag{12.19}$$

The value of I_D can be obtained from (12.14) or (12.15). The voltage at PCC, the DSTATCOM current and injected reactive power by DSTATCOM can be found as

$$\overrightarrow{V_{n+1}} = V_{n+1}\angle\theta_{n+1} \tag{12.20}$$

$$\overrightarrow{I_D} = I_D\angle\left(\theta_{n+1} + \frac{\pi}{2}\right) \tag{12.21}$$

$$Q_D = Im\left(V_{n+1}\angle\theta_{n+1}\left(I_D\angle\left(\theta_{n+1} + \frac{\pi}{2}\right)\right)^*\right) \tag{12.22}$$

12.2.3 Unified Power Quality Conditioner (UPQC)

UPQC is a powerful controller that has applied for enhancing the power quality of the electric system where it has the ability to minimize the voltage sags, balance the system, mitigate the existed harmonics and minimizing the power loss, etc.

UPQC consists of two inverters on of these inverters is connected in series with a certain transmission line while the other converter is connected in shunt to the common bus. These inverters are combined thought dc linked bus. The inverters are connected to the network by coupling transformers as shown in as shown in Fig. 12.5 [26–28]. The main purpose of the series inverter is injecting an ac series voltage to system to mitigate the supply voltage flickers or imbalance from the load and forces the shunt branch to absorb harmonics generated by the nonlinear loads. The shunt converter is employed for delivering the reactive power compensation for improving the power factor correction in addition the shunt converter is used to mitigate of current distortions and adjusting the dc bus voltage. In other words, the series converter regulates the load voltage to be balanced and sinusoidal while the shunt converter ensures the balancing of system current and become sinusoidal (harmonic free). Several types of UPQC have produced which can be classified based on the converter topology or the supply system or UPQC configuration [28].

Figure 12.5 shows the UPQC controller which included in the radial distribution system where the series controller is included between buses *n*, *n* + *1* while the shunt converter is connected at bus *n* + 1. It should highlight that the series injected

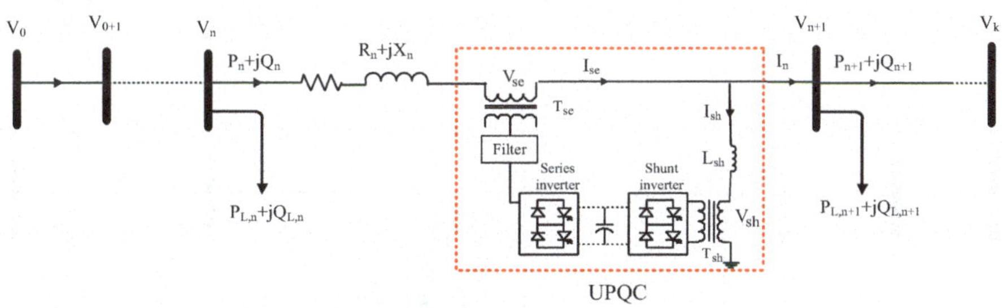

Fig. 12.5 Schematic diagram of UPQC controller

voltage is kept in quadrature with current flow. In other words, the series and shunt current are kept in quadrature with the voltage of bus $n + 1$ [29]. Referring to Fig. 12.5, the voltage at bus $n + 1$ can be given as

$$V_{n+1}\angle\theta_{n+1} = V_n\angle\theta_n - (R_n + jX_n)\left(I_n\angle\delta + I_{sh}\angle\left(\theta_{n+1} + \frac{\pi}{2}\right)\right) + V_{se}\angle\theta_{se} \tag{12.23}$$

where

V_{se}	The magnitude of the series injected voltage.
θ_{se}	The phase angle of the injected voltage.
I_n	The current flow through the transmission line.
$I_{sh}\angle\left(\theta_{n+1} + \frac{\pi}{2}\right)$	The injected current of the shunt converter

The injected current of the series converter can be found as

$$\overrightarrow{I_{se}} = \overrightarrow{I_n} + \overrightarrow{I_{sh}} \tag{12.24}$$

However, two equations are obtained by separating the real and imaginary part of (12.23). Three quantities are unknown $(V_{se}, \theta_{n+1}, I_{sh})$. For solving this problem, it is assumed that the reactive shunt power by shunt converter is represented as the negative reactive load at bus *n* + *1* as shown in Fig. 12.6 [28].

Referring to Fig. 12.6, the injected series voltage can be found as

$$V_{se}\angle\theta_{se} = V_{n+1}\angle\theta_{n+1} + Z_n\left(\acute{I}_n\angle\acute{\delta}\right) - V_n\angle\theta_n \tag{12.25}$$

where

$$\theta_{se} = \acute{\delta} + \frac{\pi}{2} \quad \acute{\delta} \le 0 \tag{12.26}$$

$$\theta_{se} = \acute{\delta} - \frac{\pi}{2} \quad \acute{\delta} > 0 \tag{12.27}$$

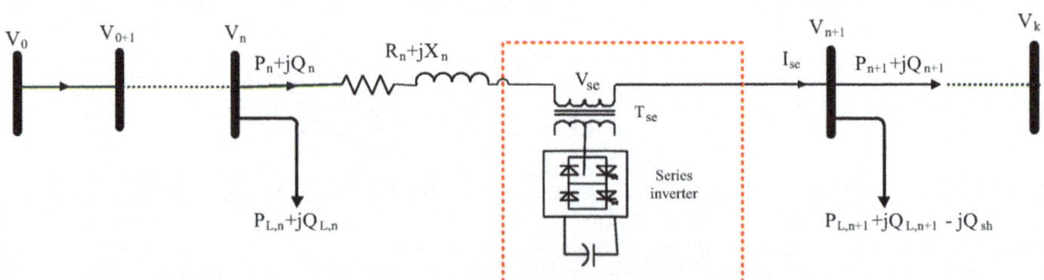

Fig. 12.6 Representation of UPQC in a distribution system

By separating the real and imaginary terms of (12.25) as

$$V_{se}\cos(\theta_{se}) = V_{n+1}\cos(\theta_{n+1}) + Re\left(Z_n\left(\acute{I}_n\angle\acute{\delta}\right)\right) - Re(V_n\angle\theta_n) \tag{12.28}$$

$$V_{se}\sin(\theta_{se}) = V_{n+1}\sin(\theta_{n+1}) + Im\left(Z_n\left(\acute{I}_n\angle\acute{\delta}\right)\right) - Im(V_n\angle\theta_n) \tag{12.29}$$

Equations (12.28) and (12.29) can be simplified as

$$V_{se}K_1 = b_3\cos(\theta_{n+1}) + b_1 \tag{12.30}$$

$$V_{se}K_2 = b_3\sin(\theta_{n+1}) + b_2 \tag{12.31}$$

where

$$x_1 = V_{se}$$

$$x_2 = \theta_{n+1}$$

$$K_1 = \cos\ (\theta_{se})$$

$$K_2 = \sin\ (\theta_{se})$$

$$b_1 = Re\left(Z_n\left(\acute{I}_n\angle\acute{\delta}\right)\right) - Re(V_n\angle\theta_n)$$

$$b_2 = Im\left(Z_n\left(\acute{I}_n\angle\acute{\delta}\right)\right) - Im(V_n\angle\theta_n)$$

$$b_3 = V_{n+1}$$

Solving (12.30) and (12.31), the value of V_{se} can be given as

$$V_{se} = \frac{-B \pm \sqrt{B^2 - 4AC}}{2A} \tag{12.32}$$

where

$$A = \frac{k_1^2 + k_2^2}{b_3}$$

$$B = -2 \times \frac{K_1b_1 + K_2b_2}{b_3}$$

$$C = \frac{b_1^2 + b_2^2}{b_3}$$

The value of θ_{n+1} can be obtained from (12.30) or (12.31) as

$$\theta_{n+1} = \cos^{-1}\left(\frac{K_1x_1 - b_1}{b_3}\right) \tag{12.33}$$

$$\theta_{n+1} = \sin^{-1}\left(\frac{K_2x_1 - b_2}{b_3}\right) \tag{12.34}$$

The reactive power of series compensator can be found as

$$Q_{series} = Im\left(V_{n+1}\angle\theta_{n+1}\left(\acute{I}_n\angle\acute{\delta}\right)^*\right) \tag{12.35}$$

12.3 Optimization Techniques

Recently, the several optimization techniques are widely applied to determine the optimal sizes and locations of compensation device in distribution networks. Variety of optimization techniques have been proposed based on nature-swarm inspired methods, human-inspired methods, physics inspired methods and evolutionary inspired algorithms. In this section, a survey including the previous techniques for solving the allocation problem of compensation devices is presented. Table 12.1 shows an overview of application the optimization techniques in radial distribution systems.

12.4 Problem Formulation

12.4.1 Capacitor Allocation Problem Formulation

The objective of optimal capacitor placement problem of the radial distribution system is minimizing the total cost including the energy loss cost along with capacitor cost. The objective function can be formulated as

$$Minimize \quad Cost = K_pP_{loss} + \sum_{i=1}^{nc} K_{c,i}Q_{c,i} \tag{12.36}$$

Table 12.1 Summary of the literature review regarding compensation devices placement problem

	Algorithm	Objective function	Controller	Sensitivity using?	Year	Refs.
Nature inspired	Ant Colony	P_{loss}	Reconfiguration and capacitor	No	2008	[30]
	Bacterial Foraging	P_{loss} and Cost	Capacitor	Yes	2015	[31]
	Particle Swarm Optimization	Cost	Capacitor	No	2010	[32]
	Cuckoo Search Algorithm	Cost	Capacitor	Yes	2013	[33]
	Artificial Bee Colony Algorithm	Cost and VSI	Capacitor	Yes	2014	[34]
	Ant Colony Optimization	P_{loss} and Cost and VD	Capacitor	Yes	2016	[35]
	Crow Search Algorithm	P_{loss} and Cost	Capacitor	No	2016	[36]
	Flower Pollination Algorithm		Capacitor		2016	[37]
	Bat and Cuckoo Search Algorithms	P_{loss} and Cost	Capacitor	No	2015	[38]
	Oppositional Krill Herd Algorithm	P_{loss}	Reconfiguration and capacitor	No	2016	[39]
	Monkey Search Optimization	P_{loss} and Cost and VD and emission	Capacitor	No	2016	[40]
	MOPSO	P_{loss} and VSI and VD and current balancing	Capacitor and DG	No	2015	[41]
	Firefly Algorithm	SAIFI and SAIDI and AENS and Cost	Capacitor	No	2014	[42]
	Particle Swarm Optimization	Cost	Capacitor	Yes	2014	[43]
	Genetic Algorithm	Loadability and total cost	DSTATCOM and DG	No	2016	[44]
	Fuzzy system and Expanded Invasive Weed Optimization	P_{loss} and VSI and VD	DSTATCOM and DG	No	2016	[45]
	Particle Swarm Optimization	P_{loss} and VD	DSTATCOM	No	2014	[46]
				No	2015	[47]

(continued)

Table 12.1 (continued)

	Algorithm	Objective function	Controller	Sensitivity using?	Year	Refs.
	Fuzzy Multiobjective Approach and Ant Colony Optimization	P_{loss} and VD and feeder load balancing	Reconfiguration and DSTATCOM and Photovoltc arry			
	Bacterial Foraging Optimization Algorithm	P_{loss} and VD and Cost	DSTATCOM and DG		2016	[48]
	Bat algorithm	P_{loss} and VSI and Cost	DSTATCOM	Yes	2015	[49]
	Improved Cat Swarm Optimization	P_{loss} and VD	DSTATCOM and DG	No		
	Cuckoo Optimization Algorithm	P_{loss} and VD and Cost and harmonic	UPQC	No	2016	[50]
	Particle Swarm Optimization	P_{loss} and VD and UPQC rating	UPQC	No	2014	[51]
	Plant Growth Simulation Algorithm	P_{loss} and Cost	Capacitor	Yes	2011	[52]
Human-inspired methods	Tabu Search	Cost	Capacitor	Yes	1996	[53]
	Improved Harmony Search	Cost	Capacitor		2016	[54]
	Harmony Search and the Particle Artificial Bee Colony	P_{loss} and VD	DG and Capacitor	Yes	2016	[55]
	Modified Cultural Algorithm	Cost	Capacitor	No	2013	[56]
	Imperialist Competitive Algorithm and Genetic Algorithm	P_{loss} and VSI and VD and load balancing	DG and Capacitor	No	2014	[57]
	Teaching Learning Based Optimization	P_{loss} and Cost	Capacitor	No	2014	[58]
	Harmony Search (HS) Algorithm	P_{loss} and VD	UPQC	No	2015	[59]

(continued)

Table 12.1 (continued)

	Algorithm	Objective function	Controller	Sensitivity using?	Year	Refs.
Physics inspired methods	Gravitational Search Algorithm	Cost	Capacitor	yes	2015	[60]
	Simulated Annealing and Greedy Search	Cost	Capacitor	No	1995	[61]
	Biogeography Based optimization (BBO)	P_{loss}	Capacitor and DG	No	2016	[62]
	Big bang-big Crunch Optimization	P_{loss}	Capacitor	No	2011	[63]
	Simulated Annealing	Cost	Capacitor	No	1990	[64]
Evolutionary inspired algorithms	Integrated Evolutionary Algorithms	Cost	Capacitor	yes	2013	[65]
	Penalty Free Genetic Algorithm	Cost	Capacitor	No	2016	[66]
	Differential Evolutionary	Cost	Capacitor	No	2015	[67]
	Fuzzy GA Genetic Algorithm	VSI and VD	Capacitor	Yes	2014	[68]
	Genetic Algorithms	Cost	Capacitor and voltage regulators	No	2012	[69]
	Genetic Algorithms	Cost	Capacitor	Yes	1994	[61]
	Immune Algorithm	Cost	DSTATCOM	No	2014	[1]
	Differential Evolution	P_{loss} and VD	DSTATCOM	No	2011	[70]
	Differential Evolution	P_{loss}	DSTATCOM	No	2016	[71]

SAIFI System average interruption frequency index
SAIDI System average interruption duration index
AENS Average energy not supplied index

where

Cost	The total cost
P_{loss}	The total active power loss (kW)
Q_c	The capacitor reactive power (kVar)
K_p	The annual cost of energy losses
K_c	The cost of capacitor per kVar

12.4.2 DSTATCOM Allocation Problem Formulation

The objective of optimal placement problem of DSTATCOM in the radial distribution system is minimizing the total loss, improving the voltage profile and enhancing the voltage stability index simultaneously as

$$f_1 = \frac{\sum_{i=1}^{nl} (P_{loss}(i))_{after\,DSTATCOM}}{\sum_{i=1}^{nl} (P_{loss}(i))_{before\,DSTATCOM}} \tag{12.37}$$

$$f_2 = \frac{\sum_{i=1}^{nb} \left(\left|V(i) - V_{ref}\right|\right)_{after\,DSTATCOM}}{\sum_{i=1}^{nl} \left(\left|V(i) - V_{ref}\right|\right)_{before\,DSTATCOM}} \tag{12.38}$$

$$f_3 = \frac{1}{\sum_{i=1}^{nb} (|VSI(i)|)_{after\,DSTATCOM}} \tag{12.39}$$

where *nl* is the number of branches in electric distribution network while *nb* is the number of buses in the network.

12.4.3 System Constraints

The required objective functions are subjected to equality and inequality constraints related to electric distribution network which can be represented as

- **Equality constraints**

The equality constraints of the system are the active and reactive power flow constraints which can be obtained as

$$P_{slack} = \sum_{i=1}^{n} P_L(i) + \sum_{j=1}^{nb} P_{loss}(j) \tag{12.40}$$

$$Q_{slack} + \sum_{i=1}^{nc} Q_c(i) = \sum_{i=1}^{n} Q_L(i) + \sum_{j=1}^{nb} Q_{loss}(j) \tag{12.41}$$

where P_{slack} and Q_{slack} are active power and reactive powers supplied from the slack bus, respectively. P_L and Q_L are the active and reactive load demands respectively. nb is the number of branches in the network while nc is the number of compensation units.

- **Inequality constraints**

I. Bus voltage constraints

$$V_{min} \le V_i \le V_{max} \tag{12.42}$$

where V_{min} and V_{max} are the minimum and the maximum allowable bus voltage limit.

II. Total reactive power constraint

Practically, the total injected reactive power using compensation devices is equal to or less than the reactive load demand.

$$\sum_{i=1}^{nc} Q_c(i) \le \sum_{i=1}^{n} Q_L(i) \tag{12.43}$$

where Q_L is the reactive load at a certain bus and Q_c is compensator reactive power.

III. Thermal limit

The current flow through network branches must be within their allowable limits as

$$I_{n,i} \le I_{max,i} \quad i = 1, 2, 3 \ldots Nb \tag{12.44}$$

Nb is the number of branches in the distribution system.

12.5 Overview of Grasshopper Optimization Algorithm (GOA)

GOA is a new optimization technique that is inspired form the movement and migration of grasshopper in natural. The adult insects of grasshopper travel together over long distance which mimics exploration of optimization technique while the

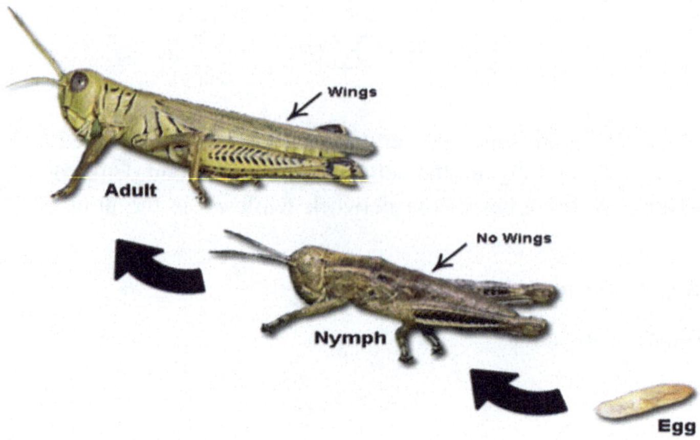

Fig. 12.7 The life cycle of grasshopper

nymphs have no wings, so it move in small area which mimics the exploitation of optimization technique [72].

Grasshoppers are harmful insects that can destroy a wide area of the agriculture and crops where the grasshoppers swarm consist of million members which can cover a wide area up to 1000 km. The life cycle of Grasshopper consists of three stages as depicted in Fig. 12.7. The grasshopper can be found in two phases. In the first phase the individual of grasshoppers avoids interaction together (solitary phase) while in the other phase (gregarious phase), grasshoppers became sociable and form a swarm. The swarm became a flying swarm depends upon environmental consideration such as air temperature, sunshine and wind speed [73].

The swarm of grasshopper moves in rolling motion where groups are formed in ground firstly by a collection of individuals of insects which move in the ground or locally and short flight then these groups became coordinated together, and the insects share a common spatial orientation. The behavior of grasshopper swarm can be summarized as

(1) The swarm flies with downwind.
(2) The grasshoppers in front of swarm settle on the ground.
(3) The settled insects start eating and resting.
(4) The swarm starts taking of gain to altitude.

The grasshopper swarm navigation behavior aligned the wind is depicted in Fig. 12.8.

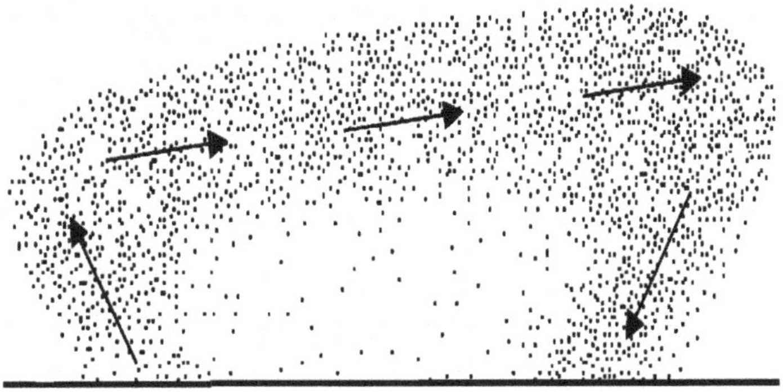

Fig. 12.8 Motion of grasshopper swarm aligned with wind

The grasshopper swarm behavior depends upon the social interaction between a grasshopper, the gravity force and the downwind advection. Hence mathematical behavior can be represented as [74]:

$$X_i = r_1 A_i + r_2 B_i + r_3 C_i \tag{12.45}$$

where

X_i	The position of ith grasshopper
A_i	The social interaction
B_i	The gravity force on the ith grasshopper
C_i	Wind advection
r_1, r_2, r_3	Random numbers

A social forced between two grasshoppers is established biologically where the repulsion forces are existed in order to prevent collisions over a short length scale and attraction force is existed for aggregation. The social interaction between grasshoppers can define as

$$A_i = \sum_{\substack{j=1 \\ i \neq j}}^{N} s\left(Dis_{ij}\right)\left(\frac{x_i - x_j}{Dis_{ij}}\right) \tag{12.46}$$

where Dis_{ij} is the distance between i and j grasshoppers that equals to $Dis_{ij} = \left|\mathrm{x_i} - \mathrm{x_j}\right|$ and the s function represents the social forces which can be represented as

$$s\left(Dis_{ij}\right) = Fe^{\frac{Dis_{ij}}{l}} - e^{Dis_{ij}} \tag{12.47}$$

where F is the intensity of attractive force and l is the attractive length scale. The swarm motion is directly affected by the gravity force which can be found as

$$B_i = -g\overrightarrow{e_g} \quad (12.48)$$

where g is the gravitational constant and $\overrightarrow{e_g}$ the unity vector towards the center of earth. The wind advection effect on the motion swarm

$$C_i = u\overrightarrow{e_w} \quad (12.49)$$

Substituting the value of A_i, B_i and C_i from (12.46), (12.48) and (12.49) in (12.45) yields

$$X_i = \sum_{\substack{j=1 \\ i \neq j}}^{N} s(Dis_{ij})\left(\frac{x_i - x_j}{Dis_{ij}}\right) - g\overrightarrow{e_g} + u\overrightarrow{e_w} \quad (12.50)$$

The previous equation is modified to be implemented for optimization problems and for enhancing the capability global searching of the algorithm it can be modified as

$$X_i^m = C\left(\sum_{\substack{j=1 \\ i \neq j}}^{N} C\left(\frac{Upper(m) - Lower(m)}{2}\right) s(Dis_{ij})\left(\frac{x_i - x_j}{Dis_{ij}}\right)\right) + P_{best}^m \quad (12.51)$$

where $Upper(.)$ and $Lower(.)$ are the upper and lower limits of the control variable, respectively. P_{best}^m is the best position (the target position). C is an adaptive coefficient that decrease linearly for enhancing the search capability of GOA which can be represented as

$$C = C_{max} - T\frac{C_{max} - C_{min}}{T_{max}} \quad (12.52)$$

where C_{max}, C_{min} are the maximum and the minimum values of C, respectivly. T and T_{max} are the current iterations and the maximum iteration, respectively.

Step 1: Determine the input data of GOA algorithm including number of the search agents (N), maximum number of iterations,C_{min}, C_{max}, F, L and the upper and lower boundaries of control variables.
Step 2: Initialize the population of GOA algorithm as

$$P_i^m = Lower(i,m) + rand * (Upper(i,m) - Lower(i,m)) \quad (12.53)$$

Step 3: Calculate the fitness functions for each search agent.
Step 4: Determine the best position (target position) in term of the best fitness function.
Step 5: Update the position of search agent according to (12.51).
Step 6: Check the boundaries of the updated agents and bring the violated variable to accepted limit.
Step 7: Calculate the fitness function for the updated positions and determine the target position.
Step 8: Repeat steps form (12.5) to (12.7) until the stopping criterion is achieved (current iteration equals to maximum iteration).
Step 9: Obtain the optimal solution by capture the target position and the related fitness function.

12.6 Numerical Examples

In this section the grasshopper optimization technique is employed to determine the optimal locations and sizes of shunt capacitors and DSTATCOM in the 69-bus radial distribution network. The line diagram of the network is shown in Fig. 12.9. The network data are given in [75] which are also tabulated in Table 12.7. A program code for optimal allocation of compensators is written using MATLAB 2009a and run on a PC with core i5 processor, 2.50 GHz and 4 GB RAM. The selected parameters of GOA technique are listed in Table 12.2. The parameters required for implementation of the proposed algorithm are adjusted by 50 times running of this algorithm. The obtained results using the GOA algorithm are compared with compared with other well-known optimization algorithms such as; Grey Wolf Optimizer (GWO) [76], Sine Cosine Algorithm (SCA) [77] and other meta-heuristics techniques. The studied cases are presented as

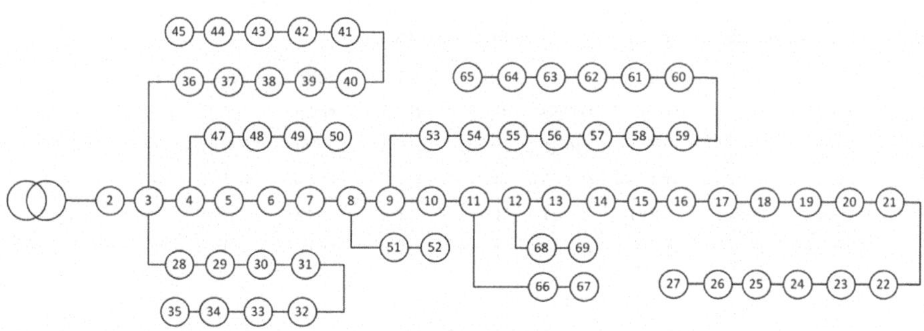

Fig. 12.9 The line diagram of the 69-bus system

Table 12.2 The selected parameters of GOA

Parameter	T_{max}	Search agents NO.	C_{max}	C_{min}	F	L
Value	100	30	1	0.00001	0.5	1.5

12.6.1 Case 1

The GOA technique is applied for optimal allocation of the capacitor in the 69-bus network to minimize the total cost as described in (12.36). The sizes of capacitors are selected to be standard with the available industrial market. The available sizes and costs of capacitors are listed in Table 12.3. The Total active and reactive load demands are 3801.89 kW and 2694.1 kVar respectively. The substation voltage is 12.66 kV and the single line diagram. The system power loss without inclusion compensation devices equal to 225 kW and the total cost for the system without any capacitor is found to be 37,800.0 $. The optimal size of capacitors, their locations and the impact of optimal placement and sizing of capacitors on the energy loss cost, capacitor cost and total cost of the system by 50 run trials are given in Table 12.4. Moreover, the best, worst and mean obtained results by GOA also are listed in Table 12.4. The power loss decreased to 145.405 MW with incorporating capacitor banks optimally using GOA. Moreover, the value of total cost is enhanced to 24,820.84 $. From Table 12.5 it can also be found that the objective value found by the GOA technique is better than those obtained by the CSA [33], DSA [78], TLBO [58], GSA [2], GWO and SCA. This demonstrates that the GOA successfully achieves better simulation results than other techniques. The voltage profiles of all system buses are enhanced significantly with incorporating capacitor banks optimally using GOA as shown in Fig. 12.10. The average computational time taken by the GOA technique and the other techniques are reported in Table 12.4. It can be obvious that GOA needs less computational time compared with other reported techniques. The convergence characteristic of the GOA, GWO

Table 12.3 Available capacitor size and related cost ($/kVar)

Size (kVar)	150	300	450	600	750	900	1050	1200	1350
Cost ($/kVar)	0.5	0.35	0.253	0.22	0.276	0.183	0.228	0.170	0.207
Size (kVar)	1500	1650	1800	1950	2100	2250	2400	2550	2700
Cost $/kVar	0.201	0.193	0.187	0.211	0.176	0.197	0.170	0.189	0.187
Size (kVar)	2850	3000	3150	3300	3450	3600	3750	3900	4050
Cost ($/kVar)	0.183	0.180	0.195	0.174	0.188	0.170	0.183	0.182	0.179

Table 12.4 Obtained results for the 69-bus test system using different optimization techniques

	Base case	CSA [33]	DSA [78]	TLBO [58]	GSA [2]	GWO	SCA	GOA
$V_{min}(p.u)$	0.9092	0.930	0.9318	0.9321	0.9519	0.93079	0.93145	0.93079
$P_{loss}(kW)$	225.00	147.95	147	146.35	145.9	145.569	145.440	145.405
Capacitor location (kVar)	–	21 (250) 62 (1200)	61 (900) 15 (450) 60 (450)	22 (300) 61 (1050) 62 (300)	26 (150) 13 (150) 15 (1050)	61 (1200) 12 (450) 26 (150)	61 (1200) 9 (450) 17 (350)	61 (1200) 12 (450) 21 (150)
Total (kVar)	–	1450	1800	1650	1350	1800	2000	1800
Capacitors cost ($)	–	291.5	392.4	446.4	451.5	392.85	440.35	392.85
Energy cost ($)	37,800.0	24,855.6	24,696	24,586.8	24,511.2	24,455.51	24,433.98	24,427.99
Total annual cost ($)	37,800.0	25,147.1	25,088.4	25,033.2	24,962.7	24,848.36	24,874.33	24,820.84
Net savings ($)	–	12,652.9	12,711.6	12,766.8	12,837.3	12,951.6	12,925.7	12,979.1
Best annual cost ($)	37,800.0	25,147.1	25,088.4	25,033.2	24,962.7	24,848.36	24,874.33	24,820.84
Worst annual cost ($)		NA	NA	NA	NA	25,093.56	25,040.17	25,040.17
Mean annual cost ($)		NA	NA	NA	NA	24,938.10	24,930.81	24,930.81
Average CPU time (S)		125.80	NA	36.87	NA	23.66	24.15	21.30

and SCA are depicted in Fig. 12.11. From the convergence graph, it may be observed that the objective value (total cost) converges and smoothly rapidly at the 15th iteration compared to GWO and SCA. This confirms the convergence reliability of the proposed GWO algorithm.

12.6.2 Case2

In this case, GOA technique is employed to determine the optimal locations and sizes of DSTATCOMs in the 69-bus network for minimizing the total loss, improving the voltage profile and enhancing the voltage stability index simultaneously as described in (12.37), (12.38) and (12.39). Hence, in this case, the objective function is a multi-objective function which can be formulated as

$$f_t = w_1 f_1 + w_2 f_2 + w_3 f_3 \tag{12.54}$$

where w_1, $w2$ and $w3$ are weighting factors. The value of any weighting factor is selected based on the relative important on its related objective function with others objective functions. The sum of the absolute values of the weight factors in (12.54) assigned to all impacts should add up to one as [79]

$$|w_1| + |w_2| + |w_3| = 1 \tag{12.55}$$

In this chapter, w_1 is set as 0.5 while w_2 and w_3 equal 0.25.It should point out that the constraint of injected reactive power of DSTATCOM is restricted as [1]

$$0 \leq Q_{STATCOM} \leq 10,000 KVAR \tag{12.56}$$

$$\sum_{i=1}^{nc} Q_{STATCOM}(i) \leq \sum_{i=1}^{n} Q_L(i) \tag{12.57}$$

In this case, three DSTATCOM devices are included in the 69-bus system. The optimal locations and sizes of DSTATCOMs that have been determined using GOA, GWO and SCA, are listed in Table 12.5. It is obvious that the power loss is reduced to 145.146 and the summation of voltage deviations is also reduced from 1.8374 to 1.3872 p.u with incorporating of the DSTATCOMs optimally using GOA. Moreover, the voltage stability is also enhanced to 62.7759 p.u with inclusion of DSTATCOMs. From Table 12.6, it is clear that the obtained results by GOA are better than those obtained by GWO and SCA.

Table 12.5 Simulation results of the 69-bus system at different loadings

Loading		Base case	GWO	SCA	GOA
100%	Minimum voltage	0.9092	0.93079	0.93145	0.93079
	Total active loss kW	225.00	145.569	145.440	145.405
	Annual cost $/year	37,800.0	24,848.36	24,874.33	24,820.84
	Location and size	–	61 (1200) 12 (450) 26 (150)	61 (1200) 9 (450) 17 (350)	61 (1200) 12 (450) 21 (150)
75%	Minimum voltage	0.93353	0.94874	0.94873	0.94874
	Total active loss kW	121.030	79.971	81.383	79.971
	Annual cost $/year	20,333.04	13,722.35	13,959.48	13,722.35
	Location and size	–	61 (900) 12 (350)	61 (900) 9 (350)	61 (900) 12 (350)
50%	Minimum voltage	0.95668	0.96569	0.96569	0.96569
	Total active loss kW	51.606	35.757	35.757	35.757
	Annual cost $/year	8669.808	6139.1694	6139.1694	6139.1694
	Location and size	–	61 (600)	61 (600)	61 (600)
Net injected kVar			Fixed 600 at bus 61 Switched 600 at bus 61 Switched 450 at bus 12 Switched 350 at bus 26	Fixed 600 at bus 61 Switched 600 at bus 61 Switched 450 at bus 9 Switched 350 at bus 17	Fixed 600 at bus 61 Switched 600 at bus 61 Switched 450 at bus 12 Switched 350 at bus 21

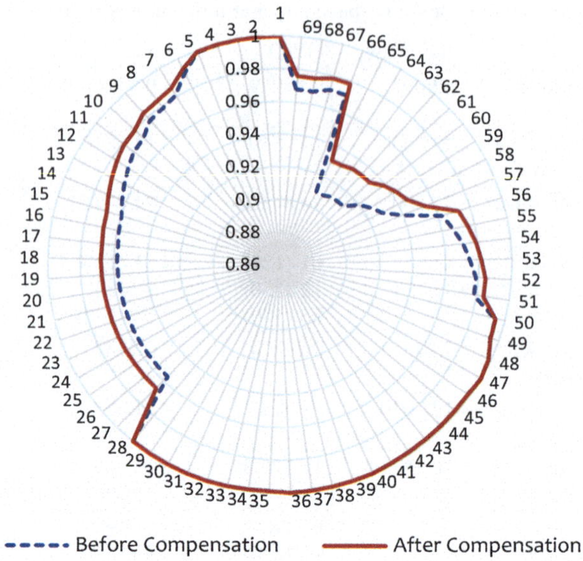

Fig. 12.10 Effect of compensation on system voltages for the 69-bus system

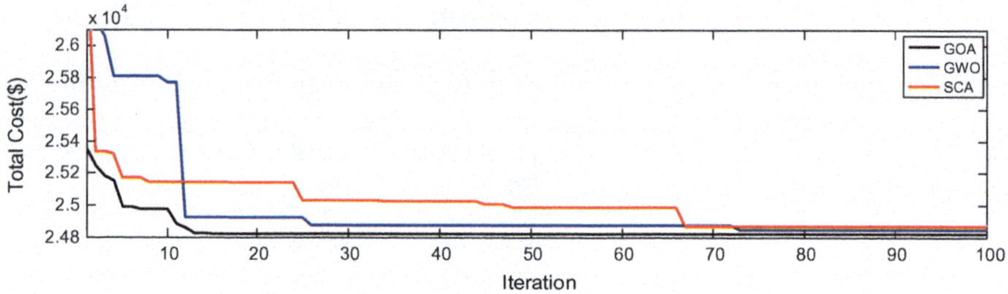

Fig. 12.11 Change of total cost with iterations for the 69-bus using GOA, GWO and SCA

Table 12.6 Obtained results for optimal allocation of DSTATCOM using different optimization techniques

	Base case	GWO	SCA	GOA
$V_{min}(p.u)$	0.90919	0.93093	0.93132	0.93121
$V_{max}(p.u)$	0.99997	0.9999	0.99998	0.99998
$VSI_{min}(p.u)$	0.6833	0.7511	0.7523	0.7520
$VSI_{max}(p.u)$	0.9999	1.0000	0.9999	0.9999
$\sum VSI$	61.2181	62.6904	62.7154	62.7759
$P_{loss}(KW)$	225.00	146.453	145.840	145.146
$VD(p.u)$	1.8374	1.4105	1.4046	1.3872
Optimal locations and size of DSTATCOM (kVar)	–	61 (1264.5) 17 (346.9973) 36 (687.7078)	12 (548.01) 61 (1245.6) 49 (562.84)	11 (374.71) 61 (1224.21) 18 (242.430)

Appendix

See Table 12.7.

Table 12.7 Data of the 69-bus test systems

S. NO.	From bus	To bus	R (Ω)	X (Ω)	PL (kW)	QL (kVar)
1	1	2	0.0005	0.0012	0	0
2	2	3	0.0005	0.0012	0	0
3	3	4	0.0015	0.0036	0	0
4	4	5	0.0251	0.0294	0	0
5	5	6	0.366	0.1864	2.60	2.20
6	6	7	0.3811	0.1941	40.40	30
7	7	8	0.0922	0.0470	75	54
8	8	9	0.0493	0.0251	30	22
9	9	10	0.819	0.2707	28	19
10	10	11	0.1872	0.0619	145	104
11	11	12	0.7114	0.2350	145	104
12	12	13	1.0300	0.3400	8	5
13	13	14	1.0440	0.3450	8	5.50
14	14	15	1.0580	0.3496	0	0
15	15	16	0.1966	0.0650	45.50	30
16	16	17	0.3744	0.1238	60	35
17	17	18	0.0047	0.0016	60	35
18	18	19	0.3276	0.1083	0	0

(continued)

Table 12.7 (continued)

S. NO.	From bus	To bus	R (Ω)	X (Ω)	PL (kW)	QL (kVar)
19	19	20	0.2106	0.0690	1	0.60
20	20	21	0.3416	0.1129	114	81
21	21	22	0.0140	0.0046	5	3.50
22	22	23	0.1591	0.0526	0	0
23	23	24	0.3463	0.1145	28	20
24	24	25	0.7488	0.2475	0	0
25	25	26	0.3089	0.1021	14	10
26	26	27	0.1732	0.0572	14	10
27	27	28	0.0044	0.0108	26	18.60
28	28	29	0.0640	0.15650	26	18.60
29	29	30	0.3978	0.1315	0	0
30	30	31	0.0702	0.0232	0	0
31	31	32	0.3510	0.1160	0	0
32	32	33	0.8390	0.2816	14	10
33	33	34	1.7080	0.5646	9.50	14
34	34	35	1.4740	0.4873	6	4
35	35	6	0.0044	0.0108	26	18.55
36	36	37	0.0640	0.1565	26	18.55
37	37	38	0.1053	0.1230	0	0
38	38	39	0.0304	0.0355	24	17
39	39	40	0.0018	0.0021	24	17
40	40	41	0.7283	0.8509	1.20	1
41	41	42	0.3100	0.3623	0	0
42	42	43	0.0410	0.0478	6	4.30
43	43	44	0.0092	0.0116	0	0
44	44	45	0.1089	0.1373	39.22	26.30
45	45	46	0.0009	0.0012	39.22	26.30
46	4	47	0.0034	0.0084	0	0
47	47	48	0.0851	0.2083	79	56.40
48	48	49	0.2898	0.7091	384.70	274.50
49	49	50	0.0822	0.2011	384.70	274.50
50	8	51	0.0928	0.0473	40.50	28.30
51	51	52	0.3319	0.1114	3.60	2.70
52	9	53	0.1740	0.0886	4.35	3.50
53	53	54	0.2030	0.1034	26.40	19
54	54	55	0.2842	0.1447	24	17.20
55	55	56	0.2813	0.1433	0	0
56	56	57	1.5900	0.5337	0	0
57	57	58	0.7837	0.2630	0	0
58	58	59	0.3042	0.1006	100	72

(continued)

Table 12.7 (continued)

S. NO.	From bus	To bus	R (Ω)	X (Ω)	PL (kW)	QL (kVar)
59	59	60	0.3861	0.1172	0	0
60	60	61	0.5075	0.2585	1244	888
61	61	62	0.0974	0.0496	32	23
62	62	63	0.1450	0.0738	0	0
63	63	64	0.7105	0.3619	227	162
64	64	65	1.0410	0.5302	59	42
65	11	66	0.2012	0.0611	18	13
66	66	67	0.0047	0.0014	18	13
67	12	68	0.7394	0.2444	28	20
68	68	69	0.0047	0.0016	28	20
Tie lines						
69	11	43	0.5	0.5	6.0	4.30
70	13	21	0.5	0.5	5.00	3.50
71	15	46	1.0	1.0	39.22	26.30
72	50	59	2.0	2.0	100.0	72
73	27	65	1.0	1.0	59.0	42.0

References

1. S.A. Taher, S.A. Afsari, Optimal location and sizing of DSTATCOM in distribution systems by immune algorithm. Int. J. Electr. Power Energy Syst. **60**, 34–44 (2014)
2. S. Ganguly, Impact of unified power-quality conditioner allocation on line loading, losses, and voltage stability of radial distribution systems. IEEE Trans. Power Delivery **29**, 1859–1867 (2014)
3. S. Devi, M. Geethanjali, Optimal location and sizing of distribution static synchronous series compensator using particle swarm optimization. Int. J. Electr. Power Energy Syst. **62**, 646–653 (2014)
4. H. Ng, M. Salama, A. Chikhani, Classification of capacitor allocation techniques. IEEE Trans. Power Delivery **15**, 387–392 (2000)
5. J. Schmill, Optimum size and location of shunt capacitors on distribution feeders. IEEE Trans. Power Appar. Syst. **84**, 825–832 (1965)
6. N. Neagle, D. Samson, Loss reduction from capacitors installed on primary feeders. Trans. Am. Inst. Electr. Eng. Part III: Power Appar. Syst. **75**, 950–959 (1956)
7. Y. Bae, Analytical method of capacitor allocation on distribution primary feeders. IEEE Trans. Power Appar. Syst. 1232–1238 (1978)
8. T.H. Fawzi, S.M. El-Sobki, M.A. Abdel-halim, New approach for the application of shunt capacitors to the primary distribution feeders. IEEE Trans. Power Appar. Syst. 10–13 (1983)
9. H. Dura, Optimum number, location, and size of shunt capacitors in radial distribution feeders a dynamic programming approach. IEEE Trans. Power Appar. Syst. 1769–1774 (1968)
10. M. Baran, F.F. Wu, Optimal sizing of capacitors placed on a radial distribution system. IEEE Trans. Power Delivery **4**, 735–743 (1989)

11. M. Ponnavsikko, K.P. Rao, Optimal choice of fixed and switched shunt capacitors on radial distributors by the method of local variations. IEEE Trans. Power Appar. Syst. 1607–1615 (1983)
12. S. Lee, J. Grainger, Optimum placement of fixed and switched capacitors on primary distribution feeders. IEEE Trans. Power Appar. Syst. 345–352 (1981)
13. K. Padiyar, *FACTS Controllers in Power Transmission and Distribution* (New Age International, 2007)
14. N.G. Hingorani, L. Gyugyi, *Understanding Facts* (IEEE press, 2000)
15. S. Kamel, F. Jurado, D. Vera, A simple implementation of power mismatch STATCOM model into current injection Newton-Raphson power-flow method. Electr. Eng. **96**, 135–144 (2014)
16. S. Kamel, F. Jurado, Z. Chen, M. Abdel-Akher, M. Ebeed, Developed generalised unified power flow controller model in the Newton-Raphson power-flow analysis using combined mismatches method. IET Gener. Transm. Distrib. **10**, 2177–2184 (2016)
17. S. Abd el-sattar, S. Kamel, M. Ebeed, Enhancing security of power systems including SSSC using moth-flame optimization algorithm, in *Power Systems Conference (MEPCON), 2016 Eighteenth International Middle East* (2016), pp. 797–802
18. M. Ebeed, S. Kamel, F. Jurado, Determination of IPFC operating constraints in power flow analysis. Int. J. Electr. Power Energy Syst. **81**, 299–307 (2016)
19. M. Chakravorty, D. Das, Voltage stability analysis of radial distribution networks. Int. J. Electr. Power Energy Syst. **23**, 129–135 (2001)
20. A. Elnady, M.M. Salama, Unified approach for mitigating voltage sag and voltage flicker using the DSTATCOM. IEEE Trans. Power Delivery **20**, 992–1000 (2005)
21. Z. Shuai, A. Luo, Z.J. Shen, W. Zhu, Z. Lv, C. Wu, A dynamic hybrid var compensator and a two-level collaborative optimization compensation method. IEEE Trans. Power Electron. **24**, 2091–2100 (2009)
22. R. Majumder, Reactive power compensation in single-phase operation of microgrid. IEEE Trans. Industr. Electron. **60**, 1403–1416 (2013)
23. R. Yan, B. Marais, T.K. Saha, Impacts of residential photovoltaic power fluctuation on on-load tap changer operation and a solution using DSTATCOM. Electr. Power Syst. Res. **111**, 185–193 (2014)
24. O.P. Mahela, A.G. Shaik, A review of distribution static compensator. Renew. Sustain. Energy Rev. **50**, 531–546 (2015)
25. M. Hosseini, H.A. Shayanfar, Modeling of series and shunt distribution FACTS devices in distribution systems load flow. J. Electr. Syst. **4**, 1–12 (2008)
26. A. Ghosh, G. Ledwich, *Power Quality Enhancement Using Custom Power Devices* (Springer Science & Business Media, 2012)
27. M.-C. Wong, C.-J. Zhan, Y.-D. Han, L.-B. Zhao, A unified approach for distribution system conditioning: distribution system unified conditioner (DS-UniCon), in *Power Engineering Society Winter Meeting, 2000. IEEE* (2000), pp. 2757–2762
28. V. Khadkikar, Enhancing electric power quality using UPQC: a comprehensive overview. IEEE Trans. Power Electron. **27**, 2284–2297 (2012)
29. M. Hosseini, H. Shayanfar, M. Fotuhi-Firuzabad, Modeling of unified power quality conditioner (UPQC) in distribution systems load flow. Energy Convers. Manag. **50**, 1578–1585 (2009)
30. C.-F. Chang, Reconfiguration and capacitor placement for loss reduction of distribution systems by ant colony search algorithm. IEEE Trans. Power Syst. **23**, 1747–1755 (2008)
31. K. Devabalaji, K. Ravi, D. Kothari, Optimal location and sizing of capacitor placement in radial distribution system using bacterial foraging optimization algorithm. Int. J. Electr. Power Energy Syst. **71**, 383–390 (2015)
32. A.A. Eajal, M. El-Hawary, Optimal capacitor placement and sizing in unbalanced distribution systems with harmonics consideration using particle swarm optimization. IEEE Trans. Power Delivery **25**, 1734–1741 (2010)

33. A.A. El-Fergany, A.Y. Abdelaziz, Cuckoo search-based algorithm for optimal shunt capacitors allocations in distribution networks. Electric Power Components and Systems **41**, 1567–1581 (2013)
34. A.A. El-Fergany, Involvement of cost savings and voltage stability indices in optimal capacitor allocation in radial distribution networks using artificial bee colony algorithm. Int. J. Electr. Power Energy Syst. **62**, 608–616 (2014)
35. A.A.A. El-Ela, R.A. El-Sehiemy, A.-M. Kinawy, M.T. Mouwafi, Optimal capacitor placement in distribution systems for power loss reduction and voltage profile improvement. IET Gener. Transm. Distrib. **10**, 1209–1221 (2016)
36. A. Askarzadeh, Capacitor placement in distribution systems for power loss reduction and voltage improvement: a new methodology. IET Gener. Transm. Distrib. **10**, 3631–3638 (2016)
37. A. Abdelaziz, E. Ali, S.A. Elazim, Flower pollination algorithm and loss sensitivity factors for optimal sizing and placement of capacitors in radial distribution systems. Int. J. Electr. Power Energy Syst. **78**, 207–214 (2016)
38. S.K. Injeti, V.K. Thunuguntla, M. Shareef, Optimal allocation of capacitor banks in radial distribution systems for minimization of real power loss and maximization of network savings using bio-inspired optimization algorithms. Int. J. Electr. Power Energy Syst. **69**, 441–455 (2015)
39. S. Sultana, P.K. Roy, Oppositional krill herd algorithm for optimal location of capacitor with reconfiguration in radial distribution system. Int. J. Electr. Power Energy Syst. **74**, 78–90 (2016)
40. F.G. Duque, L.W. de Oliveira, E.J. de Oliveira, An approach for optimal allocation of fixed and switched capacitor banks in distribution systems based on the monkey search optimization method. J. Control Autom. Electr. Syst. **27**, 212–227 (2016)
41. A. Zeinalzadeh, Y. Mohammadi, M.H. Moradi, Optimal multi objective placement and sizing of multiple DGs and shunt capacitor banks simultaneously considering load uncertainty via MOPSO approach. Int. J. Electr. Power Energy Syst. **67**, 336–349 (2015)
42. A.K. Fard, T. Niknam, Optimal stochastic capacitor placement problem from the reliability and cost views using firefly algorithm. IET Sci. Meas. Technol. **8**, 260–269 (2014)
43. A. Elsheikh, Y. Helmy, Y. Abouelseoud, A. Elsherif, Optimal capacitor placement and sizing in radial electric power systems. Alex. Eng. J. **53**, 809–816 (2014)
44. H. Karami, B. Zaker, B. Vahidi, G.B. Gharehpetian, Optimal multi-objective number, locating, and sizing of distributed generations and distributed static compensators considering loadability using the genetic algorithm. Electr. Power Compon. Syst. **44**, 2161–2171 (2016)
45. H. Bagheri Tolabi, A. Lashkar Ara, and R. Hosseini, A fuzzy-ExIWO method for optimal placement of multiple DSTATCOM/DG and tuning the DSTATCM's controller, COMPEL: Int. J. Comput. Math. Electr. Electron. Eng. **35**, 1014–1033 (2016)
46. S. Devi, M. Geethanjali, Placement and sizing of D-STATCOM using particle swarm optimization, in *Power Electronics and Renewable Energy Systems* (Springer, 2015), pp. 941–951
47. H.B. Tolabi, M.H. Ali, M. Rizwan, Simultaneous reconfiguration, optimal placement of DSTATCOM, and photovoltaic array in a distribution system based on fuzzy-ACO approach. IEEE Trans. Sustain. Energy **6**, 210–218 (2015)
48. K. Devabalaji, K. Ravi, Optimal size and siting of multiple DG and DSTATCOM in radial distribution system using bacterial foraging optimization algorithm. Ain Shams Eng. J. **7**, 959–971 (2016)
49. T. Yuvaraj, K. Ravi, K. Devabalaji, DSTATCOM allocation in distribution networks considering load variations using bat algorithm. Ain Shams Eng. J. (2015)
50. J. Sarker, S. Goswami, Optimal location of unified power quality conditioner in distribution system for power quality improvement. Int. J. Electr. Power Energy Syst. **83**, 309–324 (2016)
51. S. Ganguly, Multi-objective planning for reactive power compensation of radial distribution networks with unified power quality conditioner allocation using particle swarm optimization. IEEE Trans. Power Syst. **29**, 1801–1810 (2014)

52. R.S. Rao, S. Narasimham, M. Ramalingaraju, Optimal capacitor placement in a radial distribution system using plant growth simulation algorithm. Int. J. Electr. Power Energy Syst. **33**, 1133–1139 (2011)
53. Y.-C. Huang, H.-T. Yang, C.-L. Huang, Solving the capacitor placement problem in a radial distribution system using tabu search approach. IEEE Trans. Power Syst. **11**, 1868–1873 (1996)
54. E. Ali, S.A. Elazim, A. Abdelaziz, Improved harmony algorithm and power loss index for optimal locations and sizing of capacitors in radial distribution systems. Int. J. Electr. Power Energy Syst. **80**, 252–263 (2016)
55. K. Muthukumar, S. Jayalalitha, Optimal placement and sizing of distributed generators and shunt capacitors for power loss minimization in radial distribution networks using hybrid heuristic search optimization technique. Int. J. Electr. Power Energy Syst. **78**, 299–319 (2016)
56. V. Haldar, N. Chakraborty, Power loss minimization by optimal capacitor placement in radial distribution system using modified cultural algorithm. Int. Trans. Electr. Energy Syst. **25**, 54–71 (2015)
57. M.H. Moradi, A. Zeinalzadeh, Y. Mohammadi, M. Abedini, An efficient hybrid method for solving the optimal sitting and sizing problem of DG and shunt capacitor banks simultaneously based on imperialist competitive algorithm and genetic algorithm. Int. J. Electr. Power Energy Syst. **54**, 101–111 (2014)
58. S. Sultana, P.K. Roy, Optimal capacitor placement in radial distribution systems using teaching learning based optimization. Int. J. Electr. Power Energy Syst. **54**, 387–398 (2014)
59. V. Renu, S. Jeyadevi, Optimal design of UPQC devices in radial distribution network for voltage stability enhancement. Int. J. Appl. Eng. Res. **10** (2015)
60. Y.M. Shuaib, M.S. Kalavathi, C.C.A. Rajan, Optimal capacitor placement in radial distribution system using gravitational search algorithm. Int. J. Electr. Power Energy Syst. **64**, 384–397 (2015)
61. S. Sundhararajan, A. Pahwa, Optimal selection of capacitors for radial distribution systems using a genetic algorithm. IEEE Trans. Power Syst. **9**, 1499–1507 (1994)
62. H. Sadeghi, N. Ghaffarzadeh, A simultaneous biogeography based optimal placement of DG units and capacitor banks in distribution systems with nonlinear loads. J. Electr. Eng. **67**, 351–357 (2016)
63. M. Sedighizadeh, D. Arzaghi-Haris, Optimal allocation and sizing of capacitors to minimize the distribution line loss and to improve the voltage profile using big bang-big crunch optimization. Int. Rev. Electr. Eng. **6** (2011)
64. H.-D. Chiang, J.-C. Wang, O. Cockings, H.-D. Shin, Optimal capacitor placements in distribution systems. II. Solution algorithms and numerical results. IEEE Trans. Power Delivery **5**, 643–649 (1990)
65. A.A. El-Fergany, Optimal capacitor allocations using evolutionary algorithms. IET Gener. Transm. Distrib. **7**, 593–601 (2013)
66. J. Vuletić, M. Todorovski, Optimal capacitor placement in distorted distribution networks with different load models using penalty free genetic algorithm. Int. J. Electr. Power Energy Syst. **78**, 174–182 (2016)
67. R. Hosseinzadehdehkordi, H. Shayeghi, M. Karimi, P. Farhadi, Optimal sizing and siting of shunt capacitor banks by a new improved differential evolutionary algorithm. Int. Trans. Electr. Energy Syst. **24**, 1089–1102 (2014)
68. A.R. Abul'Wafa, Optimal capacitor placement for enhancing voltage stability in distribution systems using analytical algorithm and Fuzzy-Real Coded GA. Int. J. Electr. Power Energy Syst. **55**, 246–252 (2014)
69. I. Szuvovivski, T. Fernandes, A. Aoki, Simultaneous allocation of capacitors and voltage regulators at distribution networks using genetic algorithms and optimal power flow. Int. J. Electr. Power Energy Syst. **40**, 62–69 (2012)
70. S. Jazebi, S. Hosseinian, B. Vahidi, DSTATCOM allocation in distribution networks considering reconfiguration using differential evolution algorithm. Energy Convers. Manag. **52**, 2777–2783 (2011)

71. J. Sanam, A. Panda, S. Ganguly, Optimal phase angle injection for reactive power compensation of distribution systems with the allocation of multiple distribution STATCOM. Arab. J. Sci. Eng. 1–9 (2016)
72. S. Saremi, S. Mirjalili, A. Lewis, Grasshopper optimisation algorithm: theory and application. Adv. Eng. Softw. **105**, 30–47 (2017)
73. B. Uvarov, Grasshoppers and locusts.in *A Handbook of General Acridology Vol. 2. Behaviour, Ecology, Biogeography, Population Dynamics* (Centre for Overseas Pest Research, 1977)
74. C.M. Topaz, A.J. Bernoff, S. Logan, W. Toolson, A model for rolling swarms of locusts. Eur. Phys. J.-Spec. Top. **157**, 93–109 (2008)
75. S. Chandramohan, N. Atturulu, R.K. Devi, B. Venkatesh, Operating cost minimization of a radial distribution system in a deregulated electricity market through reconfiguration using NSGA method. Int. J. Electr. Power Energy Syst. **32**, 126–132 (2010)
76. S. Mirjalili, S.M. Mirjalili, A. Lewis, Grey wolf optimizer. Adv. Eng. Softw. **69**, 46–61 (2014)
77. S. Mirjalili, SCA: a sine cosine algorithm for solving optimization problems. Knowl.-Based Syst. **96**, 120–133 (2016)
78. M.R. Raju, K.R. Murthy, K. Ravindra, Direct search algorithm for capacitive compensation in radial distribution systems. Int. J. Electr. Power Energy Syst. **42**, 24–30 (2012)
79. E. Ali, S.A. Elazim, A. Abdelaziz, Ant lion optimization algorithm for renewable distributed generations. Energy **116**, 445–458 (2016)

Chapter 13
Optimal Allocation of Automatic Reclosers

Carlos Frederico Meschini Almeida, Gabriel Albieri Quiroga, Henrique Kagan and Nelson Kagan

Abstract This chapter presents a methodology for the allocation of Automatic Reclosers (AR) in medium voltage electric distribution networks. The methodology defines strategic positions for installing Normally Closed (NC) and Normally Opened (NO) reclosers to improve the system's performance in terms of quality of power supply. The restriction relies on the budget available for investing in purchasing and installing AR. The methodology supports power distribution planning activities, as it focusses on defining the optimal positions for installing reclosers in a large network. Due to the size of the electric distribution networks considered during planning activities, hundreds different positions for installing Normally-Opened Automatic Reclosers (NO-AR) and Normally-Closed Automatic Reclosers (NC-AR) must be assessed. To deal with the size of the problem, covering all states the network may assume and assuring the positions for installing AR were optimum ones, the proposed methodology divides this problem into three states. Through this approach, the planning engineer need to carry out several simulations in just a few minutes, evaluating the technical benefits achieved from different investment levels. Similar approaches could not be found in the current literature. The methodology was assessed considering two substations of a Brazilian electric distribution company, corresponding to twenty-five medium voltage feeders. Two analyses were carried out: the brown field analysis, where the positions of thirty new automatic reclosers were determined; and the green field analysis, where forty-five existing automatic reclosers were reallocated. The results indicate significant improvements in quality of service indices, which may reach over 30% reduction level.

C. F. M. Almeida (✉) · G. A. Quiroga · N. Kagan
Escola Politecnica, Universidade de Sao Paulo, Sao Paulo, Brazil
e-mail: cfmalmeida@usp.br

G. A. Quiroga
e-mail: gabriel.quiroga@usp.br

N. Kagan
e-mail: nelsonk@pea.usp.br

H. Kagan
Sinapsis Inovacao em Energia, Sao Paulo, Brazil
e-mail: henrique.kagan@sinapsisenergia.com

F. Shahnia et al. (eds.), *Electric Distribution Network Planning*,
Power Systems, https://doi.org/10.1007/978-981-10-7056-3_13

Keywords Electric distribution planning · Genetic algorithms · Electric distribution reliability · Automatic reclosers · Power quality

13.1 Introduction

The allocation of AR is an interesting alternative to improve the quality of supply provided by electric distribution utilities. Besides the operational flexibility provided by such equipment, these investments usually come fully in the regulatory asset base in Brazil. Other actions, such as tree pruning or change of network patterns, are alternatives that represent an increase in operational costs and may represent a regulatory risk of not being classified as prudent investment. Several Brazilian distribution companies have invested for the significant increase in the presence of AR on their networks. Such investments correspond to installation of hundreds of devices per year. Thus, a new problem arose in distribution planning: how to determine the optimal locations for the installation of hundreds of equipment in the concession area of an electric distribution company?

Allocating an AR is a very difficult task. Firstly, planning engineers obtain the data regarding the interruption occurrences recorded in the Outage Management System (OMS) database from the previous years to define where most of fault problems occur. Then, the fault occurrence information is crossed with the customer distribution throughout the electric distribution network, obtained from the Georeferenced Information System (GIS). In a very simplified way, the planning engineers try to define the strategic positions for installing AR that would not allow the regions where most of the faults occur affect the regions where there are most of the customers are located. Thus, just to locate one single AR may take up to a few hours according to the traditional approach.

Today, planning engineers need to assess different configurations the electric distribution network may assume during its operation. For each configuration, one also needs to assess the level of improvement that different sets of AR may introduce to the network performance in terms of quality of supply indices. Depending on the level of improvement of each set of AR, one can determine the investment level required to deliver the quality of supply required by the customers. It is important to clarify that these analyses cannot take much longer to be performed, because the investment plan on utility companies is normally executed in a few months of the year. So, a methodology that proposes the strategic positions for allocating AR in the networks becomes an interesting alternative.

There are several references dealing with the problem of allocating switching devices. Some references simply provide conceptual approaches for allocating switching resources [4, 6]. Others, for the sake of simplicity, focus on allocation NC switching devices only [5, 10, 11, 14, 15, 17, 18, 20, 23] or focus on the problem of allocating NO switches only [9]. Normally, the methodologies proposed in current references deal with simplified networks [2, 3, 5, 7, 8, 10–12, 14–16, 18, 23, 25]. Moreover, some methodologies address the problem of allocation of switching devices in networks by testing the installation of such equipment in every bus [7, 8], others consider every line segment as candidate position for assessing the improvement of installing a switching device [12, 25], and others have the candidate positions for the installation of the switching devices previously selected [16, 18, 21, 22]. Due to the combinatorial nature of the problem, and the vast possible candidate position available for testing the allocation of a switching device these approaches are impractical for planning investments across an entire concession area.

When dealing with electric distribution networks of considerable size, the methodologies present in the current literature tend to simplify the approach in the decision-making problem for defining the number of switching devices to be allocated. In [13], authors have defined a methodology that distributes AR in medium voltage feeders regarding their importance. Thus, no reliability simulation regarding a detailed modelling approach towards the network topology configuration and the spatial distribution of fault occurrences throughout the network is considered. Besides, the specific position where the AR should be installed is not provided either. In [20], the methodology defines the optimal position for installing AR one at a time. Thus, the optimal position for the next AR depends on the position defined for the previous one. Such approach does not allow de possibility of defining an optimum when one is trying to allocate a set of AR simultaneously.

In [21, 22, 24] authors propose an approach for allocation NC-AR and NO-AR in electric distribution networks of considerable size. These references only consider the allocation of new AR. In [21, 22] the authors proposed a methodology that simultaneously allocates a NC-AR and a NO-AR, so it would be possible to assure that the load blocks may be supplied during a contingency from another source. That approach is not always required when installing new AR in network, as the power demand may significantly vary throughout the day. In [24] the proposed methodology considers average failure rates and average failure duration times for the entire network. Such approach does not provide a realistic behavior of the network, as one region may be more prone for failure than others, due to the existence of trees, for example.

In this chapter, the methodology proposed for the optimal allocation of AR aims its application at real distribution networks, with several thousands of buses. It defines optimal alternatives for the installation of new NC-AR and NO-AR simultaneously, indicating the positions where the devices should be installed to improve the quality of electricity supply and respecting the total amount of equipment to be allocated and the total investment budget available. The methodology also considers interruption occurrences, extracted from the utility company's OMS, and the real topology, extracted from the utility company's GIS. Thus, reliability calculation is performed in a more realistic way.

An important innovation of the proposed methodology relies on the three-stage approach proposed. At the first stage, the possible states for each medium voltage feeder, depending on the allocation of NO-AR are enumerated. At the second stage, the optimum solutions for each set of NC-AR to be allocated are defined for each possible state and stored. At last, the optimum solution for each medium voltage feeder is chosen, maximizing the improvement in quality of supply regarding the level of budget available. As the set of optimum solutions are stored, the planning engineer may carry out several different simulations, considering different budget levels to evaluate the investment level required to deliver the proper quality of supply. The methodology also allows one to ignore the existing AR and reallocate them to adequate the electric distribution network towards the current condition it is subjected to.

As the proposed methodology indicates the strategic position for installing AR with respect to the spatial fault occurrence distribution and spatial customer distribution, no coordination between the protective devices is assure. Several other references that deal with the switch allocation problem do not guarantee the coordination and selectivity of the protection device either. Nevertheless, it is important to highlight that the coordination may be achieved by altering the protection parameters of the existing devices of the network and that defining the strategic positions is the more arduous task when one is trying to install new AR.

13.2 Methodology

The methodology proposed was divided into three stages. The first stage consists of enumerating the possible states a medium voltage network may assume. A state is defined by the positions for installing Normally-Opened Automatic Reclosers (NO-AR). The second stage consists of evaluating configurations for the installation of Normally-Closed Automatic Reclosers (NC-AR). Several optimization problems are solved to determine the optimal positions for installing a specific number NC-AR for each medium voltage feeder. Each set of optimization problems is related to a specific state that the medium voltage feeder may assume. The third and final stage consists on selecting a set of NC-AR and NO-AR to be installed in each of the medium voltage feeder. In this stage, the objective is to maximize the improvement in terms of quality of service that includes, in a weighted way, collective indices of quality of service. The maximum number of AR to be allocated is the main restriction of the optimization problem in this stage. It addresses the issue of budget limitations. One may carry out different simulations, considering different budget values, to evaluate if the estimated improvements would support the required investment level. Figure 13.1 shows the detailed flowchart of the methodology. Further details regarding each of the stages of the methodology are provided in the following sections.

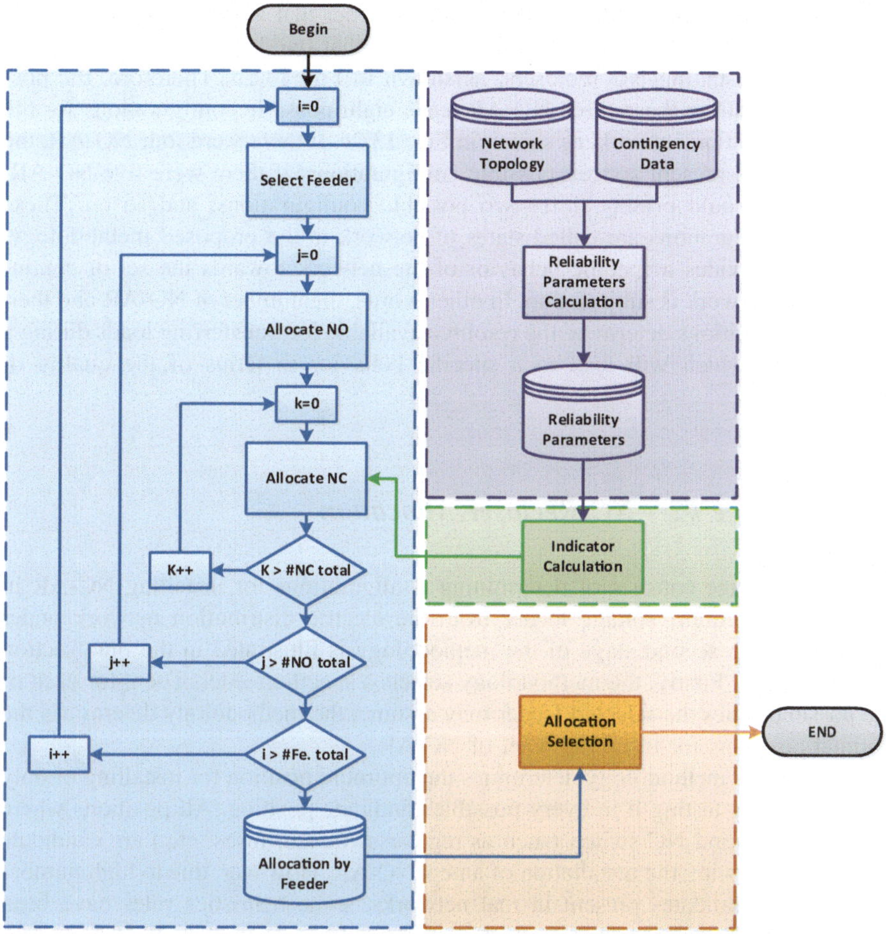

Fig. 13.1 Simplified methodology flowchart

13.2.1 Stage #1—State Enumeration—NO Recloser Allocation

As stated before, the first stage consists of enumerating configurations for the installation of AR-NO between the medium voltage feeders that belongs to the electric distribution network under analysis. At first, all positions where there is some kind of NO switch are candidate positions for testing the installation of a new NO-AR. One may also provide extra candidate position for testing NO-AR by adding a regular NO switch between two medium voltage feeders. The combination of possible positions for testing the installation of NO-AR defines a state for a specific medium voltage feeder.

For example, consider the medium voltage feeder that already has two NC-AR installed, as shown in Fig. 13.2a. Also consider that one has indicated three possibilities for installing NO reclosers, as shown in Fig. 13.2b. Therefore, the possibility of installing these three NO-AR leads eight possible configurations for the electric distribution network, as shown in Fig. 13.2c. If there were four NO-AR, the network would present sixteen possible configurations; if there were five NO-AR, the network would present thirty-two possible configurations; and so on. These possible configurations are called states of network in the proposed methodology. Each state provides a specific behavior of the network towards the set of contingencies the network is subjected to. In other words, the number of NO-AR and their installation positions determine the resource available for transferring loads during a contingency, which will lead to a specific behavior in terms of the quality of service.

13.2.2 Stage #2—NC Recloser Allocation

The second stage consists of determining configurations for installing NC-AR in each of the medium voltage feeder from the electric distribution network under evaluation. The second stage of the methodology is illustrated in the blue section from Fig. 13.1. Firstly, the methodology selects a specific feeder. Then, for each of the possible states the selected feeder may assume, the methodology determines the optimal positions for installing a set of NC-AR.

Initially, the methodology determines the optimum position for installing of only one NC-AR by testing it in every possible candidate position. All positions where there is some kind NC switch (such as regular switched, fuses, etc.) are candidate positions for testing the installation of a new NC-AR. However, due to high number of possible candidates present in real networks, some heuristics rules have been developed to determine the candidate positions for testing the effect of a new NC-AR. Some of the heuristics rules considered are:

- it is not possible to install AR after a fuse;
- it is not possible to install more than three NC-AR in series;
- it is not possible to install NC-AR next to the substation (less than 50 m, for example).

After determining the optimum position for installing only one NC-AR in the selected feeder, the methodology solves several optimization problems to determine the optimum positions for installing higher numbers of NC-AR (two NC-AR installed simultaneously, three NC-AR installed simultaneously, and so on). The process is repeated for every possible state the selected feeder may assume to determine the optimal solutions for installing a specific set of NC-AR. After concluding the optimization processes for one feeder, the methodology repeats the whole process for the next feeder that belongs to the electric distribution network

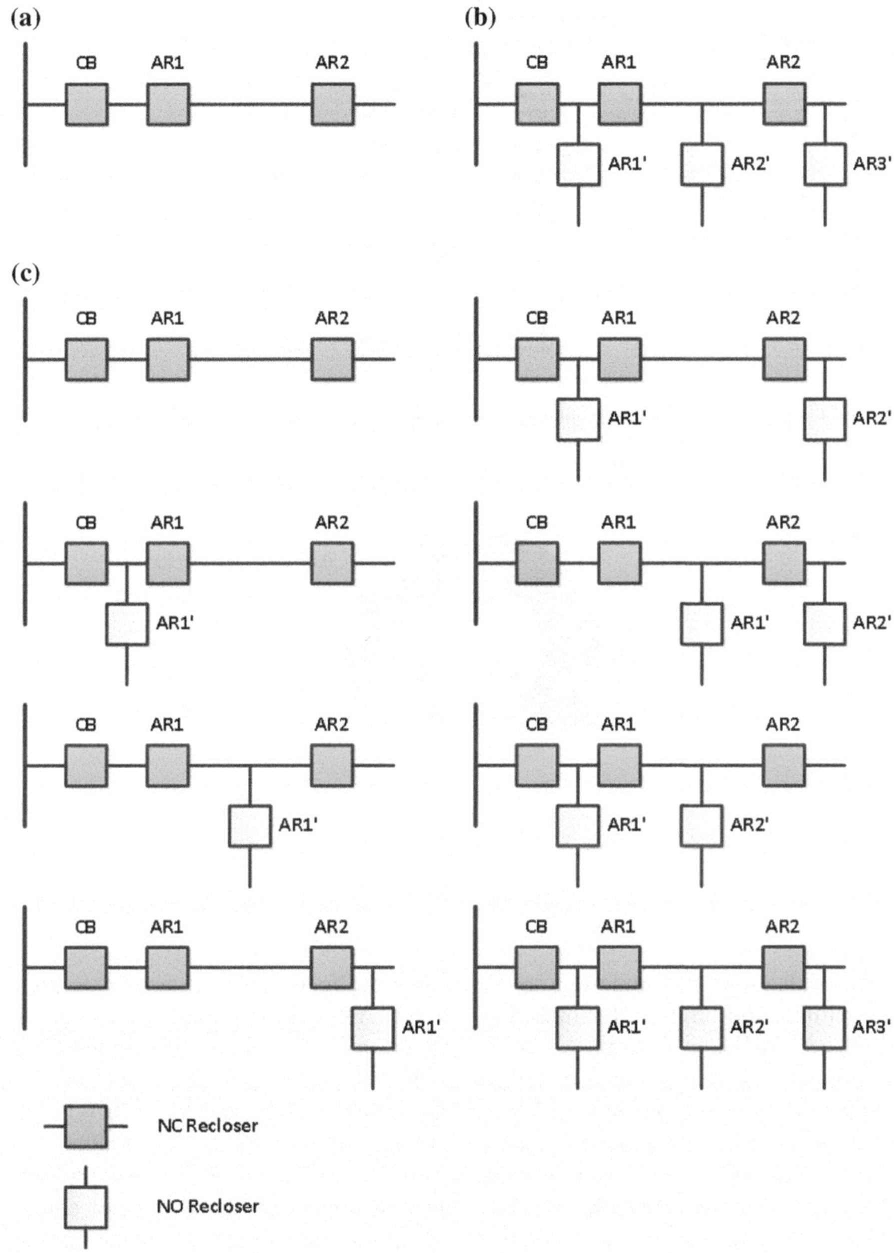

Fig. 13.2 Possible installation alternatives for NO devices

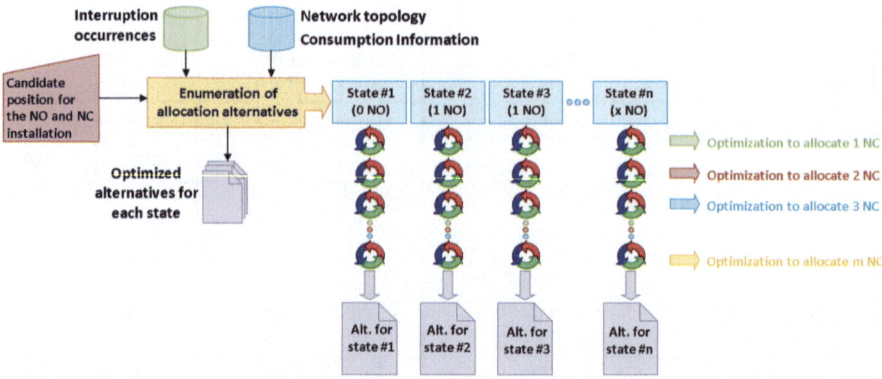

Fig. 13.3 Stage #2—NC recloser allocation methodology process

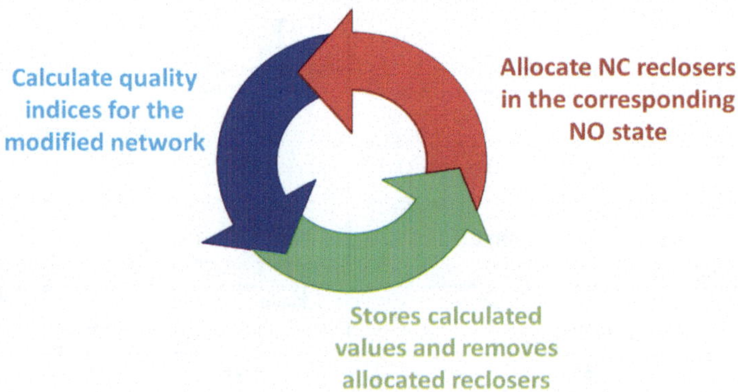

Fig. 13.4 Detail of the process of assessing the NC-AR allocation impact on the quality indices

under evaluation. It is important to clarify that the methodology starts by solving the optimization problem for installing two NC-AR and the number of NC-AR is increased until the cost of the set of NC-AR exceeds the amount of avoided ENS. To compare the investment with the avoided ENS he methodology considered a US $20,000.00 cost for the purchasing and installing a AR, 60.00 US$/MWh/year interruption cost, a 7.5% return rate and 20-year life time for the AR. Figure 13.3 illustrates the automated process regarding the second stage of the methodology. The process of assessing the NC-AR allocation impact on the quality of supply indices is detailed at Fig. 13.4. From Fig. 13.3 one may observe that the optimal results for every feeder, regarding every possible state and every possible number of NC-AR are stored in specific files.

As stated before, when the methodology is determining the optimal positions for installing two or more NC-AR, the optimum positions are determined by solving several optimization problems. As there are several candidate positions for installing NC-AR, and all possible states each medium voltage feeder may assume should be covered, several optimization problems must be solved. Using an approach based on GA becomes convenient, as, when one defines the maximum number of individual and generations to be considered, it also determines the maximum time spent for solving each optimization problem. In fact, any other metaheuristic approach could also be considered to solve the same problem. But, the focus should not be on the solution method considered, but on the methodology itself, regarding the allocation of multiple NO and NC devices, simultaneously, and on real electric distribution networks, composed by thousands of buses.

13.2.2.1 Genetic Algorithms

Genetic Algorithms (GA) optimization technique was introduced by Holland in 1975 [1] with the objective of mathematically formalizing and rigorously explain adaptation processes in natural systems and developing artificial systems that retain the original mechanisms found in natural systems. GA start from a string, an element that must have an explicit relation with the parameters of the problem. A string can be understood, from genetics, as a chromosome that presents genes (or bits), at different locus of the chromosome (positions in the string), representing several characteristics of an individual. The value of each gene, which corresponds to a specific characteristic, matches to one allele. Also, the genetic package, usually called a genotype, may have its correspondence in GA by the data structures that define the string. The interaction of this genetic package with the environment, which defines the characteristics of the individual is called a phenotype, which corresponds, in GA, in the decoding of the structure to form a possible alternative solution or possible set of solution parameters of the problem. Once the basic element of GA is established, that is a string, and its relation to the real problem, that is its coding, the mechanism of a GA is relatively simple.

String Coding Approach

The string coding approach for the Stage #2 considers integer vectors. The number of positions is defined by the number of NC-AR to be allocated. The integer inside the position is defined by the candidate position where the NC-AR should be installed. Figure 13.5 shows an example the string coding approach. Figure 13.5a shows a feeder F, with a circuit breaker CB, two NO switches and five NC switches that are the candidate position for assessing the NC-AR allocation. At its side, there is respective general string representation for the allocation of two NC-AR. As there are five candidate positions, the values in each position of the vector may vary from 0 to 4. Obviously, during the string coding process, the methodology does not allow repeated values in the string positions. Figure 13.3b indicates that the feeder performance should be assessed considering the installation of two NC-AR in the position of the switches NC2 and NC4. Figure 13.3c indicates that the feeder

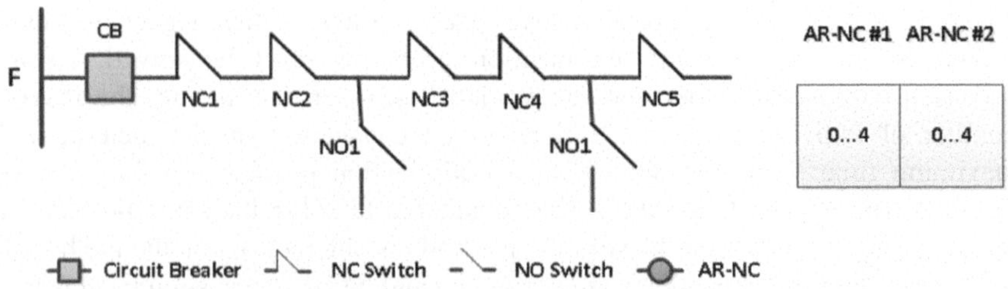

(a) Example of possible installation alternatives for two AR-NC.

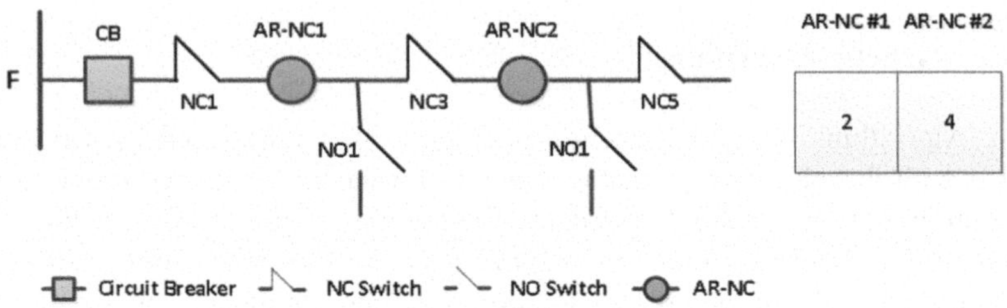

(b) Example of the installation of two AR-NC.

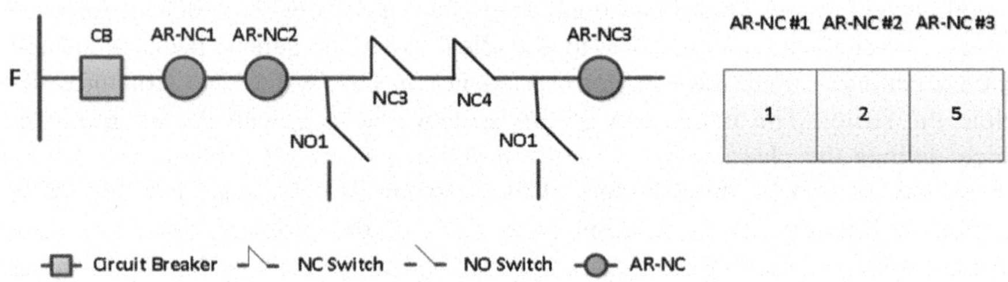

(c) Example of the installation of three AR-NC.

Fig. 13.5 String coding approach for second stage GA-based optimization problems

performance should be assessed considering the installation of three NC-AR in the position of the switches NC1, NC2 and NC4.

At this point of the methodology, the basic problem lies in determining the "best" individual or the "best fit", which is measured by the value of an evaluation function applied to every individual.

Evaluation Function

For the second stage of the methodology, the evaluation function aims at measuring the improvement of the quality of supply of the electric distribution network, by reducing the average frequency and duration of contingencies and improving revenue saving by reducing the amount of energy not supplied. Thus, for the modeling of the evaluation function, three indices are considered:

- Interrupted Customers (*IC*): total number of customers interrupted in a certain period;
- Hours of Interrupted Customers (*HIC*): total number of hours that customers were interrupted in a certain period;
- Energy Not Supplied (*ENS*): total energy not supplied due to occurrences of interruptions in a certain period.

The evaluation function considered during the several optimization problems is illustrated in Eq. (13.1).

$$max[K_{\mathrm{IC}} \cdot MI_{IC} + K_{\mathrm{HIC}} \cdot MI_{HIC} + K_{\mathrm{ENS}} \cdot MI_{ENS}] \tag{13.1}$$

where

- MI_{IC}: is the merit index for *IC* reduction;
- MI_{HIC}: is the merit index for *HIC* reduction;
- MI_{ENS}: is the merit index for *ENS* reduction;
- K_{IC}: is the weighting factor for *IC*;
- K_{HIC}: is the weighting factor for *HIC*;
- K_{ENS}: is the weighting factor for *ENS*.

The Merit Index (*MI*) formulation was based on the difference between any evaluation grade for the initial state of the medium voltage feeder (without installation of the AR-$Grade_{without}$) and the same grade for the state determined by an alternative allocation (with the installation of the AR-$Grade_{with}$). To allow comparison with impacts on different grades, the difference is then normalized by $Grade_{without}$. Equation (13.2) illustrates the generic behavior for the proposed *MI* formulation.

$$MI = \frac{Grade_{without} - Grade_{with}}{Grade_{without}} = 1 - \frac{Grade_{with}}{Grade_{without}} \tag{13.2}$$

Thus, (13.3–13.5) illustrates the specific formulation for MI_{IC}, MI_{HIC}, and MI_{ENS}:

$$MI_{IC} = 1 - \frac{IC_{with}}{IC_{without}} \tag{13.3}$$

$$MI_{HIC} = 1 - \frac{HIC_{with}}{HIC_{without}} \tag{13.4}$$

$$MI_{ENS} = 1 - \frac{ENS_{with}}{ENS_{without}} \tag{13.5}$$

The improvements in terms of each of these indices were determined for every state of every feeder that belongs to the electric distribution network under evaluation. The *IC* index is related to the SAIFI index. The *HIC* index is related to the SAIDI index. *IC* and *HIC* were considered instead of the original SAIFI and SAIDI because the methodology aims for a global improvement over the whole concession area. Using the values for SAIFI and SAIDI determined for each feeder may not lead to allocation that minimizes the quality of service indices (i.e. provides the maximum improvement in terms of quality of supply). The calculation of the indices is performed during the evaluation of every possible individual and follows the 'A Priori' Reliability Calculation Approach [19].

'A Priori' Reliability Calculation Approach

The "A Priori" Reliability Calculation Approach estimates the reliability indicators for an electric distribution network, based on the use of average reliability parameters obtained through interruptions historic data. The indicators calculated are the number of interrupted customers (*IC*), the number of hours of interrupted customers (*HIC*), and the energy not supplied (ENS) due to the interruptions. For this, it is necessary to calculate the reliability parameters for each load block and for each protection block. A load block is a set of contiguous line sections delimited by sectioning equipment. A protection block is a set of load blocks delimited only by protective equipment.

In this approach, a load block is defined by four parameters, which are the average failure rate, the average failure duration, the number of consumers and its demand, as illustrated in Fig. 13.6. In the schematic feeder presented in Fig. 13.6, there are four load blocks and two protection blocks.

The average failure rate (f^{avg}) of each load block is calculated from interruptions historic data, which may be obtained from the OMS database of electric distribution companies. Thus, the interruptions occurrences are grouped according to the protection equipment responsible for the interruption, and the number of total interruptions is divided by the total length from the corresponding protection block. Thus, the f^{avg} from Load Block #1 is the same from Load Block #2, because Load Block #1 and Load Block #2 are in the same protection block.

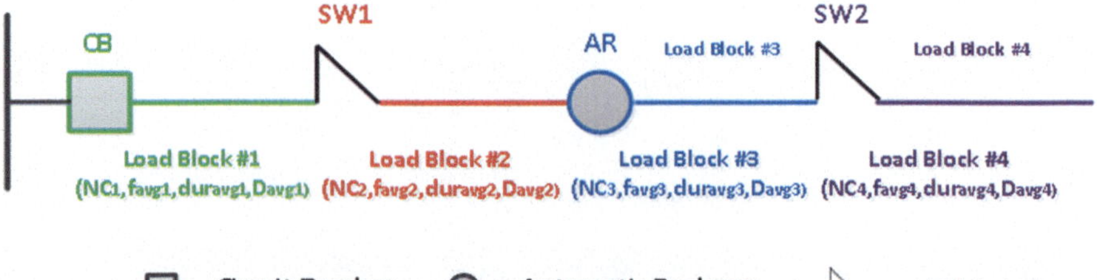

Fig. 13.6 Example of load blocks and its parameters

The average failure duration (dur^{avg}) is also calculated from interruptions historic data. In the same way, the interruptions occurrences are grouped according to the protection equipment responsible for the interruption, and dur^{avg} is the arithmetic average of duration of each interruption occurrence. Thus, the dur^{avg} from Load Block #1 is the same from Load Block #2, because Load Block #1 and Load Block #2 are in the same protection block.

Only the number of customers (*NC*) and average demand (D^{avg}) are specific for each load block. The (*NC*) is calculated from the sum of the customers that are inside the block. Typically, such information may be obtained from GIS database of electric distribution companies. The D^{avg} is calculated through the energy consumed by the customers connected to the load block. Typically, such information may be obtained from ERP (Enterprise Resource Planning) system database of electric distribution companies. With these four parameters, one may calculate the quality of supply indices as presented in the following sections.

Total Number of Customers Interrupted

The (*IC*) is calculated as illustrated in Eq. (13.6):

$$IC = \sum_{i}^{n} (NC_i \cdot f_i^{avg}) \tag{13.6}$$

where

- n: refers to the total number load blocks present the electric distribution system under analysis;
- i: refers to one specific load block;
- f_i^{avg}: refers to the average failure rate of load block i;
- NC_i: which is the number of customers that had their supply interrupted by a contingency in load block i.

Total Number Hours of Interrupted Customers

The *HIC* is calculated as illustrated in Eq. (13.7):

$$HIC = \sum_{i}^{n} (NC_i \cdot f_i^{avg} \cdot dur_i^{avg}) \tag{13.7}$$

where dur_i^{avg} is the average failure duration for a contingency in load block i (which is usually measured in hours, and considers, only interruptions that last more than 3 min, according to Brazilian regulation).

Energy Not Supplied

The *ENS* can be calculated as illustrated in Eq. (13.8):

$$ENS = \sum_{i}^{n} (D_i^{avg} \cdot f_i^{avg} \cdot Dur_i) \quad (13.8)$$

where D_i^{avg}: is the average power demand of load block *i*.

The average demand for each load block is obtained by the relation between the monthly consumption of each customers and the average number of hours in a month, as seen in Eq. (13.9):

$$D_i^{avg} = \frac{\varepsilon_i^{monthly}}{730} \quad (13.9)$$

where

- *i*: refers to one specific load block;
- $\varepsilon_i^{monthly}$: is total amount of energy consumed by the customers connected to load block *i* (which is usually measured in kWh).

For evaluating the possibility of load transference, one needs to check if capacity and voltage limits conditions would be violated. To guarantee a good computational performance and to allow a conservative approach, only one power flow calculation is carried out for the original state of the network.

Regarding loading constraints, the transference capacity for each NO switch is established through the capacity limits of the line segments that connect the NO switch to the source. Thus, the line segment with the lowest capacity located between the NO switch and the source defines the load amount that the medium voltage feeder may receive during a contingency.

Regarding voltage limit constraints, the transference capacity for each NO switch is established through the total impedance from the line segments that connect the NO switch to the source. Thus, the amount of load transferred cannot increase a voltage drop at the bus where the NO switch to values that would violate the voltage limits.

Through this approach, there is no need to perform a power flow calculation for every state the electric network may assume, during the 'A Priori Reliability Calculation Approach'. Thus, the time spent of the calculation process does not compromises the performance for the whole methodology.

It is important to clarify that the 'A Priori Reliability Calculation Approach' only considers the transference of load blocks from one feeder to another, when the load blocks are located downstream from the load block where a fault is being considered. Thus, by opening the switches to isolate the faulty load block and closing the NO switch to reestablish the supply to the downstream load blocks the radiality condition of the electric distribution network is not violated. Thus, there is no need to perform a radiality checking as commonly is done on reconfiguration problems.

In GA, the initial population of individuals, that is, an initial set of strings, is usually established randomly. Populations then evolve into generations, basically through three operators, reproduction, crossing-over and mutation. Reproduction corresponds to a process in which individuals are copied to the future generation according to their evaluation function. Crossing-over corresponds to an operator acting on a randomly chosen pair of strings. And mutation corresponds to an operator that can modify, with certain probability, the values of genes (alleles) of the strings. Based on the population from the previous generation, such operators are applied to create a new one which corresponds to the population from the next generation. In this new population, new phenotypes are then introduced, which may lead to new results in terms of the "best" individual. Figure 13.7 shows how these stages are related to each other, which illustrates the GA considered. Further details regarding GA may be found in [1]. For the second stage of the methodology, the selection operator applied was the Tournament Selection with as evaluation with three individuals, as illustrated in Fig. 13.8.

The mutation operator considered a 1% probability rate and is illustrated in Fig. 13.9. The crossing-over operator considered a 75% probability rate and is illustrated in Fig. 13.10.

13.2.3 Stage #3—Global Optimization

The third state of the methodology consists on determining which of the optimal alternatives listed at the second stage for each medium voltage feeder should be applied, to maximize the improvement in the overall quality of power supply indices of the electric distribution network. Thus, the third stage determines which state will be considered, by indicating how many NO-AR should be installed, and which alternative will be considered, by indicating how many NC-AR should be installed, for each medium voltage feeder. The main restriction in this stage relies on the budget available for investing in AR allocation, or the maximum number of AR to be installed.

To clarify the approach at this stage of the methodology, Eq. (13.10) illustrates the optimization process from a linear programming perspective. The evaluation of each state and alternative combination considers the benefits achieved in terms of reduction of the quality of supply indices and in terms of the maximum number of reclosers considered.

$$\begin{aligned}
& maximize \sum_{i=1}^{n} \frac{benef_i}{num_dev_i} \\
& subject\ to: \\
& num_dev_i \leq max_num_dev \quad max_num_dev \leq \frac{budget}{unit_cost}
\end{aligned} \tag{13.10}$$

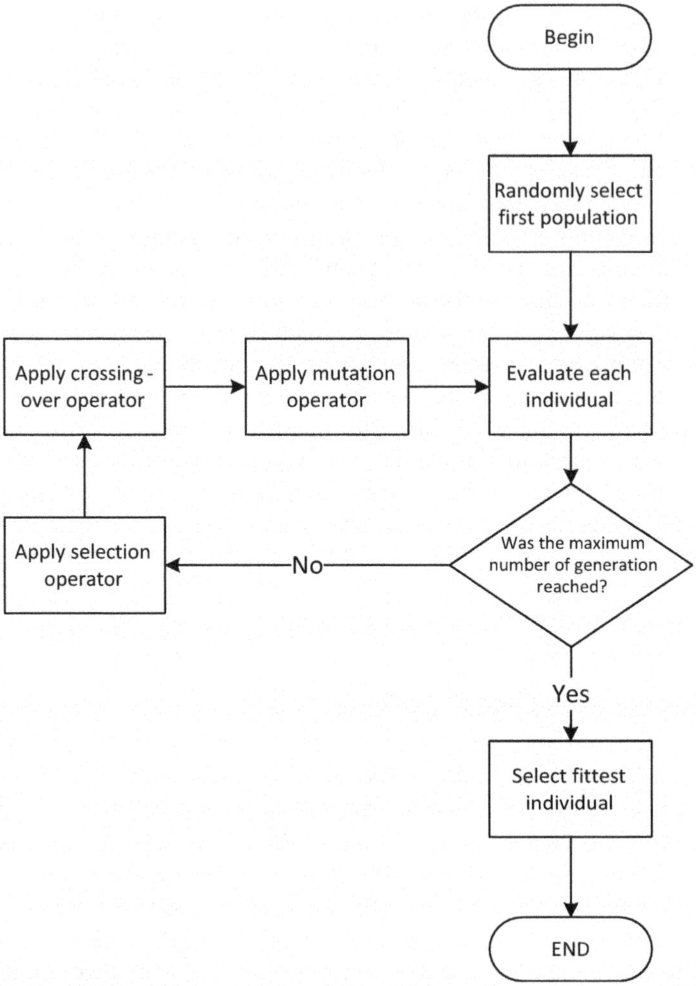

Fig. 13.7 Simplified GA flowchart

where

- *i*: refers to a specific medium voltage feeder;
- *n*: refers to the maximum number of medium voltage feeders present in the network under evaluation;
- $benef_i$: refers to the reduction on the quality of supply indices in feeder *i*;
- num_dev_i: refers to the number of reclosers to be allocated in feeder *i*;
- *max_num_dev*: is the maximum number of reclosers to be allocated in the network under evaluation;
- *budget*: the value of the budget available;
- *unit_cost*: the unit cost of purchasing and installing one recloser.

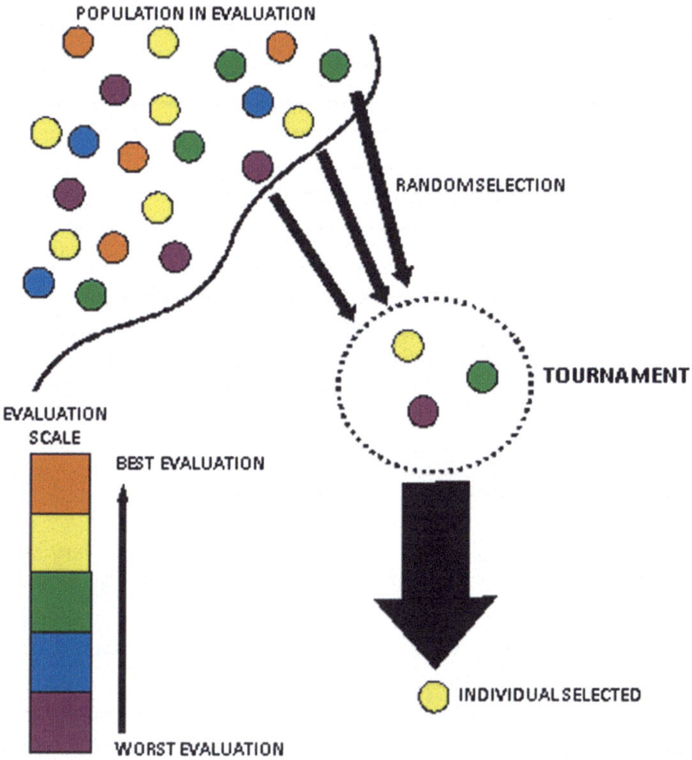

Fig. 13.8 Tournament selection

1	4	0
[Possible values: 0, 1, 2, 3]	[Possible values: 0, 1, 2, 3, 4, 5]	[Possible values: 0, 1, 2]

(a) Original String (third position selected for mutation)

1	4	2
[Possible values: 0, 1, 2, 3]	[Possible values: 0, 1, 2, 3, 4, 5]	[Possible values: 0, 1, 2]

(b) New String (after mutation execution)

Fig. 13.9 Mutation operator

Any other optimization technique could also be considered to solve the optimization problem described through Eq. (13.10). Due to its ease of implementation, GA were also considered to solve the optimization problem regarding this stage of the methodology.

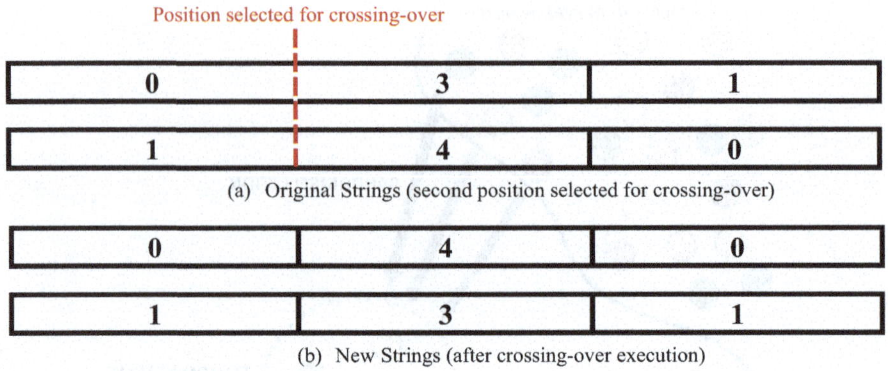

Fig. 13.10 Crossing-over operator

13.2.3.1 Genetic Algorithms

The GA process considered at this stage of the methodology was very similar to the one considered for the second stage of the methodology. At this stage, the selection, mutation and crossing-over operator were the same as the ones considered in the second stage. The probability rates for the mutation and crossing-over operators were also 1 and 75%, respectively.

The main difference from the approach considered for the second stage is that, instead of solving multiple optimization problems, only one optimization problem is now solved for the whole electric distribution network. As the NO-AR determines the state for the network, one should be able to identify the corresponding state for each feeder, to locate the correct optimal solutions regarding the allocation of NC-AR. Such problem was addressed through the string coding approach considered.

String Coding Approach

The coding for the proposed solution alternatives is a string with two types positions. The binary positions at the beginning of the string correspond to the NO-AR. They determine the installation (unit value) or not (null value) of an NO-AR at a specific candidate position at the electric network. The integer positions in the string correspond to a specific medium voltage feeder present in the network under evaluation. Figure 13.11 shows an example of the string coding approach. Figure 13.11 shows four feeders and the candidate positions for installing AR, defined by NC and NO switches.

One may or may not allocate an NO-AR in each NO switch present in Fig. 13.11. In that way, its coding is represented by a binary position. Since that are four NO candidate switches in the example, the string code must have four binary positions.

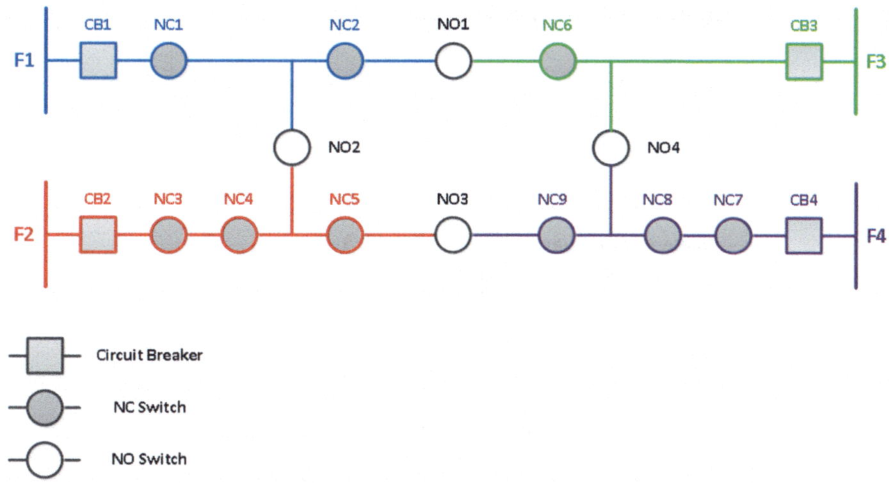

Fig. 13.11 Feeders and its candidate positions for installing AR

For each medium voltage feeder, there are many optimal alternatives for allocating NC-AR. Each of these alternatives correspond to a specific state of the electric distribution network and was previously determined at the second stage of the methodology. Thus, there are several sets of optimal solution for each feeder. Each set of optimal solutions lists the optimal solutions for allocating one NC-AR, two NC-AR, three NC-AR and so on, until the *ENS* avoided by the installation of the NC-AR does not exceeds the unit cost of the AR. Each set of optimal solutions also corresponds to a specific state of the medium voltage feeder. And the state of the medium voltage feeder is determined by the NO-AR considered. For example, in Fig. 13.11, F1 has two candidate positions for installing NC-AR, which leads to four possible alternatives for NC-AR installations: the allocation of no NC-AR, the allocation of an NC-AR at the position of switch NC1, the allocation of an NC-AR at the position of switch NC2 and the allocation of two NC-AR, one at the position of switch NC1 and another one at the position of switch NC2. In this way, an integer coding is convenient. Since there are four medium voltage feeders in the example, the string code must have four integer positions, one for each medium voltage feeder. The generic string coding for the network illustrated in Fig. 13.11 is detailed in Fig. 13.12.

To illustrate it better, Fig. 13.13 shows a coded string for a viable solution.

By assessing the string shown in Fig. 13.13, one can identify that this solution indicates the allocation of two NO-AR at the positions of switches NO1 and NO2, the allocation of two NC-AR in feeder F1, at the position of switches NC1 and NC2, the allocation of one NC-AR in feeder F3, at the position of switch NC6, and the allocation of two NC-AR in feeder F4, at the positions of switches NC8 and NC9. The solution represented by this coding can be seen in Fig. 13.14.

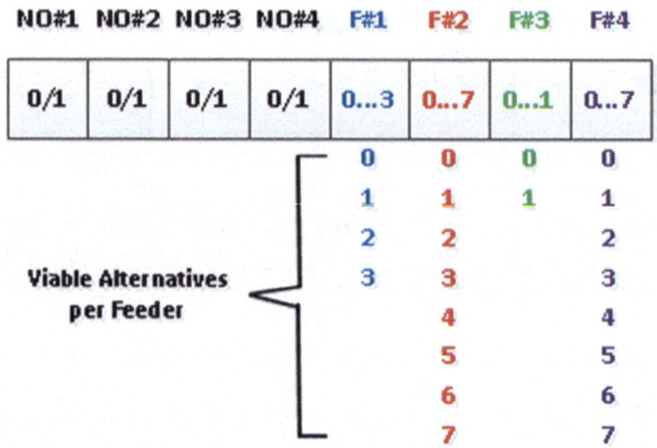

Fig. 13.12 String considered to code all possible AR allocation positions

NO#1	NO#2	NO#3	NO#4	F#1	F#2	F#3	F#4
1	1	0	0	3	0	1	6

Fig. 13.13 Example of viable solution defined by a string

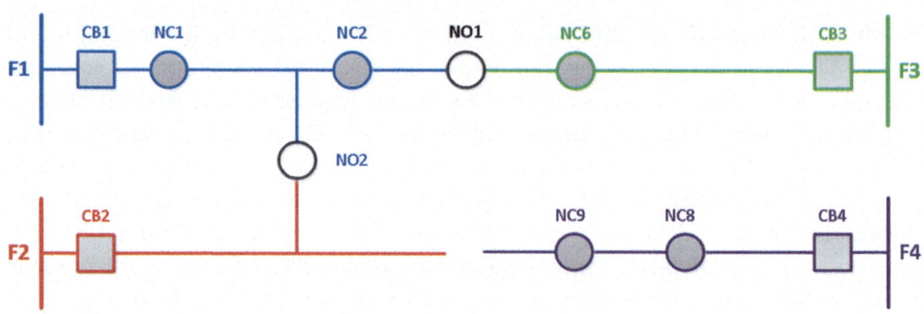

Fig. 13.14 Example of a possible solution

13.2.3.2 Evaluation Function

For the third stage of the methodology, the evaluation function also aims at measuring the improvement of quality of supply of electric distribution network, by reducing the average frequency and duration of contingencies and improving revenue saving by reducing the amount of energy not supplied. Thus, for the modeling of the evaluation function, *IC*, *HIC* and *ENS* indices were also considered.

The evaluation function for evaluating every possible solution during the several optimization problems is illustrated in Eq. (13.11).

$$max[K_{IC} \cdot f_{IC} + K_{HIC} \cdot f_{HIC} + K_{ENS} \cdot f_{ENS}] \tag{13.11}$$

where f_{IC}: is a function that provides the overall reduction in terms of the *IC* for the whole network; f_{HIC}: is a function that provides the overall reduction in terms of the *HIC* for the whole network; f_{ENS}: is a function that provides the overall reduction in terms of the *ENS* for the whole network, which are formulated as

$$f_{IC} = 1 - \frac{\sum_i^n IC_{with}(feeder_i, NO_1, \ldots, NO_k, Alternative\#)}{\sum_i^n IC_{without}(feeder_i)} \tag{13.12}$$

where

- $feeder_i$: refers to a specific medium voltage feeder present in the network under evaluation;
- $NO_1, \ldots, NO_k$: refer to the candidate positions for installing NO-AR, which defines the state $feeder_i$ is subjected to;
- *Alternative#*: refers to the combination of NC-AR to be installed in $feeder_i$;
- $IC_{with}(feeder_i, NO_1, \ldots, NO_k, Alternative\#)$: is a function that returns the corresponding value of *IC* for $feeder_i$, depending on the state defined by the combination of values on $NO_1, \ldots, NO_k$, and on the combination of NC-AR to be installed;
- $IC_{without}(feeder_i)$: is a function that returns the original value of *IC* for $feeder_i$, without the installation of the AR.

$$f_{HIC} = 1 - \frac{\sum_i^n HIC_{with}(feeder_i, NO_1, \ldots, NO_k, Alternative\#)}{\sum_i^n HIC_{without}(feeder_i)} \tag{13.13}$$

where

- $HIC_{with}(feeder_i, NO_1, \ldots, NO_k, Alternative\#)$: is a function that returns the corresponding value of *HIC* for $feeder_i$, depending on the state defined by the combination of values on $NO_1, \ldots, NO_k$, and on the combination of NC-AR to be installed;
- $HIC_{without}(feeder_i)$: is a function that returns the original value of *HIC* for $feeder_i$, without the installation of the AR.

$$f_{ENS} = 1 - \frac{\sum_i^n ENS_{with}(feeder_i, NO_1, \ldots, NO_k, Alternative\#)}{\sum_i^n ENS_{without}(feeder_i)} \tag{13.14}$$

where

- $ENS_{with}(feeder_i, NO_1, \ldots, NO_k, Alternative\#)$: is a function that returns the corresponding value of *ENS* for $feeder_i$, depending on the state defined by the combination of values on $NO_1, \ldots, NO_k$, and on the combination of NC-AR to be installed;
- $ENS_{without}(feeder_i)$: is a function that returns the original value of *ENS* for $feeder_i$, without the installation of the AR.

13.3 Results

The simulations were performed considering interruptions occurrences referring to the years of 2012, 2013, 2014 and part of 2015 (until February 19, 2015) together, corresponding to 327,472 records for the whole utility concession area. Such occurrences were extracted from the OMS database of the Brazilian utility company. Each occurrence describes the protection equipment that have operated, the time the interruption started and the time the service was restored. With this information, it is possible to calculate the average duration time for repairing fault at each load block and the average failure rate for eat load block.

For the second stage, all GA-based optimizations considered 50 generations with 100 individuals each. For the third stage, all GA-based optimizations considered 200 generations with 500 individuals each. The simulation time spent in both analysis was around 15 min. All simulations were executed in a virtual machine, considering 2 processors and 8 GB of RAM memory.

Another relevant premise for the studies carried out was the consideration of AR already previously installed or not. That is, the studies can be divided into:

- Brown field: considers the effects of existing reclosers and intends to carry out the installation of new AR;
- Green field: when the existing reclosers are disregarded and it is intended to reallocate existing AR.

Figure 13.15 illustrates the pilot area where the proposed methodology was applied. The pilot area for application of the methodology is composed by two real substations (SED) from a Brazilian utility. Some of the characteristics of the network under evaluation are:

- The first SED is composed of nine feeders, one of which is a distress feeder;
- The second SED consists of sixteen feeders, 2 of which are distress feeders;
- There are 14,513 buses;
- There are 13,803 segments of line;
- There are 17 AR installed in the feeders served by first SED, of which 4 are NO type;

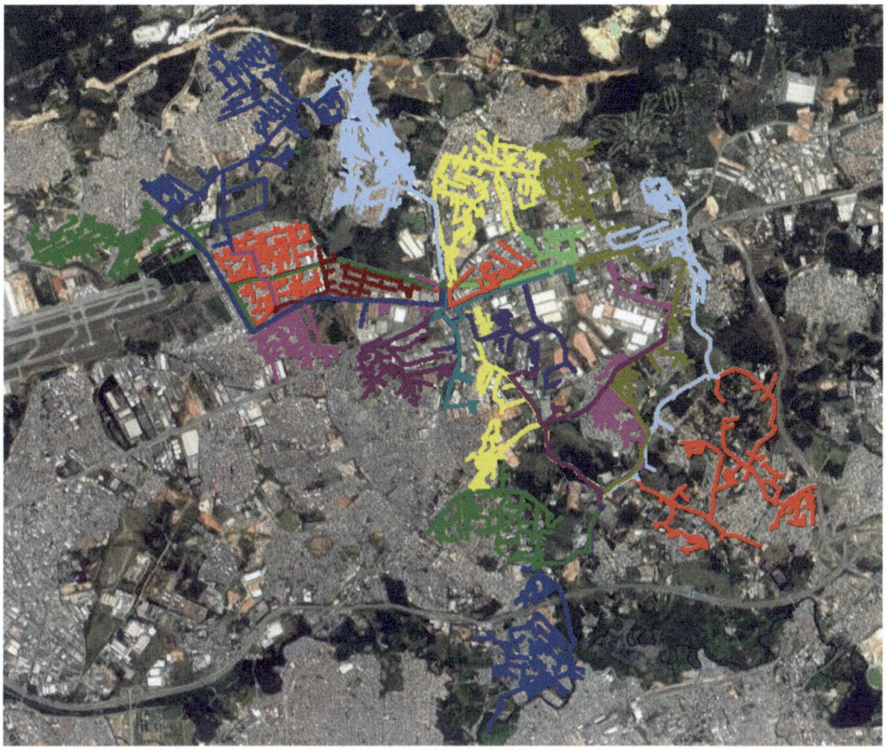

Fig. 13.15 Pilot Area

- There are 28 AR installed in the feeders served by second SED, of which 8 are NO type;
- There are 765 switches installed (among regular switches and fuses);
- There are 99,946 customers;
- The original HIC value is 779,727.13 h/year;
- The original IC value is 641,862.52 customers/year;
- The original ENS value is 304.65 MWh/year.

13.3.1 Brown Field Analysis

The second stage of the methodology determine 261 possible positions for installing NC reclosers, and 35 possible positions for installing NO reclosers. Table 13.1 details the results found for the brown field analysis.

As one can observe, regardless of the limitations imposed, the most significant reduction in terms of CIH is achieved by installing 20 new NO reclosers and 48

Table 13.1 Brown field simulation

	Best HIC	Best IC	Best ENS	Allocation of 10 new reclosers	Allocation of 20 new reclosers	Allocation of 30 new reclosers
# of NO reclosers	20	15	20	2	5	8
# of NC reclosers	48	34	47	8	15	22
Benefit in HIC reduction	19.23%	–	–	10.16%	15.00%	16.44%
Benefit in IC reduction	–	19.25%	–	10.38%	16.24%	17.84%
Benefit ENS reduction	–	–	25.61%	9.52%	16.18%	21.38%

new NC reclosers. In terms of CI reduction, the best result is achieved by installing 15 new NO reclosers and 34 new NC reclosers. And in terms of ENS reduction, the best result is achieved by installing 20 new NO reclosers and 47 new NC reclosers. These results are obtained by combining the most significant reduction achieved for each feeder at the second stage of the methodology. Thus, the GA approach for global optimization had not been used yet.

The global optimization simulations consider constraints for the allocation of ten, twenty and thirty reclosers. In addition, the analysis considers the best results for each collective index, regardless of the limitations imposed. Thus, for the simulation for allocation ten reclosers, the methodology proposed the installation of 2 NO reclosers and 8 NC reclosers. For the simulation for allocation twenty reclosers, the methodology proposed the installation of 5 NO reclosers and 15 NC reclosers. And for the simulation for allocation thirty reclosers, the methodology proposed the installation of 8 NO reclosers and 22 NC reclosers.

13.3.2 Green Field Analysis

The second stage of the methodology determine 286 possible positions for installing NC reclosers, and 38 possible positions for installing NO reclosers. Table 13.2 details the results found for the green field analysis.

As one can observe, regardless of the limitations imposed, the most significant reduction in terms of *HIC* is achieved by installing 26 new NO reclosers and 61 new NC reclosers. In terms of *IC* reduction, the best result is achieved by installing

Table 13.2 Green field simulation

	Best HIC	Best IC	Best ENS	Reallocation of 45 existing reclosers
# of NO reclosers	26	25	28	9
# of NC reclosers	61	56	61	36
Benefit in HIC reduction	37.46%	–	–	34.44%
Benefit in IC reduction	–	36.12%	–	33.29%
Benefit in ENS reduction	–	–	41.83%	32.73%

25 new NO reclosers and 56 new NC reclosers. And in terms of *ENS* reduction, the best result is achieved by installing 28 new NO reclosers and 61 new NC reclosers.

The global optimization simulations consider constraints for the reallocation of forty-five reclosers. Thus, the simulation for allocation ten reclosers, the methodology proposed the installation of 9 NO reclosers and 36 NC reclosers.

13.3.3 General Remarks

A comparison of the results obtained from both analysis was presented in Fig. 13.16. Observing Fig. 13.16 it is possible to identify that the allocation of new reclosers improves the performance indices of service quality. It is also noticed that the improvement tends to be less intense with the increase of the number of devices allocated, leading to a saturation of the benefit coming from the allocation of the reclosers.

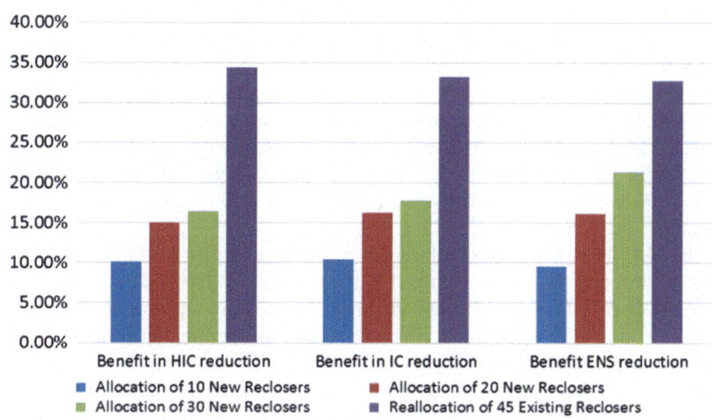

Fig. 13.16 Histogram comparing both analysis

Another interesting analysis is comparing the results obtained from the brown field analysis for allocating 20 new reclosers, which results the results obtained with the analysis for allocating 45 existing reclosers in total and the green field simulation for 45 reclosers, which also results in 45 reclosers in total. Observing these results, one can see the relevance of the methodology, since the relocation results are superior to the benefits of installing new reclosers. Thus, it is assumed that the methodology can present superior results when compared to the current methodologies of allocation used by most of the utility companies.

References

1. D.E. Goldberg, *Genetic Algorithms in Search, Optimization, and Machine Learning* (Addison-Wesley, 1989)
2. G. Levitin, S. Mazal-Tov, D. Elmakis, Genetic algorithm for optimal sectionalizing in radial distribution systems with alternative supply. Electr. Power Syst. Res. **35**(3) (1995)
3. R. Billinton, S. Jonnavithula, Optimal switching device placement in radial distribution systems. IEEE Trans. Power Deliv. **11**(3) (1996)
4. F. Soudi, K. Tomsovic, Optimal distribution protection design: quality of solution and computational analysis. Int. J. Electr. Power Energy Syst. **21**(5) (1999)
5. G. Celli, F. Pilo, Optimal sectionalizing switches allocation in distribution networks. IEEE Trans. Power Deliv. **14**(3) (1999)
6. F. Soudi, K. Tomsovic, Optimal trade-offs in distribution protection design. IEEE Trans. Power Deliv. **16**(2) (2001)
7. J.H. Teng, C.N. Lu, Feeder-switch relocation for customer interruption cost minimization. IEEE Trans. Power Deliv. **17**(1) (2002)
8. J.H. Teng, Y.H. Liu, A novel ACS-based optimum switch relocation method. IEEE Trans. Power Syst. **18**(1) (2003)
9. M.R. Haghifam, Optimal allocation of tie points in radial distribution systems using a genetic algorithm. Eur. Trans. Electr. Energy Syst. (2004)
10. L.G.W. Silva, R.A.F. Pereira, J.R.S. Mantovani, Allocation of protective devices in distribution circuits using nonlinear programming models and genetic algorithms. Electr. Power Syst. Res. **69**(1) (2004)
11. D.H. Popovic, J.A. Greatbanks, M. Begovic, A. Pregelj, Placement of distributed generators and reclosers for distribution network security and reliability. Int. J. Electr. Power Energy Syst. **27**(5–6) (2005)
12. C.S. Chen, C.H. Lin, H.J. Chuang, C.S. Li, M.Y. Huang, C.W. Huang, Optimal placement of line switches for distribution automation systems using immune algorithm. IEEE Trans. Power Syst. **21**(3) (2006)
13. V.C. Zamborlini, D.R. Trindade, E. Zambon, B.B. Garcia, E.F Azeredo, *Otimização da Alocação de Religadores em Larga Escala; II CBEE - Congresso Brasileiro de Eficiência Energética* (Vitória/ES, Brazil, 2007)
14. A. Moradi, M.F. Firuzabad, Optimal switch placement in distribution systems using trinary particle swarm optimization algorithm. IEEE Trans. Power Deliv. **23**(1) (2008)
15. L.G.W. Silva, R.A.F. Pereira, J.R. Abbad, J.R.S. Mantovani, Optimised placement of control and protective devices in electric distribution systems through reactive Tabu search algorithm. Electr. Power Syst. Res. **78**(3) (2008)
16. A. Helseth, A.T. Holen, Impact of energy end use and customer interruption cost on optimal allocation of switchgear in constrained distribution networks. IEEE Trans. Power Deliv. **23**(3) (2008)

17. H. Falaghi, M.R. Haghifam, C. Singh, Ant colony optimization-based method for placement of sectionalizing switches in distribution networks using a fuzzy multiobjective approach. IEEE Trans. Power Deliv. **24**(1) (2009)
18. W. Tippachon, D. Rerkpreedapong, Multiobjective optimal placement of switches and protective devices in electric power distribution systems using ant colony optimization. Electr. Power Syst. Res. **79**(7) (2009)
19. N. Kagan, C.C.B. Oliveira, E.J. Robba, *Introdução aos Sistemas de Distribuição de Energia Elétrica. 2ª Edição* (Editora Edgard Blucher, 2010)
20. C.C.B. Oliveira, D. Takahata, M. Maia, *Metodologia de Alocação Otimizada de Dispositivos de Proteção em Alimentadores Baseada no Desempenho Máximo do Alimentador (DMA). IX CBQEE - Conferência Brasileira sobre Qualidade da Energia Elétrica* (Cuiabá/MT, Brazil, 2011)
21. D.P. Bernardon, M. Sperandio, V.J. Garcia, J. Russia, L.N. Canhab, A.R. Abaideb, E.F.B. Daza, Methodology for allocation of remotely controlled switches in distribution networks based on a fuzzy multi-criteria decision-making algorithm. Electr. Power Syst. Res. **81**(2) (2011)
22. D.P. Bernardon, M. Sperandio, V.J. Garcia, L.N. Canha, A.R. Abaide, E.F.B. Daza, AHP decision-making algorithm to allocate remotely controlled switches in distribution networks. IEEE Trans. Power Deliv. **26**(3) (2011)
23. A.A. Jahromi, M.F. Firuzabad, M. Parvania, M. Mosleh, Optimized sectionalizing switch placement strategy in distribution systems. IEEE Trans. Power Deliv. **27**(1) (2012)
24. L.S. Assis, J.F.V. González, F.L. Usberti, C. Lyra, C. Cavellucci, F.J. Von Zuben, Switch allocation problems in power distribution systems. IEEE Trans. Power Syst. **30**(1) (2015)
25. J.C. López, J.F. Franco, M.J. Rider, Optimisation-based switch allocation to improve energy losses and service restoration in radial electrical distribution systems. IET Gener. Transm. Distrib. **10**(11) (2016)

The manufacturer's authorised representative in the EU is Springer
Nature Customer Service Centre GmbH, Europaplatz 3, 69115 Heidelberg,
Germany. If you have any concerns regarding our products, please
contact ProductSafety@springernature.com

Printed and bound by CPI Group (UK) Ltd, Croydon, CR0 4YY
25/09/2026
02222853-0005